ADVANCES IN CHEMICAL PHYSICS

VOLUME LVIII

Advances in
CHEMICAL PHYSICS

EDITED BY

I. PRIGOGINE

University of Brussels
Brussels, Belgium
and
University of Texas
Austin, Texas

AND

STUART A. RICE

Department of Chemistry
and
The James Franck Institute
The University of Chicago
Chicago, Illinois

VOLUME LVIII

AN INTERSCIENCE® PUBLICATION
JOHN WILEY & SONS
NEW YORK · CHICHESTER · BRISBANE · TORONTO · SINGAPORE

An Interscience® Publication

Library of Congress Catalog Number: 58-9935

ISBN 0-471-81349-4

Printed in the United States of America

10 9 8 7 6 5 4 3 2 1

CONTRIBUTORS TO VOLUME LVIII

R. STEPHEN BERRY, The Department of Chemistry and the James Franck Institute, The University of Chicago, Chicago, Illinois

THOMAS KEYES, Department of Chemistry, Boston University, Boston, Massachusetts

W. G. MCMILLAN, Department of Chemistry and Biochemistry, University of California, Los Angeles, California

GUILHEM MARC, Department of Chemistry and Biochemistry, University of California, Los Angeles, California

ANDREW J. MASTERS, Royal Signals and Radar Establishment, Malvern, United Kingdom

GRIGORY A. NATANSON, Department of Chemistry and the James Franck Institute, The University of Chicago, Chicago, Illinois

JOSEPH N. WEBER, The Department of Chemistry and the James Franck Institute, The University of Chicago, Chicago, Illinois

INTRODUCTION

Few of us can any longer keep up with the flood of scientific literature, even in specialized subfields. Any attempt to do more and be broadly educated with respect to a large domain of science has the appearance of tilting at windmills. Yet the synthesis of ideas drawn from different subjects into new, powerful, general concepts is as valuable as ever, and the desire to remain educated persists in all scientists. This series, *Advances in Chemical Physics*, is devoted to helping the reader obtain general information about a wide variety of topics in chemical physics, which field we interpret very broadly. Our intent is to have experts present comprehensive analyses of subjects of interest and to encourage the expression of individual points of view. We hope that this approach to the presentation of an overview of a subject will both stimulate new research and serve as a personalized learning text for beginners in a field.

ILYA PRIGOGINE

STUART A. RICE

CONTENTS

ADVANCES IN CHEMICAL PHYSICS

VOLUME LVIII

TAGGED-PARTICLE MOTION IN DENSE MEDIA: DYNAMICS BEYOND THE BOLTZMANN EQUATION

THOMAS KEYES

Department of Chemistry
Boston University
Boston, Massachusetts 02215

ANDREW J. MASTERS

Royal Signals and Radar Establishment
Malvern, United Kingdom, WR14 3PS

CONTENTS

I. INTRODUCTION

Tagged particle motion in equilibrium may be characterized by the velocity correlation function,

$$C(t) = \langle \mathbf{v}_1(t) \cdot \mathbf{v}_1(0) \rangle \tag{1.1}$$

where $\langle\ \rangle$ denotes an equilibrium ensemble average; the self-diffusion coefficient, D, is given by the time integral

$$D = d^{-1} \int_0^\infty dt\, C(t), \tag{1.2}$$

where d is the dimension. The quantities $C(t)$ and D have been studied extensively, both theoretically and via computer simulation. Some of the basic facts are as follows. For a particle small compared to the mean free path (l) of a dilute gas, C and D are accurately given by the[1] Lorentz–Boltzmann (LB) theory, to be discussed later. The resulting expression depends upon the potentials of interaction. In the simple case of hard spheres, to an excellent approximation, $C(t)$ decays exponentially, and[1]

$$D_{\text{LB}} = \frac{3}{8\rho\sigma_{12}^2}\left(\frac{k_{\text{B}}T}{2\pi\mu}\right)^{1/2} \tag{1.3}$$

where μ is the reduced mass, $\mu = m_1 m/(m_1 + m)$, m_1 and m being the tagged-particle and bath-particle masses, respectively; $\sigma_{12} = R_1 + a$, R_1 and a being the tagged- and bath-particle radii; ρ is the density of the gas; k_B is Boltzmann's constant; and T is the temperature. For other potentials possessing a harshly repulsive core and a reasonably short-ranged attractive region, $C(t)$ and D should be qualitatively similar to the hard-sphere form. The kinetic theory discussed in this review will be almost completely confined to the case of smooth hard spheres. The problems to be discussed are sufficiently difficult that consideration of the extra complications caused by more realistic potentials would be premature.

Even if the gas remains dilute, particles large compared to the mean free path will see the gas as a hydrodynamic continuum, and will thus, if sufficiently massive, obey the Stokes–Einstein (SE) law for D,

$$D_{\text{SE}} = \frac{k_{\text{B}}T}{c\pi\eta R_1} \tag{1.4}$$

where η is the shear viscosity of the bath. The coefficient c equals 4 for slip boundary conditions (appropriate for smooth spheres) and 6 for stick (appropriate for rough potentials). If the gas is made dense, further complications ensue, even for particles the same size as the bath particles. Alder, Gass, and Wainwright have shown,[2] in a computer simulation, that the diffusion constant of a tagged member of a pure hard-sphere liquid obeys the slip Stokes–Einstein law. They also found that $C(t)$ was highly nonexponential, becoming negative after a few collision times and decaying as a power law, $t^{-d/2}$, for very long times. The ratio of the self-diffusion constant to its LB value, as a function of density, was seen to first increase, reaching a peak at a reduced density, $\rho^* = \rho a^3$, of ~ 0.5, and to sharply decrease at liquid density, corresponding to $D_{LB} \gg D_{SE}$.

None of the behavior discussed in the paragraph above fits with the Lorentz–Boltzmann theory. On the other hand, until fairly recently, this theory and its similar[1] Enskog extension constituted the only detailed, microscopic theory of self-diffusion.

For the calculation of D and $C(t)$ for a small particle in a more concentrated gas, Enskog's kinetic theory has proved extremely successful. This theory takes into account the equilibrium structure of the gas around the tagged particle, and the result is to divide the right-hand side of Eq. (1.3) by a factor of the radial distribution function at contact, g_{12}^{+}; $C(t)$ retains its exponential form. Thus, although this is an improvement upon the Lorentz–Boltzmann theory, it cannot account for the facts listed above, and a more complete theory—especially, one which can give the Stokes–Einstein law when appropriate—is needed. Of course, the ultimate goal is the construction of a theory correct for all sizes, masses, and densities. On the other hand, it has not been obvious until recently, even in the continuum limit, how to carry out any microscopic derivation of the SE law, so this has also been an important challenge. Considerable effort has been expended on these problems over the past ten or fifteen years. The research splits into two main lines. One approach retains the apparatus of kinetic theory, and aims to derive kinetic descriptions more complete than that of LB. The other invokes generalized hydrodynamics in several forms, in particular, mode coupling. In this review, we will discuss the various methods which have been used, the relations which exist among the methods, the progress which has been made, and the problems which remain unsolved.

II. KINETIC THEORY

A. Background

The velocity correlation function $C(t)$ is a moment of the tagged-particle phase-space correlation function $f(\mathbf{r}, \mathbf{v}, \mathbf{v}', t)$:

$$C(t) = \int d\mathbf{r}\, d\mathbf{v}\, d\mathbf{v}' (\mathbf{v}\cdot\mathbf{v}') f(\mathbf{r}, \mathbf{v}, \mathbf{v}', t) \tag{2.1}$$

$$f(\mathbf{r}, \mathbf{v}, \mathbf{v}', t) = \left\langle \delta(\mathbf{r} - \mathbf{r}_1(t))\, \delta(\mathbf{v} - \mathbf{v}_1(t))\, \delta(\mathbf{r}_1(0))\, \delta(\mathbf{v}' - \mathbf{v}_1(0)) \right\rangle \tag{2.2}$$

where $\mathbf{r}_1$ and $\mathbf{v}_1$ denote the position and velocity of the tagged particle. Kinetic theory may be discussed in the terms of nonequilibrium distribution functions or equilibrium correlation functions; the two approaches are equivalent if the only nonequilibrium situations considered are very near equilibrium. We will use the correlation-function language. In this case, f is the fundamental quantity, and is analogous to the tagged-particle distribution function in the other method.

The Fourier–Laplace transform of the LB equation for f is

$$(z+i\mathbf{k}\cdot\mathbf{v}-\rho\Lambda_{\mathrm{LB}}(\mathbf{v}))f(\mathbf{k},\mathbf{v},\mathbf{v}',z)=f(\mathbf{k},\mathbf{v},\mathbf{v}',t=0) \tag{2.3}$$

where z is the frequency, $\mathbf{k}$ the wavevector, and Λ_{LB} the linearized LB collision operator. For hard spheres, on which we focus,[1]

$$\Lambda_{\mathrm{LB}}(\mathbf{v}_1)g(\mathbf{v}_1)=\int d2\,\varphi_0(v_2)T(12)g(\mathbf{v}_1) \tag{2.4}$$

where "2" is shorthand for $\mathbf{r}_2$ and $\mathbf{v}_2$, φ_0 is the Maxwellian, and T is the binary collision operator,

$$T(12)=\sigma_{12}^2\int d\hat{\boldsymbol{\sigma}}\,W(\mathbf{v}_{12}\cdot\hat{\boldsymbol{\sigma}})|\mathbf{v}_{12}\cdot\hat{\boldsymbol{\sigma}}|\delta(\mathbf{r}_{12}-\sigma_{12}\hat{\boldsymbol{\sigma}})(1-b_{12}) \tag{2.5}$$

Here W is the Heaviside function, "12" denotes a relative coordinate, $\hat{\boldsymbol{\sigma}}$ is the perihelion vector, and b_{12} is an operator which replaces precollision with postcollision (12) velocities $\mathbf{v}^*$:

$$\begin{aligned}\mathbf{v}_1^*&=b_{12}\mathbf{v}_1=\mathbf{v}_1-\left(\frac{2\mu}{m_1}\right)(\hat{\boldsymbol{\sigma}}\cdot\mathbf{v}_{12})\hat{\boldsymbol{\sigma}}\\ \mathbf{v}_2^*&=b_{12}\mathbf{v}_2=\mathbf{v}_2+\left(\frac{2\mu}{m_2}\right)(\hat{\boldsymbol{\sigma}}\cdot\mathbf{v}_{12})\hat{\boldsymbol{\sigma}}\end{aligned} \tag{2.6}$$

Formal solution of Eq. (2.3), with use of Eq. (2.2), yields

$$\begin{aligned}C_{\mathbf{k}}(z)&\equiv\int d\mathbf{v}\,d\mathbf{v}'\,\mathbf{v}\cdot\mathbf{v}'f(\mathbf{k},\mathbf{v},\mathbf{v}',z)\\ &=\int d\mathbf{v}\,\varphi_0(v)\mathbf{v}\cdot[z+i\mathbf{k}\cdot\mathbf{v}-\rho\Lambda_{\mathrm{LB}}(\mathbf{v})]^{-1}\mathbf{v}\end{aligned} \tag{2.7}$$

where we have used the zero-time result,

$$f(\mathbf{k},\mathbf{v},\mathbf{v}',t=0)=\varphi_0(v)\delta(\mathbf{v}-\mathbf{v}') \tag{2.8}$$

The Fourier–Laplace-transformed quantity $C_{\mathbf{k}}(z)$ is closely related to the generalized **k**- and z-dependent diffusion constant,

$$\begin{aligned} D &= d^{-1} \lim_{k,z\to 0} C_{\mathbf{k}}(z) \\ C(z) &= \lim_{k\to 0} C_{\mathbf{k}}(z) \end{aligned} \tag{2.9}$$

In general, the diffusion constant must be evaluated with use of a "projected" time evolution, but this is negligible for $k \to 0$. The inverse in Eq. (2.7) may be evaluated via standard kinetic-theory techniques, giving rise to the expression previously given for D, Eq. (1.3), in the so-called first Sonine polynomial approximation.

Of course, the LB equation was originally written down from phenomenological arguments. The most important of these is that, in moving through the gas, the tagged particle undergoes *uncorrelated binary collisions* only, that is, it collides with particles it has never encountered before and will never see again. Cohen,[3] Dorfman and Cohen,[4] Kawasaki and Oppenheim,[5] Zwanzig,[6] and others formulated a systematic density expansion of the true, exact kinetic operator K, defined so that, if Λ_{LB} is replaced with K in Eq. (2.7), the true, exact $C_{\mathbf{k}}(z)$ results, at any density:

$$C_{\mathbf{k}}(z) \equiv \int d\mathbf{v}\,\varphi_0(\mathbf{v})\mathbf{v}\cdot[z + i\mathbf{k}\cdot\mathbf{v} - \rho K_{\mathbf{k}}(\mathbf{v}, z)]^{-1}\mathbf{v} \tag{2.10}$$

They showed that

$$K = \Lambda_{\text{LB}} + \sum_{n=1}^{\infty} \rho^n B_n \tag{2.11}$$

with formal expressions for the B's, and it appeared[7] that a better approximation to K than Λ_{LB} might be obtained from Eq. (2.11), that is, by the inclusion of a few more terms or by partial summation of the series. It must be pointed out that, at the time, attention was not sharply focused on the defects of LB theory mentioned in the introduction, but rather on the more general aspects of developing a systematic kinetic theory.

It was soon found, however, that the expansion, Eq. (2.11), did not exist, that is, some of the B's diverge. An excellent review, up to and including this point, is given in Ref. 7. Kawasaki and Oppenheim[5] attempted to deal with the problem of the nonexistence of the density expansion by summing the most strongly divergent terms, the "rings." The hope was that an infinite number of individually infinite terms would sum to a finite result; such was indeed the case. Concomitantly, the density dependence of K was seen to be nonanalytic.

In a ring, the tagged particle collides with a gas particle, undergoes a finite number of uncorrelated binary collisions, and finally recollides with the original collision partner. As will be seen, recollisions—absent from LB theory—are the single most important aspect of tagged-particle motion at high density. Roughly, individual rings diverge because, at long times, the finite number of intermediate binary collisions allows too much velocity correlation to persist. The ring operator R resulting from the Kawasaki–Oppenheim resummation has the form of a single ring, with the intermediate propagation given by the LB equation, viz,

$$R(1,z)=\int d2\,\varphi_0(v_2)\bar{T}(12)G(12,z)T(12) \tag{2.12}$$

where G is the intermediate tagged-particle–bath-particle propagator,

$$G(12,z)=\left[z+\mathbf{v}_1\cdot\nabla_1+\mathbf{v}_2\cdot\nabla_2-\rho\Lambda_{\mathrm{LB}}(\mathbf{v}_1)-\rho\Lambda_{\mathrm{B}}(\mathbf{v}_2)\right]^{-1} \tag{2.13}$$

the ordinary (not tagged-particle) Boltzmann operator is Λ_{B}, and $\bar{T}$ is a different version of the binary collision operator,

$$\bar{T}(12)=\sigma_{12}^2\int d\hat{\boldsymbol{\sigma}}\,W(\mathbf{v}_{12}\cdot\hat{\boldsymbol{\sigma}})|\mathbf{v}_{12}\cdot\hat{\boldsymbol{\sigma}}|\left[\delta(\mathbf{r}_{12}-\sigma_{12}\hat{\boldsymbol{\sigma}})-\delta(\mathbf{r}_{12}+\sigma_{12}\hat{\boldsymbol{\sigma}})b_{12}\right] \tag{2.14}$$

The two T's correspond to the original collision and the recollision, and the inverse in between describes intermediate propagation of the tagged and bath particles by LB and Boltzmann dynamics. We have written R in position space, not **k**-space, so $\mathbf{v}\cdot\nabla$ appears instead of $i\mathbf{k}\cdot\mathbf{v}$. Because the T operators are not diagonal in **k**, Fourier transformation of Eq. (2.12) yields

$$\begin{aligned} R_{\mathbf{k}}(\mathbf{v}_1,z)&=\int d\mathbf{k}'\int d\mathbf{v}_2\varphi_0(v_2)\bar{T}_{\mathbf{k}-\mathbf{k}'}(\mathbf{v}_1,\mathbf{v}_2)G_{\mathbf{k},\mathbf{k}'}(\mathbf{v}_1,\mathbf{v}_2)T_{\mathbf{k}'-\mathbf{k}}(\mathbf{v}_1,\mathbf{v}_2)\\ G_{\mathbf{k},\mathbf{k}'}(\mathbf{v}_1,\mathbf{v}_2)&=\left[z+i(\mathbf{k}+\mathbf{k}')\cdot\mathbf{v}_1-i\mathbf{k}'\cdot\mathbf{v}_2+\rho\Lambda_{\mathrm{LB}}(\mathbf{v}_1)+\rho\Lambda_{\mathrm{B}}(\mathbf{v}_2)\right]^{-1} \end{aligned} \tag{2.15}$$

where

$$\begin{aligned} \bar{T}_{\mathbf{k}'}(\mathbf{v}_1,\mathbf{v}_2)&=\sigma_{12}^2\int d\hat{\boldsymbol{\sigma}}\,\theta(\mathbf{v}_{12}\cdot\hat{\boldsymbol{\sigma}})|\mathbf{v}_{12}\cdot\hat{\boldsymbol{\sigma}}|\left(e^{i\mathbf{k}'\cdot\hat{\boldsymbol{\sigma}}\sigma_{12}}-e^{-i\mathbf{k}'\cdot\hat{\boldsymbol{\sigma}}\sigma_{12}}b_{12}\right)\\ T_{\mathbf{k}'}(\mathbf{v}_1,\mathbf{v}_2)&=\sigma_{12}^2\int d\hat{\boldsymbol{\sigma}}\,\theta(\mathbf{v}_{12}\cdot\hat{\boldsymbol{\sigma}})|\mathbf{v}_{12}\cdot\hat{\boldsymbol{\sigma}}|e^{i\mathbf{k}'\cdot\hat{\boldsymbol{\sigma}}\sigma_{12}}(1-b_{12}) \end{aligned} \tag{2.16}$$

With the ring operator in hand, it was natural to investigate the ring approximation, $K \cong \Lambda_{LB} + R \equiv K^R$. Dorfman and Cohen[4,8] and Dufty[9] showed that such a theory was capable of reproducing the "long-time tails" on $C(t)$. The calculation proceeds as follows. At sufficiently low density, R should be a small perturbation on Λ_{LB}. Thus, in the ring approximation,

$$G \sim G_{LB} + \rho G_{LB} R G_{LB} \tag{2.17}$$

where G denotes the inverse in Eq. (2.10) with $K = K^R$, and so on. We then obtain

$$\begin{aligned} C_{\mathbf{k}}(z) &= C_{LB,\mathbf{k}}(z) + \rho \int d\mathbf{k}'\, d\mathbf{v}_1\, d\mathbf{v}_2\, \varphi_0(v_1)\varphi_0(v_2) \\ &\times \big\{ \mathbf{v}_1 \cdot G_{LB,\mathbf{k}}(\mathbf{v}_1, z) \bar{T}_{\mathbf{k}'}(\mathbf{v}_1, \mathbf{v}_2) \\ &\quad \times [z + i(\mathbf{k}+\mathbf{k}')\cdot\mathbf{v}_1 - i\mathbf{k}'\cdot\mathbf{v}_2 - \rho\Lambda_{LB}(\mathbf{v}_1) - \rho\Lambda_B(\mathbf{v}_2)]^{-1} \\ &\quad \times T_{-\mathbf{k}'}(\mathbf{v}_1, \mathbf{v}_2) G_{LB,\mathbf{k}}(\mathbf{v}_1, z)\mathbf{v}_1 \big\} \end{aligned} \tag{2.18}$$

Evaluation of the RHS of Eq. (2.18) is an exercise in mathematical methods of kinetic theory. The inverses are represented[8-10] in terms of complete sets of basis functions in the velocity,

$$\begin{aligned} C_{\mathbf{k}}(z) &= C_{LB,\mathbf{k}}(z) + \rho \int d\mathbf{k}' (\mathbf{v}_1 \cdot | G_{LB,\mathbf{k}}(z) | f_i)(f_i | \bar{T}_{\mathbf{k}'} | f_j) \\ &\times (f_j | [z + i(\mathbf{k}+\mathbf{k}')\cdot\mathbf{v}_1 - i\mathbf{k}'\cdot\mathbf{v}_2 - \rho\Lambda_{LB}(\mathbf{v}_1) - \rho\Lambda_B(\mathbf{v}_2)]^{-1} | f_l) \\ &\times (f_l | T_{-\mathbf{k}'} | f_m)(f_m | G_{LB,\mathbf{k}}(z) | \mathbf{v}_1) \end{aligned} \tag{2.19}$$

where ( ) denotes $\int dv_1\, dv_2\, \varphi_0(v_1)\varphi_0(v_2) \cdots$, and the summation convention has been used. The f's are product tagged–bath-particle basis functions, for example 11, $1\mathbf{v}_2$, $\mathbf{v}_1 1$, and so on; thus, a complete set of tagged or bath (not product) functions appears when the bath or tagged velocity enters as "1." Using the fact that G_{LB} is, to an excellent approximation, diagonal on $\mathbf{v}_1$, which matrix element further gives C_{LB}, we have

$$\begin{aligned} C_{\mathbf{k}}(z) &= C_{\mathbf{k},LB}(z) + \rho C_{\mathbf{k},LB}(z) \Big[\int d\mathbf{k}' (\mathbf{v}_1 \cdot | \bar{T}_{\mathbf{k}'} | f_j) \\ &\times (f_j | [z + i(\mathbf{k}+\mathbf{k}')\cdot\mathbf{v}_1 - i\mathbf{k}'\cdot\mathbf{v}_2 - \rho\Lambda_{LB}(\mathbf{v}_1) - \rho\Lambda_B(\mathbf{v}_2)]^{-1} | f_l) \\ &\times (f_l | T_{-\mathbf{k}'} | \mathbf{v}_1) \Big] C_{\mathbf{k},LB}(z) \end{aligned} \tag{2.20}$$

The sum of matrix elements must now be terminated. While it is not obvious *a priori* which f's are important for all k, z, it is clear that at small k, z the hydrodynamic modes of the intermediate propagator—11, $1\mathbf{v}_2$, $1\mathbf{v}_2^2$—will be very important. For these special modes, the matrix elements of the intermediate propagator have the form

$$\left(f\middle|[\ ]^{-1}\middle|f\right) \sim \left[z+|\mathbf{k}+\mathbf{k}'|^2\lambda_T + k'^2\lambda_B\right]^{-1} \tag{2.21}$$

where λ_T (λ_B) is a transport coefficient for tagged (bath) particles. For small k, z, this inverse will be very large in the small-k' part of the k' integral, and the hydrodynamic modes might be expected to dominate the small-z, long-time behavior of $C(t)$. Dorfman and Cohen[8] and Dufty[9] found that the most important bath-particle modes at long times are the "shear modes" associated with shear momentum-density flow and characterized by the transport coefficient η:

$$\mathbf{s}(\mathbf{k}) = (1-\hat{\mathbf{k}}\hat{\mathbf{k}})\cdot\mathbf{v}_2. \tag{2.22}$$

The only tagged-particle hydrodynamic mode is 1, with transport coefficient D, so

$$C_{\mathbf{k}}(z) \cong C_{\mathbf{k},\mathrm{LB}}(z)+C_{\mathbf{k},\mathrm{LB}}(z)\rho\left[\int d\mathbf{k}'(\mathbf{v}_1\cdot\bar{T}_{\mathbf{k}'}\mathbf{s}_{\mathbf{k}'}) \times(\mathbf{s}_{-\mathbf{k}'}T_{-\mathbf{k}'}\cdot\mathbf{v}_1)\left[z+|k+k'|^2D+k'^2\eta\right]^{-1}\right]C_{\mathbf{k},\mathrm{LB}}(z) \tag{2.23}$$

The long time behavior of $C(t)$ is obtained from the small-z dependence of Eq. (2.23). Since

$$C_{\mathbf{k},\mathrm{LB}}(z) \cong \left[z+\rho(\mathbf{v}_1\cdot\Lambda_{\mathrm{LB}}\mathbf{v}_1)(\mathbf{v}_1\cdot\mathbf{v}_1)^{-1}\right]^{-1}$$

this quantity may be treated as a constant. The behavior of the integral at small z comes from the small-k' regime, which further depends only on the small-k' form of (v_1Ts) (k'-independent). Evaluating the matrix elements, doing the integral, and Laplace-inverting, one finally obtains

$$\lim_{t\to\infty} C(t) = \alpha^{(d)}t^{-d/2} \tag{2.24}$$

where

$$\begin{aligned}\alpha^{(2)} &= \left[4\pi\rho\left(D+\frac{\eta}{\rho m}\right)\right]^{-1}\\ \alpha^{(3)} &= \frac{1}{4\rho}\left[\pi\left(D+\frac{\eta}{\rho m}\right)\right]^{-3/2}\end{aligned} \tag{2.25}$$

In a ring theory, D and η, the shear viscosity, must be calculated in the low-density Boltzmann approximation. Dorfman and Cohen actually went beyond the ring approximation and obtained a theory which assigns Enskog values to these quantities, and is in good agreement with the simulation at low to moderate density. Obviously, then, the use of Boltzmann expressions for D and η is accurate at low density. It is interesting to note that if η in Eq. (2.25) were assigned its true value and D neglected, the tail would be identical to that found via a simple hydrodynamic argument by[11] Alder and Wainwright.

Thus, the ring approximation, resulting from resummation of the most divergent terms in the density expansion, also produces correct non-LB long-time dynamics. It was clear that the approximation contained some of the essential features of the sought-after microscopic theory of diffusion. Less obvious was what might be missing. Since the tail was found to be given accurately even at liquid density by the expression obtained by substituting true transport coefficients, perhaps a ring operator with "true" tagged and bath particle kinetic operators in the intermediate propagator was required. But could this be the whole story?

A particular problem involved the inclusion of equilibrium structural information. As mentioned earlier, in the Enskog LB theory, the dynamical events retained are still uncorrelated binary collisions, but the collision rate is modified through the equilibrium tagged-particle–bath-particle pair distribution function; an early attempt to incorporate this effect into a ring theory was in fact made in the first calculations of the "tails," as just discussed. The original ring operator, on the other hand, contains recollisions, but was derived with neglect of all equilibrium structure. Derivation of a ring operator, with consistent inclusion of fluid structure, has been a difficult problem, and is obviously important at high or moderate density; as many "ring operators" will exist as there are schemes for incorporating the structure.

An alternative to pursuit of the high-density ring operator is to consider a model for the bath where structure is unimportant. Such a model, the Lorentz gas, occupies a prominent place in kinetic theory. A point particle moves in an array of fixed, randomly placed smooth spheres. The scatterers are allowed to overlap—they do not see each other—and thus have no equilibrium structure at all. Despite this simplicity, the dynamics of the Lorentz gas is, interestingly, not unlike that of a real gas. Using the argument just sketched, Ernst and Weyland showed[12] that $C(t)$ should have negative $t^{-(d/2+1)}$ "tails." With a computer simulation, Bruin[13] and Alder and Alley[14] showed that $C(t)$ became negative after a few collisions, again as for the real gas. It seemed reasonable, then, to test the ring approximation with the Lorentz gas. If the original ring approximation could explain the non-LB behavior of the Lorentz gas, it might follow that a ring theory, with equilibrium structure included, would accurately describe a real gas.

Weyland[15] carried out calculations of $C(t)$ for the Lorentz gas. This work is significant in several respects. First, he was able to avoid the cumbersome basis-set expansions endemic to kinetic theory. Second, he also considered the "superring"—referred to here as a repeated-ring operator—in which chains of rings are included; this operator, as will be seen later on, is essential to the attainment of our goals. Third, he brought up the point that, at high enough density, D must vanish because the overlapping scatterers divide free space into finite islands; $D = 0$ at all densities in one dimension. Thus, another key test of a high-density kinetic theory is prediction of localization of the particle in the Lorentz gas. Weyland showed that the repeated-ring approximation (RRA) predicted $D = 0$ at all densities for $d = 1$, while the ring approximation (RA) did not; this was probably the first demonstration of the importance of the repeated rings. He also found $C(t)$ to be negative in the RA at intermediate times in $d = 2, 3$, with the importance of the negative region increasing with time, in good qualitative agreement with the simulation. This work strengthened the case for believing that rings were essential at high density and suggested that repeated rings should also be kept; it left open the question of what additional collisions were important, even in the absence of static structure.

Returning now to the inclusion of structure for a real gas, a careful attack on this problem was made[16] by Dorfman and Cohen (DC), who worked out the density expansion of K in a way which explicitly demonstrated both the density dependence due to "pure" dynamics and that due to equilibrium structure. They performed an extended version of the Kawaski–Oppenheim resummation, in which equilibrium structure was included consistently to each order in the density. They also summed the terms which cause the Enskog correction to LB theory. In their theory, an "Enskog ring" theory, LB operators are everywhere replaced by Enskog operators. Thus,

$$R^{\mathrm{DC}}_{\mathbf{k},\mathbf{k}'}(\mathbf{v}_1,\mathbf{v}_2,z) = (g^{+}_{12})^2 \int d\mathbf{v}_2\, \varphi_0 \overline{T}_{\mathbf{k}'}(\mathbf{v}_1,\mathbf{v}_2)\, G^{\mathrm{DC}}_{\mathbf{k},\mathbf{k}'}(\mathbf{v}_1,\mathbf{v}_2,z)\, T_{-\mathbf{k}'}(\mathbf{v}_1,\mathbf{v}_2)$$

$$G^{\mathrm{DC}}_{\mathbf{k},\mathbf{k}'}(\mathbf{v}_1,\mathbf{v}_2,z) = \big[z + i(\mathbf{k}+\mathbf{k}')\cdot\mathbf{v}_1 - i\mathbf{k}'\cdot\mathbf{v}_2 - \rho g^{+}_{12}\Lambda_{\mathrm{LB}}(\mathbf{v}_1) - \rho g^{+}_{22}\Lambda_{\mathrm{B},-\mathbf{k}'}(\mathbf{v}_2) - \rho A_{-\mathbf{k}'}(\mathbf{v}_2)\big]^{-1} \tag{2.26}$$

where $\Lambda_{\mathrm{B},\mathbf{k}}$ is a Boltzmann operator that takes the finite size of bath particles into account,

$$\Lambda_{\mathrm{B},\mathbf{k}} = \int d\mathbf{v}_3\, \varphi_0(v_3)\big[T_0(\mathbf{v}_2,\mathbf{v}_3) + T_{\mathbf{k}}(\mathbf{v}_2,\mathbf{v}_3) P_{23}\big] \tag{2.27}$$

particle 3 is a bath particle, and P_{23} exchanges the labels 2 and 3. The tagged–bath and bath–bath radial distribution functions at contact are denoted g_{12}^{+} and g_{22}^{+}, respectively, and A is the "mean field" operator which enters the Enskog theory of the bath. Dorfman and Cohen found that use of a mixed representation of the dynamics via T and $\overline{T}$ allowed a compact representation of source collision sequences which would be an infinite sum of terms otherwise. All this is related to the treatment of excluded volume (i.e., structure) and is a crucial feature of the higher-density ring operator. With this careful Enskog-ring theory, Dorfman and Cohen verified their earlier result that Enskog transport coefficients entered the coefficients of the tails at moderate density, in accord with the idea that the true tails are simply obtained by using the true transport coefficients.

A different approach to the derivation of a high-density kinetic operator was given[17] by Mazenko, based upon a truncation scheme for the BBGKY hierarchy written for phase-space correlation functions. He derived a ring operator which differed from that of DC via the presence of one, as opposed to two, factors of g_{12}^{+}. This, by itself, would give a tail coefficient $\alpha_{\mathrm{M}} = (g_{12}^{+})^{-1}\alpha_{\mathrm{DC}}$, based upon the approximation, Eq. (2.19), of expanding the ring propagator about the LB propagator and stopping with the term containing a simple ring operator. However, Mazenko, following the original tail calculation of Dufty, tried to avoid this approximation and therefore finally concluded

$$\alpha_{\mathrm{M}} = \frac{D}{D_E}(g_{12}^{+})^{-1}\alpha_{\mathrm{DC}} \tag{2.28}$$

where D is the full ring diffusion constant and D_E is the Enskog diffusion constant. The reason for the difference was unclear. Resibois and Lebowitz[18] gave another hierarchy-truncation scheme. They obtained an Enskog-ring theory closely related to that of Mazenko, and retained some nonring three-body collisions in an attempt to obtain the exact t^2 term in the time expansion of $C(t)$. Their prediction for the tail coefficient α was in accord with that of Mazenko. Résibois,[19] Futardo, Yip, and Mazenko,[20] and Mehaffey, Desai and Kapral[21] used the modified ring operators to obtain $C(t)$ and D for all times and densities, the goal being reproduction of the Alder–Gass–Wainwright[2] (AGW) simulation at moderate or (better) liquid densities. The numerical analysis begins as in the tail calculation, but the approximation in Eq. (2.19) is avoided by writing, with an assumption of the diagonality of the full kinetic operator on $\mathbf{v}$,

$$C(t) = \left[\left(\mathbf{v}_1\cdot\middle| z - \rho\Lambda_{\mathrm{LB}}(\mathbf{v}_1) - R(\mathbf{v}_1, z)\middle|\mathbf{v}_1\right)\right]^{-1}\langle\mathbf{v}_1\cdot\mathbf{v}_1\rangle^2 \tag{2.29}$$

with the matrix element of R to be evaluated in the same way in each case.

Evaluation of the matrix element was, of course, the key problem. A truncation must be made of the infinite set of tagged-bath particle basis functions displayed in Eq. (2.19). Obviously, the hydrodynamic functions must be kept for correct long-time behavior, but other basis functions might be necessary at shorter times. One reason to expect this is as follows. It is rigorously true that $\bar{T}G^{(12)0}T = 0$, where $G^{(12)0}$ is the tagged-bath free-streaming propagator. Since $G^{(12)} \to G^{(12)0}$ as $t \to 0$, $\bar{T}G^{(12)}T$ must vanish as $t \to 0$. This property is not preserved, however, by the hydrodynamic approximation to $G^{(12)}$.

In the approach[17] of Mazenko, this difficulty is avoided by replacing $G^{(12)}$ with $G^{(12)} - G^{(12)0}$. Formally, only zero is being subtracted, but the new term will contribute when only finite numbers of basis functions are kept. Mazenko further introduced a projection-operator method in which the basis functions are divided into hydrodynamic and nonhydrodynamic, and the effect of the nonhydrodynamic ones, in a calculation where only the hydrodynamic ones are considered *explicitly*, enters via a memory function. Both FYM[20] and MDK[21] employed these methods to obtain $C(t)$ and D for a full range of times and densities; the latter authors considered the more complicated case of rough spheres. Although they both state that they use Mazenko's ring operator, they are referring[22] to a later version in which the missing g_{12}^{+} just discussed has been found, and so it may also be said that type use the Dorfman–Cohen ring operator, which is now believed to have the correct number of g_{12}^{+}'s. Their results were in qualitative agreement with the simulations for $C(t)$ and $D(\rho)$. The predicted amplitudes of the negative portion of $C(t)$ were too large. Résibois[19] also used the equation[18] due to him and Lebowitz to calculate $C(t)$. He did not try to treat nonhydrodynamic effects, so his calculations are less sophisticated mathematically than those of the other authors; but again, the results were qualitatively reasonable.

Several important conclusions may be drawn from the numerical work. The RA allows a qualitative explanation of the AGW simulation. The initial enhancement of D/D_{LB} at low density is associated with the positive tail on $C(t)$, which arises, in the basis-set approach, from the hydrodynamic shear mode. The sharp decrease of D/D_{LB} at liquid density is due to the intermediate-time negativity of $C(t)$, which in turn arises from the density mode.

An attractive physical picture of diffusion may be constructed from these facts. At short times, if the density is high enough, the particle rattles around in a cage of neighbors. This causes reversal of the original velocity [negative $C(t)$]. Physically, the cage is a density fluctuation; mathematically, it is a consequence of the density mode (basis function). At long times, the motion is dominated by the reaction back on the particle of the slowly varying shear velocity field stirred up by the particle; this tends to preserve the original velocity.

The question of what important collisions might be missing remained, however. Furthermore, it was unclear that the numerical methods just described were giving the correct $C(t)$ for a *given* RA. The quantity $\bar{T}(G^{(12)} - G^{(12)0})T$ was very sensitive to the approximation used for the nonhydrodynamic basis functions at short times. Since the approximations considered were crude, the resulting $C(t)$'s could not really be trusted.

During all this work, little attention had been paid to another essential test of a complete theory of diffusion—successful prediction of the size dependence for a given density. It is easily seen, however, that consideration of size dependence could be more instructive then testing the theory against the AGW simulation. Even in a low-density gas, obeying the Boltzmann equation, a massive tagged particle large compared to the mean free path must obey the SE law. This provides a test which does not involve the uncertainties of static structure. And this test is very important. It was a great surprise that $D = D_{SE}$ for a particle of a pure liquid. It would be an even greater surprise if this result could be reproduced by a theory which could not produce the SE law under conditions where it is *supposed* to hold.

It was shown[23] by one of us that the RA not only fails the SE "test," but it becomes unphysical as R is increased. At liquid density, the RA is highly unstable with respect to changes in R, and any good prediction of D must be a fluke; that is, if $D(R)$ is obtained accurately, D for a slightly different R will be given very badly. The reason for this is very simple. In Eq. (2.29), the LB term in the inverse is $\propto R^2$. For a large massive particle, the ring term is positive and $\propto R^3/\eta$. Thus, $D \propto (R^2 - R^3/\eta)^{-1}$, and is ill behaved at large R. The RA is *not acceptable* as the basis of a comprehensive microscopic theory of self-diffusion.

On the other hand, the RA has been seen to have many necessary features. What operator retains these while correcting the size-dependence difficulties? The answer to this question was contained in the work of Dorfman and several coworkers. Ernst and Dorfman, in a paper[24] on the kinetic theory of collective motion, introduced the repeated-ring operator RR. We will discuss their work as if it had been for tagged-particle motion, as this extension is trivial.

In the absence of static structure, RR is obtained from R by adding to $\bar{T}$ to the denominator of $G^{(12)}(12)$ (in real space, not to the Fourier transform). Physically, a repeated ring is a string of rings; similarly, it seems natural, in manipulating RR, to expand $G^{(12)}$ in a series containing increasing numbers of $\bar{T}$'s. If the tagged particle propagator is then expanded about G_{LB}, terms proliferate; each term in that expansion is itself an infinite sum.

In the language of Dorfman et al., multiple R's appearing due to expansion of the tagged-particle propagator are "iterated rings," while the repeated rings arise from the expansion of RR. Making any progress with the joint sum seems a formidable task. A further complication is that, upon Fourier transformation of the RR sum, multiple wavevector integrals result. We have

$$\begin{aligned}\mathrm{RR}_{\mathbf{k}}(\mathbf{v}_1, z) = \int d\mathbf{v}_2 \varphi_0(v_2) \\ \times \Big\{ \int d\mathbf{k}' \, \overline{T}_{\mathbf{k}'} G^{R(12)}_{\mathbf{k},\mathbf{k}'} T_{-\mathbf{k}'} + \iint d\mathbf{k}' \, d\mathbf{k}'' \, \overline{T}_{\mathbf{k}'} G^{R(12)}_{\mathbf{k},\mathbf{k}'} \overline{T}_{\mathbf{k}''-\mathbf{k}'} G^{R(12)}_{\mathbf{k},\mathbf{k}''} T_{-\mathbf{k}''} \\ + \iiint d\mathbf{k}' \, d\mathbf{k}'' \, d\mathbf{k}''' \, \overline{T}_{\mathbf{k}'} G^{R(12)}_{\mathbf{k},\mathbf{k}'} \overline{T}_{\mathbf{k}''-\mathbf{k}'} \\ \times G^{R(12)}_{\mathbf{k},\mathbf{k}''} \overline{T}_{\mathbf{k}'''-\mathbf{k}'''} G^{R(12)}_{\mathbf{k},\mathbf{k}'''} T_{-\mathbf{k}'''} + \cdots \Big\} \qquad (2.30)\end{aligned}$$

where $G^{R(12)}_{\mathbf{k},\mathbf{k}'}$ is the ring tagged–bath propagator (for the k dependence), and the first term on the RHS is $R_{\mathbf{k}}$.

Ernst and Dorfman[24] carried out an approximate summation of the series by cutting off the wavevector integrals at an upper limit, assuming that all the T's could then be replaced by $T_{k=0}$, and by inserting the hydrodynamic tagged–bath particle velocity basis functions. This was sufficient for their calculation, which focused on the hydrodynamic, small-k regime. Their results, translated to the tagged-particle case, give the same answer as would be found by using a single iteration of the RA; that is, the repeated rings and iterated rings cancel. Within the RA, iterating once is actually superior to a more complete analysis. The original paper of Ernst and Dorfman does not show how to perform the summation when the k dependence of the T's is included, which is essential for evaluation of $C(t)$ or D. This problem was addressed[25,26] in the work of Dorfman, van Beijern, and McClure, who studied the drag on a fixed sphere in a flowing Boltzmann gas via kinetic theory. At the microscopic level, the gas and sphere interact via a $\overline{T}$, and the drag is expressed as an infinite series involving more and more T's; this series is equivalent to that generated by expanding the repeated-ring operator. Projecting onto the hydrodynamic modes of the gas, but with no use of a cutoff wavevectors, these authors obtained[25] Stokes's law with a rather mysterious 5π replacing the usual coefficient 4π appropriate for slip boundary conditions (expected for smooth spheres). Mehaffey and Cukier[27] proposed an Enskog RRA, which adds $g^+_{12}\overline{T}(12)$ to the inverse of DC's Enskog ring. Applying the same methods used in the drag calculation to the repeated-ring sum, they obtained

$$D = D_E + \frac{k_B T}{5\pi \eta_E \sigma_{12}} \tag{2.31}$$

where η_E is the Enskog shear viscosity; since $\sigma_{12} \to R$ and $D_E \propto R^{-2}$ for large R, the RRA is thus seen, except for the 5π, to have the correct large-R limit; it is strongly indicated that some version of the RRA could be the kinetic equation we are seeking. Mehaffey and Cukier went on[28,29] to attempt to justify their proposed Enskog RRA, based upon the approach of Mazenko, and to calculate $D(\rho)$ for the hard-sphere liquid; their results agreed qualitatively with the AGW simulation.

The question of the 5π remained; actually the answer to this puzzle was already apparent from the work[26] of Dorfman, van Beijern, and McClure, who had shown how to obtain 4π for the drag. The key point involved the expression used for the hydrodynamic modes. It had almost always been stated that the shear modes are $\mathbf{s} = (\mathbf{1} - \hat{\mathbf{k}}\hat{\mathbf{k}})\cdot\mathbf{v}_2$, that is, the matrix elements $(sG_{\text{LB}}s)$ vary in time as $\exp(-k^2\eta_{\text{LB}}t)$. This statement is not true, however, for to be consistent with a hydrodynamic decay rate $\propto k^2$, s should be considered to $O(k)$. This makes no difference in calculations of the long-time tail, but van Beijern and Dorfman showed that s must be kept to $O(k)$ in order to obtain the correct drag; evaluation of the sum for the drag was greatly complicated by this fact, as is the repeated-ring sum. Eventually, Cukier et al.[30] and Mercer and Keyes[31 32] showed how to apply van Beijern and Dorfman's summation to self-diffusion in a low density (not Enskog) RRA, and thus to obtain a kinetic theory with the correct large-R limit.

At this point, an Enskog RRA seemed the kinetic theory of choice, but much work still had to be done. It was still not known how to get 4π in the SE law in an Enskog RRA. While one Enskog RRA had been given, its validity was not established. Quantitative agreement with the AGW simulation had not been achieved, nor had the theory been tested for predictions of the localization phenomenon in the Lorentz gas. Finally, accurate, reliable methods for calculating $C(t)$ and $D(\rho)$ from the equations needed to be developed. Research on these problems is discussed in the next section.

B. Recent Work

Consider now the problem of obtaining accurate expressions for $D(\rho)$ and $C(t)$ given some RA or RRA. The basis-set methods are not totally satisfactory, since their predictions for $C(t)$ at short times depend sensitively upon the nonhydrodynamic basis functions, which must be treated with approximations of unknown validity. This point is discussed[10] in a paper by one of us (T.K.) and Mercer, where we suggest some improvements and use the new techniques to carry out a fairly successful calculation of the Lorentz gas. More recently,[33] we (A.M. and T.K.) have given a variational method for solving the RRA, which eliminates the use of basis functions altogether and seems very powerful.

We start with the RRA (low-density form) formulated as two coupled equations in real (not Fourier) space,

$$[z - \rho\Lambda_{\text{LB}}(\mathbf{v}_1)]\,\boldsymbol{\phi}_D(\mathbf{v}_1, z) = \mathbf{v}_1\varphi_0(v_1)\langle \mathbf{v}_1\cdot\mathbf{v}_1\rangle^{-1} + \rho\int d\mathbf{r}_{12}\,d\mathbf{v}_2\,\varphi_0(v_2)\bar{T}(12)\boldsymbol{\theta}'(\mathbf{v}_1,\mathbf{v}_2,\mathbf{r}_{12}, z) \tag{2.32a}$$

$$[z + \mathbf{v}_1\cdot\nabla_1 + \mathbf{v}_2\cdot\nabla_2 - \rho\Lambda_{\text{LB}}(\mathbf{v}_1) - \rho\Lambda_{\text{B}}(\mathbf{v}_2)]\,\boldsymbol{\theta}'(\mathbf{v}_1,\mathbf{v}_2,\mathbf{r}_{12}, z) = \bar{T}(12)\boldsymbol{\theta}'(\mathbf{v}_1,\mathbf{v}_2,\mathbf{r}_{12}, z) + T(12)\boldsymbol{\phi}_D(\mathbf{v}_1, z) \tag{2.32b}$$

where

$$\int d\mathbf{v}_1\,\varphi_0(v_1)\mathbf{v}_1\cdot\boldsymbol{\phi}_D(\mathbf{v}_1, z) = C_{k=0}(z) \tag{2.33}$$

Our method can be used to get C_k at finite k by considering $\phi(v_1, r_1, z)$, but this generalization will not be discussed here. Formal solution of Eq. (2.32b) for $\boldsymbol{\theta}'(\boldsymbol{\phi})$, substitution into Eq. (2.32a), introduction of Fourier representations, and use of Eq. (2.33) shows that these equations really are equivalent to the RRA as presented in the last section.

The LHS of the second RRA equation describes the motion of $\boldsymbol{\theta}'$ via independent Lorentz–Boltzmann and Boltzmann equations for $\mathbf{v}_1$ and $\mathbf{v}_2$, respectively. The RHS is confined, because of the T, to $\delta(r_{12} - \sigma)$, that is, it is confined to the "collision sphere." The function $\boldsymbol{\theta}'$ is expected to vanish inside the collision sphere, so the $\mathbf{v}\cdot\nabla$ operators on the LHS will also produce surface delta functions. Once these are isolated, the equation may[26] be divided into smooth and delta-function parts, which are equated separately, due to the completely different spatial dependence. The equation for the δ's gives boundary conditions on the equation for the smooth parts, which, as was just mentioned, are Boltzmann-like; no rings or the like are involved.

To be specific, we write[26]

$$\boldsymbol{\theta}' = \boldsymbol{\theta} W(|r_{12}|) \tag{2.34}$$

where

$$W(x) = \begin{cases} 0, & x < \sigma_{12} \\ 1, & x \geq \sigma_{12} \end{cases}$$

and $\boldsymbol{\theta}$ is assumed to be smoothly varying. Proceeding as described, we find

$$W(z + \mathbf{v}_1\cdot\nabla_1 + \mathbf{v}_2\cdot\nabla_2 - \rho\Lambda_{\text{LB}}(\mathbf{v}_1) - \rho\Lambda_{\text{B}}(\mathbf{v}_2))\boldsymbol{\theta} = 0 \tag{2.35a}$$

$$T(\boldsymbol{\theta} + \boldsymbol{\phi}_D) = 0 \tag{2.35b}$$

The result of acting on W with ∇ combines with $\bar{T}\boldsymbol{\theta}$ to produce $T\boldsymbol{\theta}$. Equation (2.35b) is an inhomogeneous boundary condition on θ at $r_{12}=\sigma_{12}$, with $\boldsymbol{\phi}_D$ determining the inhomogeneity. The above approach is based upon that used by van Beijern and Dorfman in the drag calculation.

The point of these manipulations is that we may now take advantage of a large body of knowledge about solving Boltzmann equations in the presence of boundaries. We chose a variational method due to Cercignani et al.,[34] which requires first making[35] a BGK approximation on the Boltzmann operators. We are now pursuing other methods for more complicated RRAs, but only the Cercignani method will be discussed here.

The basic ideas may be illustrated via the Lorentz gas, for which everything depending on $\mathbf{v}_2$ is removed from Eqs. (2.32)–(2.35). The BGK form of Λ_{LB} is

$$\rho\Lambda_{\mathrm{LB}}(\mathbf{v}_1)G(\mathbf{v}_1) = -\nu\left[G(\mathbf{v}_1)-\varphi_0(v_1)\int d\mathbf{v}\,G(\mathbf{v})\right] \tag{2.36}$$

where ν is the friction constant, $\nu=\rho\pi a^2 v_0$. The magnitude of the tagged-particle velocity, v_0, is conserved in the Lorentz gas, and φ_0 is not a Maxwellian, but

$$\varphi_0(\mathbf{v}_1)=\frac{\delta(|\mathbf{v}_1|-v_0)}{4\pi v_0^2} \tag{2.37}$$

Usually, expressions like Eq. (2.36) and its well-known analogs are approximate, but the BGK "approximation" is in fact exact for the Lorentz gas when $d=3$. It is also true, because $\boldsymbol{\phi}_D$ is a vector and $v=v_0$, that we may write

$$\boldsymbol{\phi}_D=\mathbf{v}_1\varphi_0(\mathbf{v}_1)B(z) \tag{2.38}$$

Using Eq. (2.36), Eq. (2.35a) becomes

$$(z+\mathbf{v}_1\cdot\nabla_1+\nu)\boldsymbol{\theta}(\mathbf{v}_1,\mathbf{r}_{12},z)=\nu\varphi_0(\mathbf{v}_1)\mathbf{n}(\mathbf{r}_{12},z) \tag{2.39}$$

where $\mathbf{n}$ is the moment,

$$\mathbf{n}=\int d\mathbf{v}_1\,\boldsymbol{\theta} \tag{2.40}$$

Integration of Eq. (2.39) along characteristic paths which lie along a line in the $\mathbf{v}$ direction for a given $\mathbf{v}$, with incorporation of the boundary condition and Eq. (2.38), yields $\boldsymbol{\theta}$ in terms of $\mathbf{n}$ and B, which is converted into a closed equation for $\mathbf{n}$. Formal solution gives $\mathbf{n}(B)$, which gives $\boldsymbol{\theta}(B)$, which is substituted into the first RR equation. Thus, $C(t)$ is finally calculated from Eq.

(2.33), and the result is[33]

$$C(z)=\left(z+\nu-\frac{\nu\rho^*[\mathbf{g},\mathbf{S}]}{4\pi^2}\right)^{-1} \tag{2.41}$$

where [,] denotes a dot product and a spatial integral over the volume outside the collision sphere, **g** is the moment **n** in units of $v_0 B(\omega)/2\pi$, **S** is a "source" in the equation obeyed by **g**,

$$\mathbf{g}(\mathbf{x},z)=\hat{A}\mathbf{g}(\mathbf{x},z)-\mathbf{S}(\mathbf{x},z) \tag{2.42}$$

and $\hat{A}$ is an integral operator; $\hat{A}$ and **S** are given in Ref. 33.

Equation (2.41) is a formal solution of the RRA–BGK equations, in which the burden of solution has been shifted to obtaining the function **g**. Since **n** is the zeroth moment of $\boldsymbol{\theta}$, it is a number density, and **g** is a number density, in the presence of a spherical surface source (inhomogeneous boundary condition), measured in units of the strength of the source. Of course, **g** is a vector, while densities are normally scalars. The vector nature of **g** just arises from the fact that the source ($\propto \hat{r}$) is a vector; each component of **g** is an ordinary density, determined by that component of the source. It is significant that **g** does not depend on **v**, so it is much simpler than $\boldsymbol{\theta}$. Furthermore, with Eq. (2.42), it is easy to show that [**g**, **S**] is the stationary value of a functional

$$J(\tilde{\mathbf{g}})=[\tilde{\mathbf{g}},\tilde{\mathbf{g}}-\hat{A}\tilde{\mathbf{g}}+2\mathbf{S}] \tag{2.43}$$

Thus,

$$C(z)=\left(z+\nu-\frac{\nu\rho^*}{4\pi^2}\operatorname{Stat}J\right)^{-1} \tag{2.44}$$

and the velocity correlation is to be found by choosing trial functions $\tilde{\mathbf{g}}$ and varying them until J becomes stationary.

If the vector density obeys a diffusion equation, then the steady-state solution with a point source $\propto \hat{r}$ at the origin is $\propto \hat{r}/r^2$. Thus, a simple hydrodynamic trial function at $z=0$ is $\mathbf{g}(\mathbf{r}, z=0)=\alpha\hat{r}/r^2$, with α to be varied; hydrodynamic solutions from which **g** may be guessed at finite z are obtained by solving the frequency-dependent diffusion equation. If only D is required, only $z=0$ need be considered. With these trial functions, we calculated $C(t)$ and D at low to moderate density. At low density, agreement with the simulations is very good for the relevant difference, $C(t)-C_{\mathrm{LB}}(t)$, but at moderate density it is worse. As will be discussed shortly, the RRA is expected to break down at moderate density in the LG. The key point here, however, is that we believe that our results are the true RRA results—that the numerical analysis is accurate. Use of more flexible trial functions led to changes of D/D_{LB} in the third decimal place only. It seems that the variational method avoids the aforementioned difficulties in basis-set expansions.

In the limit of high density, the Lorentz-gas RRA yields

$$D = \frac{3D_{\mathrm{LB}}}{2\pi\rho^*} \tag{2.45}$$

which is incorrect; for $\rho > \rho_c$, the percolation density, D must vanish identically. Thus, the RRA is inadequate and a better kinetic operator must be found. The nature of the required operator is suggested by the mode-coupling work[36] of Goetze et al. These authors have shown that "localization" may be obtained by "self-consistent mode coupling," in which the correlation functions on the RHS of the equations are themselves given consistently with the mode-coupling theory. This approach is discussed at some length in the following sections. In terms of kinetic theory, it may be regarded as a "self-consistent ring" theory. In the RA, propagation between recollisions is treated as Boltzmann-like; at high density this is surely a poor approximation, so in the self-consistent RA (SCRA) this propagation is described by the "true" tagged-particle kinetic operator. In order to solve these equations numerically, Goetze et al. employed a "super-BGK" approximation for this true operator—an approximation of unknown validity, but one for which several plausibility arguments can[37] be given. Repeated-ring collisions were not included in this theory, however, so we[37] investigated a self-consistent repeated-ring approximation (SCRRA), obtained by adding a T to Goetze's SCRA operator. Our theory predicts $D = 0$ for $\rho > \rho_c$, with $\rho_c^* = 0$, π^{-1}, and $3/2\pi$ for $d = 1$–3. Agreement with the best computer simulation available, that of Alder and Alley[14] for $d = 2$, is extremely good. Thus, keeping the warnings about the super-BGK in mind, we may say that the SCRRA gives a good theory of localization.

The need for a self-consistent theory did not show up when we studied the real gas. This is because there, the propagation between recollisions involves both tagged- and bath-particle motion, with the dominant effects usually coming from the bath, so self-consistency in the tagged-particle motion is irrelevant. However, a real gas should start to act like a Lorentz gas, from the viewpoint of the tagged particle, as the bath-particle mass becomes very large. We may thus conclude that the Enskog RRA, which we have suggested as the desired-tagged particle kinetic equation, must be made self-consistent if it is to describe the motion of light particles.

Even if the gas is left dilute, it is clear that the SCRRA must show a rich mass dependence, in addition to the size dependence already discussed for a massive particle. We[38] set up the variational principle for a real gas SCRRA in order to study the combined size and mass dependence of D. For a given tagged-particle/bath-particle mass ratio, we calculated D/D_{LB} as a function of R_1. As expected, D/D_{LB} smoothly increases from unity to $D_{SE}/D_{LB} \propto R_1$ when R_1 exceeds the mean free path for a massive particle. For mass ratios less than unity, however, D/D_{LB} decreases sharply, reaches a minimum, and then increases again. The initial decrease is apparently associated with an incomplete version of the localization phenomena. Cages may be present instantaneously, but they break up due to the motion of the gas particles, and D does not vanish; the larger the particle, the easier it is to cage. Even larger light particles diffuse as rigid bubbles, carried along by velocity fluctuations in the gas, and the D associated with this motion must decrease more slowly with increasing R_1 than D_{LB} does. The basic physical picture of the size and mass dependence which emerges is very simple. A massive particle pushes the bath particles out of the way, with a transition from D_{LB} to D_{SE} at $R_1 \cong l$. A light particle cannot push, however, and can be trapped, which becomes more likely the larger the particle. The R_1 dependence of the caging dominates the R_1 dependence of D for light particles.

It would, of course, be desirable to test these predictions. The only data of which we are aware on the size and mass dependence are those[39] of Alder, Alley, and Dymond, who performed computer simulations for four sizes and four masses. They considered a gas density which is too high for use of the ordinary RRA ($\cong \frac{1}{3}$ close packed), but we tried to test our theory against the simulation in the hope of, at least, obtaining the correct trends. The agreement was quite good, considering the density. Most importantly, Alder, Alley, and Dymond (AAD) found D/D_{LB} to increase with R_1 for unit mass ratio (the heaviest tagged particle they studied) and to decrease with increasing R_1 for the three smaller ratios, in qualitative agreement with the physical arguments given, and with the numerical results of our theory. The findings of AAD had not been previously explained.

Returning to the problem of static structure, let us recall that the proposed[28,29] Enskog RRA (ERRA) of Cukier and Mehaffey had not been shown to produce the correct Enskog SE law. The question of finding the correct ERRA therefore remained open. This problem we reexamined by Sung and Dahler,[40] and by ourselves,[41] at about the same time and with very similar methods.

In the original derivation[24] of Ernst and Dorfman, the RRA is derived by discarding the "triply connected" part of the three-body distribution function in the second equation of the BBGKY hierarchy. So the essence of the RRA is neglect of triple correlations. To obtain an Enskog RRA, we must discard static and dynamic triple correlations in a consistent way. One way to do this is via the Mori[42] formalism discussed in Section III. If we chose one- and two-body phase-space densities as variables, then all three-body effects will be in the memory function. Thus, an ERRA may be derived[40,41] by discarding the memory function.

More specifically, the variables are

$$A(\bar{\mathbf{v}}_1) = \delta(\bar{\mathbf{v}}_1 - \mathbf{v}_1) \tag{2.46a}$$

$$B(\bar{\mathbf{v}}_1, \bar{\mathbf{v}}_2, \bar{\mathbf{r}}_{12}) = \delta(\bar{\mathbf{v}}_1 - \mathbf{v}_1)\left[\sum_{i>1} \delta(\bar{\mathbf{v}}_2 - \mathbf{v}_i)\delta(\bar{\mathbf{r}}_{12} - \mathbf{r}_{1i}) - \rho G(\bar{r}_{12})\varphi_0(\mathbf{v}_2)\right] \tag{2.46b}$$

so A is the distribution of tagged-particle velocities, and B is the product of the tagged-particle velocity distribution with the joint distribution of bath-particle velocity and tagged–bath-particle separation, minus its average. Straightforward, if messy, application of Mori's method gives equations for the correlation functions of A and B. We discard the memory function, note that

$$\int d\bar{\mathbf{v}}\bar{\mathbf{v}}\langle A(\bar{1})A(\bar{1}')\rangle = \boldsymbol{\phi}_D(\bar{\mathbf{v}}_1) \tag{2.47}$$

and that the quantity $\boldsymbol{\theta}$ can similarly be expressed in terms of $\langle AB\rangle$; the resulting RR equations are

$$z\boldsymbol{\phi}_D(\mathbf{v}_1, z) - \rho g_{12}^{+}\Lambda_{\mathrm{LB}}(\mathbf{v}_1)\boldsymbol{\phi}_D(\mathbf{v}_1, z) - \rho g_{12}^{+}\int d2\,\varphi_0(v_2)\bar{T}(12)\boldsymbol{\theta}(\mathbf{v}_1, \mathbf{v}_2, \mathbf{r}_{12}, z) = \mathbf{v}_1 \tag{2.48a}$$

$$\begin{aligned}
&zG(r_{12})\boldsymbol{\theta}(12) + z\rho\int d3\,\varphi_0(v_3)[G(123) - G(12)G(13)]\,\boldsymbol{\theta}(13) \\
&\quad - G(12)(\mathbf{v}_1\cdot\nabla_1 + \mathbf{v}_2\cdot\nabla_2)\boldsymbol{\theta}(12) \\
&\quad - \rho\int d3\,\varphi_0(v_3)G(123)T(23)(1 + P_{23})\boldsymbol{\theta}(12) \\
&\quad - \rho\int d3\,\varphi_0(v_3)\boldsymbol{\theta}(13)[G(12)G(13)\nabla_3 W(13) - G(123)\nabla_3 W(123)]\cdot\mathbf{v}_3 \\
&\quad - \rho\int d3\,\varphi_0(v_3)[G(123) - G(12)G(13)]\mathbf{v}_1\nabla_1\cdot\boldsymbol{\theta}(13) \\
&\quad - \rho\int d3\,d4\,\varphi_0(v_3)\varphi_0(v_4)\left[G(1234) - \frac{G(124)G(134)}{G(14)}\right]T(14)\boldsymbol{\theta}(13) \\
&= 0, \qquad r_{12} > \sigma_{12}
\end{aligned} \tag{2.48b}$$

$$\begin{aligned}
&g_{12}^{+}T(12)[\boldsymbol{\phi}_D + \boldsymbol{\theta}(12)] + \rho\int d3\,\varphi_0(v_3)[G(123) - G(12)G(13)]\,T(12)\boldsymbol{\theta}(13) \\
&= 0, \qquad r_{12} = \sigma_{12}.
\end{aligned} \tag{2.48c}$$

The Enskog RRA is extensively discussed in Ref. 41. For large r_{12}, the second equation reduces to an Enskog kinetic equation; for smaller values of r_{12}, the equilibrium tagged–bath-particle structure enters in a complicated way, both via the tagged–bath pair distribution function and via the tagged–bath–bath three-body function. The region where Eq. (2.48b) does not have its asymptotic form constitutes a natural "boundary layer" around the tagged particle. On the other hand, the first ERRA equation differs from its low-density form only by the presence of two g_{12}^{+}'s—one which converts Λ_{LB} into an Enskog operator, and one in the $\boldsymbol{\phi}_D - \boldsymbol{\theta}$ coupling.

These equations are obviously not easy to deal with; however, the long-time and Brownian limits have proved tractable. We have shown[41] that the tail coefficient, $\alpha^{(d)}$ in Eq. (2.24), is just the low-density result, Eq. (2.25), with η_E and D_E appearing. This is in agreement with the conclusion of DC, and provides further support for the—perhaps surprising—idea that the low-density tail coefficient need only be modified at high density via the transport coefficients. Some care is needed in obtaining this result, if one is to avoid spurious factors of g_{12}^{+}.

Both Sung and Dahler[40] (SD) and we[41] have considered the Brownian limit. Using the idea that, in the limit, $\boldsymbol{\theta}$ should be given by its hydrodynamic form, we look for Chapman–Enskog normal solutions of Eq. (2.48b). The associated fluid fields (the hydrodynamic variables) are shown to obey Enskog hydrodynamic equations with slip boundary conditions. SD state that the slip Stokes law then follows, if one calculates the force on the tagged particle with these equations in the presence of externally imposed fluid flow; this is indeed correct. SD do not actually discuss tagged-particle motion, that is, they do not try to substitute the solution of the second ERRA into the first to obtain $C(t)$. Although it might seem obvious from the hydrodynamic behavior of $\boldsymbol{\theta}$ that the ERRA gives $D = D_{SE}$ for a Brownian particle, it is in fact very tricky to obtain the SE law from the RRA given the normal form of $\boldsymbol{\theta}$. This may be immediately seen by noting that any explicit dependence of $\boldsymbol{\phi}_D$ on fluid transport coefficients must come from $\boldsymbol{\theta}$. Now, the Boltzmann normal $\boldsymbol{\theta}$ contains the Boltzmann shear viscosity. The Enskog normal $\boldsymbol{\theta}$, however, *does not* contain η_E, but only some of its "pieces" or components. Direct substitution of this $\boldsymbol{\theta}$ into Eq. (2.48a) gives a SE law with the wrong expression for η_E, and with a spurious g_{12}^{+}. In order to avoid these difficulties, one must note[41] that the normal $\boldsymbol{\theta}$ is only the true $\boldsymbol{\theta}$ outside the boundary layer, while Eq. (2.48a) requires $\boldsymbol{\theta}$ right on the collision sphere. We have shown[41] how to reexpress Eq. (2.48a), in the Brownian limit, in terms of $\boldsymbol{\theta}$ evaluated just outside the boundary layer. Use of the normal θ in this equation yields the correct slip SE law, with 4π and η_E.

Very recently,[43] we showed that for the Brownian and long-time limits, keeping the memory function which was discarded to obtain Eq. (2.48b)—

leading to a formally exact kinetic equation—only changes the transport coefficients in the SE law and tail coefficients to their true values. Thus, at last, we have a kinetic-theory calculation which reproduces the results or ordinary hydrodynamics (the hydrodynamic tail calculations will be discussed shortly) when appropriate. It remains to be seen if the formally exact equation will be tractable outside these limits.

This concludes the kinetic-theory section of the review. It is quite conceivable that a self-consistent Enskog RRA—a self-consistent version of Eqs. (2.48)—will accurately describe self-diffusion in almost any system of interest. The outstanding current problem in this field, in our view, is solving the ERRA for a pure dense fluid, to make comparison with the AGW simulation, and to try to understand *why* the slip SE law seems to work so well here. It must be noted that this fact *cannot* arise simply because the "hydrodynamic limit" for D, mysteriously, persists down to molecular dimensions. The analytic form of the hydrodynamic solution contains $\sigma_{12} = R_1 + a$ in the denominator. For a Brownian particle, $\sigma \cong R_1$, but for a pure fluid, $\sigma = 2R_1$, giving a SE law with 8π. Thus, oddly, some nonhydrodynamic effects must be involved in the apparent accuracy of the slip SE law for atoms. If a simple physical principle is involved here, use of the ERRA may help to decipher that principle.

Of next importance is a more careful treatment of the localization phenomena, especially one that avoids the "super-BGK" approximation. And, of course, this entire discussion has been about smooth hard spheres; more work needs to be done in expanding these methods to include more realistic potentials.

III. MODE COUPLING

A. Background

The basic campaign plan for the kinetic-theory approach was to begin with an exact, low-density theory and then to extend it to cover the cases of larger tagged particles and higher-density fluids. Then in order to regain the SE relation in the Brownian limit, the kinetic operators had to be projected onto the fluid hydrodynamic modes. In this section we shall outline a different approach to the problem of self-diffusion. By means of Mori's generalized Langevin equation,[42] the fluid fluctuations are immediately described in hydrodynamical terms, and the hope is that the SE result will be a natural limit, easy to obtain. Then the theory would have to be extended to deal with smaller particles—in a sense the reverse of the kinetic-theory philosophy.

Let us consider the tagged particle moving in the fluid, be it gas or liquid, and let us take the potentials of interaction to be continuous, central, short-ranged, and possessing a harshly repulsive core. Then the equation of motion of the particle is given by[42]

$$m_1 \frac{d\mathbf{v}_1(t)}{dt} = -\int_0^t ds\, \nu(t-s)\mathbf{v}_1(s) + \mathbf{F}_1(t) \tag{3.1}$$

where

$$\mathbf{F}_1(t) = \exp(Q_1 iLt)\, Q_1 iL\mathbf{v}_1(0) \tag{3.2}$$

and

$$\nu(t) = \left(m_1^2 \langle \mathbf{v}_1 \cdot \mathbf{v}_1 \rangle\right)^{-1} \langle \mathbf{F}_1(t) \cdot \mathbf{F}_1 \rangle \tag{3.3}$$

Here iL is the Liouville operator and Q the Mori projection operator that projects a variable orthogonal to $\mathbf{v}_1(0)$. $\mathbf{F}_1(t)$ is clearly orthogonal to $\mathbf{v}_1(0)$ and is generally called the random force term. If we now form the scalar product of the above equation with $\mathbf{v}_1(0)$, and then take the Laplace transform, we obtain an expression for $C(z)$ in the form

$$C(z) = \frac{\langle \mathbf{v}_1 \cdot \mathbf{v}_1 \rangle}{z + \nu(z)} \tag{3.4}$$

where $\nu(z)$, the Laplace transform of $\nu(t)$, is the frequency-dependent friction coefficient. The problem now is how to calculate it.

Firstly let us consider the Brownian limit. Suppose that at time $t = 0$ the system is displaced slightly away from equilibrium, and consider the non-equilibrium average of $\mathbf{v}_1(t)$, denoted by $\langle \mathbf{v}_1(t) \rangle_{\text{NE}}$. Equation 3.1 then yields

$$m_1 \frac{d}{dt} \langle \mathbf{v}_1(t) \rangle_{\text{NE}} = -\int_0^t ds\, \nu(t-s) \langle \mathbf{v}_1(s) \rangle_{\text{NE}} \tag{3.5}$$

It would thus seem reasonable to identify $\nu(t)$ as the drag coefficient on our particle as it moves through the fluid with a time-dependent velocity $\langle \mathbf{v}_1(t) \rangle_{\text{NE}}$. This may be calculated by using the classical equations of hydrodynamics supplemented by either slip or stick boundary conditions for the fluid flow at the surface of the sphere. The results for $\nu(z)$ for motion in an isothermal, compressible fluid for both types of boundary conditions are given by Zwanzig and Bixon[44] and Metiu et al.[45]

The general results of substituting this form for $\nu(z)$ into Eq. (3.34) are as follows. Setting $z = 0$ leads back to the SE relation. Secondly, for small z, $\nu(z)$ is of the form

$$\nu(z) = \nu + \nu^{(1)} z^{1/2} + \cdots$$

The nonanalytic $z^{1/2}$ term leads to a $t^{-3/2}$ tail in $C(t)$, that is,

$$\lim_{t\to\infty} C(t) = \frac{2k_B T}{\rho m_1}\left(\frac{4\pi\eta t}{\rho m_1}\right)^{-3/2} \tag{3.6}$$

This result was obtained even more simply, via a purely hydrodynamic argument, by Alder and Wainwright,[11] and does not depend on the type of boundary conditions used to obtain $\nu(z)$; we also note that it is independent of both the size and the mass of the tagged particle. Lastly we note that this hydrodynamic calculation incorrectly predicts $C(t)$ to be cusped at $t = 0$. This is essentially because hydrodynamic equations are only valid on a time scale long compared to a molecular collision time, so the very short-time behavior of $C(t)$ is not expected to be accurately reproduced.

As stated in the introduction, this hydrodynamic theory has proved very successful in understanding Brownian motion. More surprisingly, though, this approach has been applied remarkably successfully to the calculation of $C(t)$ for molecular self-diffusion.[44] By using frequency-dependent transport coefficients in the hydrodynamic equations and by taking the effective radius of the tagged particle (an argon atom in this case) to be such that the SE relation gave the correct diffusion constant (this effective radius, for slip boundary conditions, coming out fairly close to the true atomic radius), they were able to use the hydrodynamic form of $\nu(z)$ to reproduce fairly quantitatively the features of the velocity correlation function as calculated by computer simulation. We believe it is fair to say, though, that the theoretical reasons for this success are still very poorly understood, and a true understanding could only come from first-principle, molecular theories, to which we shall now turn our attention.

From Eq. (3.3) we have a microscopic prescription for $\nu(t)$ and hence $\nu(z)$. We thus ought to be able to start from this equation, obtain the hydrodynamic results in the Brownian limit, and perhaps develop an approximate theory for molecular diffusion. Straightaway, however, there would appear to be a problem. Looking at Eq. (3.3), one might have imagined that, as $\nu(t)$ is a force correlation function, it would decay rapidly in time, on a time scale characteristic of interparticle collisions. If this were accepted, then the conclusion would be that $C(t)$ should be the sum of a few exponentially decaying functions, with no power-law long-time dependence. Indeed, until the computer simulations[2] of Alder et al., this was the generally prevailing view.

To put this more formally, the existence of the long-time tail shows that even for a Brownian particle, the random force is not a true "fast" variable, but somehow contains "slow" components. Furthermore, the success of the SE relation and also the purely hydrodynamic explanation of the long-time tail given by Alder and Wainwright[11] suggest that in some way the tagged particle motion must couple to the hydrodynamic modes of the fluid—that is, it must couple to fluctuations of the conserved variables of the fluid, which are the number density, the momentum density, and the energy density. At long wavelengths these fluctuations decay very slowly. The problem then is how to explicitly couple these slow, fluid fluctuations to the tagged-particle variables, as by symmetry they cannot be coupled directly. The solution to this problem, within the framework of the Mori formalism, was given by Kawasaki[46] in his mode-coupling theory. For a full discussion we refer to the original papers and also the review by T. Keyes.[47] Essentially the point is that quantities such as $\mathbf{v}_1$ and $\mathbf{F}_1$ can couple to certain products of tagged-particle and fluid-particle functions. Thus they may couple to $n_{\mathbf{k}}n^1_{-\mathbf{k}}$, $\mathbf{p}_{\mathbf{k}}n^1_{-\mathbf{k}}$, and $e_{\mathbf{k}}n^1_{-\mathbf{k}}$ for all values of the wavevector $\mathbf{k}$, where

$$n_{\mathbf{k}} = \sum_{i=1}^{N} e^{i\mathbf{k}\cdot\mathbf{r}_i} \tag{3.7}$$

$$\mathbf{p}_{\mathbf{k}} = \sum_{i=1}^{N} \mathbf{p}_i e^{i\mathbf{k}\cdot\mathbf{r}_i} \tag{3.8}$$

$$e_{\mathbf{k}} = \sum_{i=1}^{N} \left(\tfrac{1}{2} m_i v_i^2 + \sum_{j>i} U(ij) \right) e^{i\mathbf{k}\cdot\mathbf{r}_i} \tag{3.9}$$

$$n^1_{\mathbf{k}} = e^{i\mathbf{k}\cdot\mathbf{r}_1} \tag{3.10}$$

are the number, momentum, energy, and tagged-particle-number densities respectively. In these definitions $\mathbf{p}_i$ is the momentum of particle i, m_i is the mass of particle i (m for a fluid particle, m_1 for the tagged particle) and $U(ij)$ is the potential of interaction between particles i and j. The products $n_{\mathbf{k}}n^1_{-\mathbf{k}}$ and so on are called bilinear variables, and for small k they decay very slowly. In a similar vein we may construct trilinear variables, and so on, to which the tagged-particle variables may couple.

We now may follow the methods of Ernst et al.[48] and Keyes and Oppenheim[49] to obtain expressions for the long-time behavior of $C(t)$. We project $\mathbf{v}_1$ onto the "slow" bilinear variables, that is, the bilinear variables with $k < k_c$, where k_c is a cutoff wavevector of the order of an inverse molecular radius or inverse particle radius. This yields

$$P_B\mathbf{v}_1 = \sum_{\mathbf{k}_1,\mathbf{k}_2}^{k_c} \left\langle \mathbf{v}_1 \cdot \mathbf{p}_{\mathbf{k}_1} n^1_{-\mathbf{k}_1} \right\rangle \chi^{-1}_{\mathbf{k}_1,\mathbf{k}_2} \mathbf{p}_{\mathbf{k}_2} n^1_{-\mathbf{k}_2} \tag{3.11}$$

where P_B means the projection onto the slow bilinear modes. The wavevector sum is over all wavevectors such that the wavelengths fit into the sample, and $\chi^{-1}_{\mathbf{k}_1,\mathbf{k}_2}$ is the inverse of

$$\chi_{\mathbf{k}_1,\mathbf{k}_2}=\left\langle \mathbf{p}_{\mathbf{k}_1}n^1_{-\mathbf{k}_1}\cdot\mathbf{p}_{\mathbf{k}_2}n^1_{-\mathbf{k}_2}\right\rangle \tag{3.12}$$

We may thus use Eq. (3.11) to obtain an expression for $C(t)$ involving correlation functions, both static and time-dependent, of bilinear variables. In order to make further progress, a Gaussian or factorization approximation is generally made. Thus we write

$$\begin{aligned}
&\left\langle \left(\mathbf{p}_{\mathbf{k}_1}(t)n^1_{-\mathbf{k}_1}(t)\right)\cdot\left(\mathbf{p}_{\mathbf{k}_2}(0)n^1_{-\mathbf{k}_2}(0)\right)\right\rangle \\
&\quad\cong\left\langle \mathbf{p}_{\mathbf{k}_1}(t)\cdot\mathbf{p}_{\mathbf{k}_2}\right\rangle\left\langle n^1_{-\mathbf{k}_1}(t)n^1_{-\mathbf{k}_2}\right\rangle+\left\langle \mathbf{p}_{\mathbf{k}_1}(t)n^1_{-\mathbf{k}_2}\right\rangle\cdot\left\langle n^1_{-\mathbf{k}_1}(t)\mathbf{p}_{\mathbf{k}_2}\right\rangle \\
&\quad=\delta_{\mathbf{k}_2,-\mathbf{k}_2}\left\langle \mathbf{p}_{\mathbf{k}_1}(t)\cdot\mathbf{p}_{-\mathbf{k}_1}\right\rangle\left\langle n^1_{-\mathbf{k}_1}(t)n^1_{\mathbf{k}_1}\right\rangle+O\left(\frac{1}{N}\right)
\end{aligned} \tag{3.13}$$

where N is the number of particles in the system. Thus, after factorization, the correlation function depends on only one wavevector rather than two. If a similar approximation is made for the static correlation function, we obtain for the bilinear contribution to $C(t)$, denoted by $C_{\mathrm{bil}}(t)$, the expression

$$\begin{aligned}
C_{\mathrm{bil}}(t)&\cong\sum_{\mathbf{k}}^{k_c}\frac{\left\langle \mathbf{v}_1\cdot\mathbf{p}_{\mathbf{k}}n^1_{-\mathbf{k}}\right\rangle^2\left\langle \mathbf{p}_{\mathbf{k}}(t)\cdot\mathbf{p}_{-\mathbf{k}}\right\rangle\left\langle n^1_{-\mathbf{k}}(t)n^1_{\mathbf{k}}\right\rangle}{\left\langle \mathbf{p}_{\mathbf{k}}\cdot\mathbf{p}_{-\mathbf{k}}\right\rangle^2\left\langle n^1_{\mathbf{k}}n^1_{-\mathbf{k}}\right\rangle^2} \\
&=\frac{1}{m^2N^2}\sum_{\mathbf{k}}^{k_c}\left\langle \mathbf{p}_{\mathbf{k}}(t)\cdot\mathbf{p}_{-\mathbf{k}}\right\rangle\left\langle n^1_{-\mathbf{k}}(t)n^1_{\mathbf{k}}\right\rangle
\end{aligned} \tag{3.14}$$

It is to be expected that the long-time behavior will arise only from $C_{\mathrm{bil}}(t)$, as the remaining contributions to $C(t)$ should decay on a "fast" time scale. Of course this may be questioned, for it might be wondered if these apparently "fast" variables contained projections onto "slow" trilinear modes, but the work of Ernst et al.[48] suggests that trilinear and higher-order modes do not, in fact, affect the asymptotic long-time behavior of $C(t)$ in three dimensions. Thus, in order to calculate the long-time tail from Eq. (3.14) we need the small-k, long-time form of these correlation functions. It is easily shown that only the transverse part of the momentum density contributes to the dominant long-time tail—the longitudinal components only contribute to a weaker $t^{-5/2}$ behavior. Thus we have, for small k the results (in $d=3$)

$$\left\langle \mathbf{p}_{\mathbf{k}}(t)\cdot\mathbf{p}_{-\mathbf{k}}\right\rangle=2Nmk_BTe^{-k^2\eta t/\rho m}+\text{longitudinal term} \tag{3.15a}$$

and

$$\left\langle n_{\mathbf{k}}^{1}(t) n_{-\mathbf{k}}^{1}\right\rangle = e^{-k^2 Dt} \tag{3.15b}$$

Substitution of the above into Eq. (3.14), followed by converting the **k** sum into a **k** integral, yields the result

$$\begin{aligned}\lim_{t\to\infty} C(t) &= \frac{2k_{\mathrm{B}}T}{(2\pi)^3 m\rho}\int_0^{k_c} d\mathbf{k} \exp\left\{-k^2\left(D+\frac{\eta}{\rho m}\right)t\right\} \\ &= \frac{2k_{\mathrm{B}}T}{\rho m}\left[4\pi\left(D+\frac{\eta}{\rho m}\right)t\right]^{-3/2}\end{aligned} \tag{3.16}$$

This is identical to the hydrodynamic result, Eq. (3.6), except that the factor $\eta/\rho m_1$ there is replaced by $D + \eta/\rho m$. In the Brownian limit, though, $D \ll \eta/\rho m$, whereupon the two expressions become the same. Before discussing further the approximations that have gone into this theoy, let us recall that Eq. (3.16) is identical to the results obtained from low-density kinetic theories except that full fluid transport coefficients appear instead of low-density Boltzmann or Enskog ones. Also it is possible to make some progress towards obtaining the SE relation in the Brownian-particle limit. To do this we may follow the methods of Keyes and Oppenheim.[49] In the above work we see that the size of the tagged particle never makes an appearance—it is as if we were dealing with point particles. In order to introduce the tagged particles's finite size, they redefined the tagged-particle density as

$$\tilde{n}_{\mathbf{k}}^{1} = \int d\mathbf{r}\, e^{i\mathbf{k}\cdot\mathbf{r}} H(\mathbf{r}-\mathbf{r}_1) \tag{3.17}$$

where

$$H(\mathbf{x}) = \begin{cases} \left(\frac{4}{3}\pi R_1^3\right)^{-1}, & |x| < R_1 \\ 0, & |x| \geq R_1 \end{cases}$$

On carrying out the **r** integration in Eq. (3.17) we find

$$\tilde{\mathbf{n}}_{\mathbf{k}}^{1} = \frac{j_1(kR_1)}{3kR_1} n_{\mathbf{k}}^{1} \tag{3.18}$$

where $j_1(x)$ is a spherical Bessel function. Thus the procedure is to replace $e^{i\mathbf{k}\cdot\mathbf{r}_1}$ by $\tilde{n}_{\mathbf{k}}^{1}$ in all the definitions in Eqs. (3.7)–(3.10). It was shown that the result of this modification—after making the factorization approximation in some, but not all, of the bilinear correlation functions—was to replace Eq. (3.14) by

$$C_{\text{bil}}(t) = \frac{1}{m^2\rho N}\int_0^{k_c} d\mathbf{k}\langle \mathbf{p}_{\mathbf{k}}(t)\cdot\mathbf{p}_{-\mathbf{k}}\rangle\langle n^1_{-\mathbf{k}}(t)n^1_{\mathbf{k}}\rangle\left\{\frac{j_1(kR_1)}{kR_1}\right\}^2 \tag{3.19}$$

The effect of the factor in the curly braces is to restrict the values of k to those $\lesssim 1/R_1$, so this does the job of the cutoff wavevector k_c. We may thus extend the upper limit of the k integral to infinity, still effectively only counting bilinear modes with values of $k < O(R_1^{-1})$. The values of the correlation functions are given by Eqs. (3.15a, b). By substituting these results into Eq. (3.19) and using the fact that $D \ll \eta/\rho m$, we obtain an expression for the bilinear contribution to the diffusion constant, D_{bil}, in the form

$$D_{\text{bil}} = \frac{k_B T}{5\pi\eta R_1} \tag{3.20}$$

Keyes and Oppenheim argued that for a large particle, the contributions of the "fast" modes to D would be negligible, so in this Brownian limit $D = D_{\text{bil}}$. Clearly this is very similar to the SE relation, the only difference being the presence of the 5 as opposed to the 4 corresponding to slip boundary conditions or the 6 corresponding to stick.

Finally, Keyes and Oppenheim[50] showed that by using the free-streaming ideal-gas form for $\langle \mathbf{p}_{\mathbf{k}}(t)\cdot\mathbf{p}_{-\mathbf{k}}\rangle$ and $\langle n^1_{\mathbf{k}}(t)n^1_{-\mathbf{k}}\rangle$ in Eq. (3.19), they could obtain an expression for D_{bil} appropriate for a small massive particle diffusing in a dilute gas. Thus they found

$$D_{\text{bil}} \propto \left(\frac{k_B T}{m}\right)^{1/2}\frac{1}{\rho R_1^2}$$

though the constant of proportionality was not in agreement with the exact LB result.

So, in conclusion, it would seem that this mode-coupling approach can account qualitatively for many of the observed features of $C(t)$ and D, and it seems to be quantitatively most accurate for the long-time tail behavior. However, it is clear that many problems remain. On the theoretical side, the factorization approximation is not well justified. Also, in the Brownian-particle limit, one might feel that R_1 should appear in the equations naturally, as a result of the intermolecular forces, rather than having to be "put in by hand," as is done in Eq. (3.17). On the numerical side, it is disappointing that the "true" SE relation was not obtained, and also, it would appear that Eq. (3.19), in the Brownian limit, would give neither the correct zero-time value of $C(t)$, nor the expected intermediate-time exponential behavior. That is, we do not regain an expression like Eq. (3.4).

A different approach is that of Bedeaux and Mazur[51]—a method somewhat in between a microscopic mode-coupling and a pure hydrodynamical theory. They couple the diffusion of a tagged particle to convective currents in the fluid, obtaining the result

$$D = D_0 + D_c + \frac{k_B T k_c}{3\pi^2(\eta + \rho m D_0)} \tag{3.21}$$

Here D_0 is a "bare" diffusion constant (that is, the diffusion constant were there no coupling to convective currents); k_c is a cutoff wavevector, of order $1/R_1$ for a large particle; and D_c is, for large particles, a small negative term proportional to the compressibility. Upon modifying the theory to deal with a large tagged particle and allowing for the effect of the stick boundary conditions imposed at the particle's surface upon the fluctuations of the velocity fields in the fluid, the SE "stick" result was recovered. However, the assumption of boundary conditions and the presence in Eq. (3.21) of an unknown, bare diffusion constant means that this cannot form the basis of a purely microscopic theory of diffusion.

B. Recent Work

In order to improve the Keyes–Oppenheim method so as to obtain better zero- and intermediate-time behavior of $C(t)$, and also to avoid any feeling of arbitrariness introduced by the presence of the function $H(\mathbf{r}-\mathbf{r}_1)$ in Eq. (3.17), it is possible to couple the bilinear variables to $\dot{\mathbf{v}}_1(t)$ or $\ddot{\mathbf{v}}_1(t)$. Coupling to the latter (that is, to the time derivative of the force acting on the tagged particle) forms the basis of a quite comprehensive molecular theory of diffusion due to Bosse et al.[52]

Let us do Mori theory on the two variables $\mathbf{v}_1$ and $(1/m_1)\mathbf{F}_1$. This leads to the result

$$C(z) = \langle \mathbf{v}_1 \cdot \mathbf{v}_1 \rangle \left[z + \frac{\langle \mathbf{F}_1 \cdot \mathbf{F}_1 \rangle}{m_1^2 \langle \mathbf{v}_1 \cdot \mathbf{v}_1 \rangle} [z + \xi(z)]^{-1} \right]^{-1} \tag{3.22}$$

where

$$\xi(z) = \frac{\langle [z - Q_2 iL]^{-1} Q_2 iL \mathbf{F}_1 \cdot Q_2 iL \mathbf{F}_1 \rangle}{\langle \mathbf{F}_1 \cdot \mathbf{F}_1 \rangle} \tag{3.23}$$

and Q_2 projects a variable orthogonally to $\mathbf{v}_1$ and $\mathbf{F}_1$.

We now express $Q_2 iL\mathbf{F}_1$ in terms of bilinear variables. The choice of modes differs slightly from what went before. Bosse et al. used product modes of $n^1_{\mathbf{k}}$ with the conserved densities of the fluid particles only—that is, the sums in Eqs. (3.7)–(3.9) did not include $i=1$. Then $p_{\mathbf{k}} n^1_{-\mathbf{k}}$ is orthogonal to $\mathbf{v}_1$, which simplifies the calculation. Projection onto these bilinear modes for all values of $\mathbf{k}$ yields

$$P_{\text{bil}}(Q_2 iLF_1^{\alpha}) = 3 \sum_{\mathbf{k}_1,\mathbf{k}_2} \left\langle (Q_2 iLF_1^{\alpha}) p^{\beta}_{\mathbf{k}_1} n^1_{-\mathbf{k}_1} \chi^{-1}_{\mathbf{k}_1,\mathbf{k}_2} p^{\beta}_{\mathbf{k}_2} n^1_{-\mathbf{k}_2} \right\rangle \tag{3.24}$$

where the Greek superscripts denote a component of a vector and the summation convention is used. Then the approximation is to replace $Q_2 iL\mathbf{F}_1$ by $P_{\text{bil}}[Q_2 iL\mathbf{F}_1]$ in Eq. (3.23) (an approximation that becomes exact in the high-mass limit but is less good for $m_1 \lesssim m$), to drop the operator Q_2 in the factor $[z - Q_2 iL]^{-1}$, and then to carry out a factorization approximation on time correlation functions of bilinear variables. Upon taking the inverse Laplace transform of Eq. (3.23), this all leads to

$$\xi(t) = \sum_{\mathbf{k}} \left(\frac{\langle (Q_2 iLF_1^{\alpha}) p^{\beta}_{\mathbf{k}} n^1_{-\mathbf{k}} \rangle \langle p^{\gamma}_{\mathbf{k}} n^1_{-\mathbf{k}} (Q_2 iLF_1^{\alpha}) \rangle}{\langle \mathbf{p}_{\mathbf{k}} \cdot \mathbf{p}_{-\mathbf{k}} \rangle^2 \langle \mathbf{F}_1 \cdot \mathbf{F}_1 \rangle} \right) \times \langle p^{\beta}_{\mathbf{k}}(t) p^{\gamma}_{-\mathbf{k}} \rangle \langle n^1_{\mathbf{k}}(t) n^1_{-\mathbf{k}} \rangle \tag{3.25}$$

The static correlation functions required are easily evaluated for a given system using, say, computer simulation data, whereas Bosse et al. had already calculated approximations to the time correlation functions needed. Thus carrying out the $\mathbf{k}$ integration yielded an estimate for $\xi(t)$ and hence $C(t)$.

One of the advantages of this method is that the short-time behavior of $C(t)$ is given correctly—that is,

$$C(t) = \langle \mathbf{v}_1 \cdot \mathbf{v}_1 \rangle - \frac{1}{2m_1^2} \langle \mathbf{F}_1 \cdot \mathbf{F}_1 \rangle t^2 + \cdots$$

Furthermore, the role played by $H(\mathbf{r})$ in the Keyes–Oppenheim theory is here played by the static correlation function $\langle [Q_2 iL\mathbf{F}_1] \cdot \mathbf{p}_{\mathbf{k}} n^1_{-\mathbf{k}} \rangle$, which, indeed, means that the size of the particle appears naturally in the theory, instead of having to be introduced by hand. It is clear that what must physically link the tagged particle's motion to that of the surrounding fluid particles is the mutual forces. This is explicit in this theory, whereas in the earlier mode-coupling theories the physics behind the coupling of tagged particle to bilinear variables was less evident.

Let us now consider the Brownian limit of Eq. (3.24). We have the identity

$$\left\langle \left(Q_2 iLF_1^{\alpha}\right) p_{\mathbf{k}}^{\beta} n_{-\mathbf{k}}^1 \right\rangle = k_B T\rho \int d\mathbf{r}\, G(r)\left(\nabla^{\alpha}\nabla^{\beta}U(r)\right)e^{i\mathbf{k}\cdot\mathbf{r}} \tag{3.26}$$

The bulk of this integral comes from the steeply repulsive portion of $U(r)$, and because near there $G(r)$ is a sharply increasing function of r and $\nabla\nabla U(r)$ is a sharply decreasing function of r, it is a good approximation to write[53]

$$k_B T\rho G(r)\nabla^{\alpha}\nabla^{\beta}U(r) \cong \delta(r - R_1)\hat{r}_{\alpha}\hat{r}_{\beta}\langle \mathbf{F}_1\cdot\mathbf{F}_1\rangle \tag{3.27}$$

Really the delta function is best centered at the first maximum of $G(r)$, but in the Brownian limit, this is just R_1 to an excellent approximation. Substitution of this approximation into Eq. (3.26) yields

$$\left\langle \left(Q_2 iLF_1^{\alpha}\right) p_{\mathbf{k}}^{\beta} n_{-\mathbf{k}}^1 \right\rangle = \langle \mathbf{F}_1\cdot\mathbf{F}_1\rangle \left\{ (\mathbf{1} - \hat{\mathbf{k}}\hat{\mathbf{k}})_{\alpha\beta} \frac{j_1(kR_1)}{kR_1} + (\hat{\mathbf{k}}\hat{\mathbf{k}})_{\alpha\beta}\left(\frac{j_0(kR_1)}{3} - 2\frac{j_2(kR_1)}{3} \right) \right\} \tag{3.28}$$

Then if the result is substituted into Eq. (3.24) and we use Eq. (3.15a) for $\langle p_{\mathbf{k}}^{\alpha}(t) p_{-\mathbf{k}}^{\beta}\rangle$, we obtain a diffusion constant given by Eq. (3.20)—the 5π appearing once again. In order to obtain this limit we did have to make the approximation given in Eq. (3.27), but the small errors introduced there cannot be sufficient to convert the 5π into a 4π.

Thus it must again be concluded that in spite of the many attractive features of this theory, it still does not give the correct SE limit. On the other hand, just as in the previous mode-coupling theories, the long-time tail behavior is given by Eq. (3.16).

This Brownian-limit calculation just described is similar in many ways to an earlier calculation of Michaels and Oppenheim,[54] when they considered a fixed sphere and wanted to calculate $\langle \mathbf{F}_1(t)\cdot\mathbf{F}_1\rangle_0$, where $\mathbf{F}_1$ is the fluctuating force exerted on the sphere by the fluid. One would expect that this quantity, in the high-mass, low-frequency limit, would be related to $\nu(z)$ by

$$\nu(z) = \frac{\langle \mathbf{F}_1(z)\cdot\mathbf{F}_1\rangle_0}{3m_1 k_B T}$$

the Einstein relation. $\mathbf{F}_1$ was coupled to the same bilinear variables as used by Bosse et al. (i.e. $\mathbf{p}_{\mathbf{k}} n^1_{-\mathbf{k}}$, with $i=1$ excluded in the sum definiting $p_{\mathbf{k}}$), and obtained, at zero frequency, the slip-boundary Stokes result (i.e. $\nu = 4\pi\eta R_1$). The methods used can be criticized, however, on the grounds that insufficient care was taken in distinguishing transverse modes from longitudinal modes. Thus firstly, the calculation assumed three transverse modes, whereas there should only have been two. Secondly, as is stressed by Bosse et al. and as is shown explicitly in Eq. (3.28), the vertex function [i.e., the k-dependent function that multiplies the time correlation functions in (e.g.) Eqs. (3.19) and (3.25)] is different for transverse modes than for longitudinal modes. The vertex function used by Michaels and Opperheim is essentially the trace of Eq. (3.28), whereas the correct vertex function, linked to the transverse modes, is the coefficient of $(\mathbf{1}-\hat{\mathbf{k}}\hat{\mathbf{k}})_{\alpha\beta}$. When these changes are made, the friction coefficient, as before, is given by $\nu = 5\pi\eta R_1$. Similar objections may be leveled against the calculation of Tokuyama and Oppenheim[55] —again, if transverse modes are carefully distinguished from longitudinal modes, $\nu = 5\pi\eta R_1$ appears as opposed to the $4\pi\eta R_1$ quoted in the article.

Thus all the microscopic mode-coupling theories so far described yield the correct functional form of the SE relation, but an incorrect coefficient. However, it is worth pointing out that the following calculation, which would appear to be making approximations no worse than what has gone before, does not even get the correct form for D in the Brownian limit. The previous calculations essentially coupled either $\mathbf{v}_1$ or $Q_2 iL\mathbf{F}_1$ to $\mathbf{p}_{\mathbf{k}} n^1_{-\mathbf{k}}$. It is an equally plausible approach to couple $\mathbf{F}_1$ to $n_{\mathbf{k}} n^1_{-\mathbf{k}} - \langle n_{\mathbf{k}} n^1_{-\mathbf{k}} \rangle$. However, after using the standard factorization approximations, this leads to the result

$$\nu(t) = \frac{1}{3m_1 k_B T} \sum_{\mathbf{k}} \frac{\left|\langle \mathbf{F}_1 n_{\mathbf{k}} n^1_{-\mathbf{k}} \rangle\right|^2}{\langle n_{\mathbf{k}} n_{-\mathbf{k}} \rangle^2} \langle n_{\mathbf{k}}(t) n_{-\mathbf{k}} \rangle \langle n^1_{-\mathbf{k}}(t) n^1_{\mathbf{k}} \rangle \qquad (3.29)$$

which says that the diffusion constant is determined by the decay of the fluid number-density fluctuation. Use of the hydrodynamic form of $\langle n_{\mathbf{k}}(t) n_{-\mathbf{k}} \rangle$ does not lead to a SE result and does not give a long-time tail in agreement with Eq. (3.16).

So, it would seem that there are serious deficiencies in the methods so far described. None of them get the correct Brownian limit, and one seems to get different answers depending on whether $\mathbf{v}_1$, $\mathbf{F}_1$, or $Q_2 iL\mathbf{F}_1$ is coupled to the bilinear variables. We note, though, that what all these approaches have in common is the use of a factorization approximation to simplify the calculation. It would therefore seem quite possible that it was this that was causing all the problems. This point was the subject of some work by Masters and Madden.[56] They, like Bosse et al.,[52] started from Eqs. (3.22) and (3.23). Then the correlation function $\xi(z)$ was analyzed using the bilinear modes

$Q_2(n_{\mathbf{k}}n^1_{-\mathbf{k}} - \langle n_{\mathbf{k}}n^1_{-\mathbf{k}}\rangle)$ and $\mathbf{p}_{\mathbf{k}}n^1_{-\mathbf{k}}$—again with $i=1$ excluded from the sums in Eqs. (3.7) and (3.9). In the high-mass limit, $Q_2 iL\mathbf{F}_1$ is given exactly as a sum over $\mathbf{p}_{\mathbf{k}}n^1_{-\mathbf{k}}$:

$$Q_2 iLF_1^\alpha = \sum_{\mathbf{k}_1,\mathbf{k}_2} \left\langle (Q_2 iLF_1^\alpha)\, p^\beta_{\mathbf{k}_1} n^1_{-\mathbf{k}_1} \right\rangle \chi^{-1}_{\mathbf{k}_1,\mathbf{k}_2}\, p^\beta_{\mathbf{k}_2} n^1_{-\mathbf{k}_2} \tag{3.30}$$

Then, by a straightforward application of Mori theory, we obtain

$$\xi(z) = \frac{1}{\langle \mathbf{F}_1 \cdot \mathbf{F}_1 \rangle} \sum_{\mathbf{k}_1,\mathbf{k}_2} \left\langle (Q_2 iLF_1^\alpha)\, p^\beta_{\mathbf{k}_1} n^1_{-\mathbf{k}_1} \right\rangle \mathrm{RR}_{p^\beta p^\gamma}(\mathbf{k}_1,\mathbf{k}_2,z) \times \left\langle p^\gamma_{\mathbf{k}_2} n^1_{-\mathbf{k}_2} (Q_2 iLF_1^\alpha) \right\rangle \tag{3.31}$$

where

$$\sum_{b,\mathbf{k}_2} \mathrm{RR}_{ab}(\mathbf{k}_1,\mathbf{k}_2,z)\mathrm{RR}^{-1}_{bc}(\mathbf{k}_2,\mathbf{k}_3,z) = \delta_{ac}\delta_{\mathbf{k}_1,-\mathbf{k}_3} \tag{3.32}$$

and

$$\mathrm{RR}^{-1}_{bc}(\mathbf{k}_2,\mathbf{k}_3,z) = z\left\langle b_{\mathbf{k}_2} c_{\mathbf{k}_3} \right\rangle - \left\langle (QiLb_{\mathbf{k}_2})\, c_{\mathbf{k}_3} \right\rangle + \left\langle [z - QiL]^{-1} QiLb_{\mathbf{k}_2} (QiLc_{\mathbf{k}_3}) \right\rangle \tag{3.33}$$

The variables a, b, and c refer either to $Q_2(n_{\mathbf{k}}n^1_{-\mathbf{k}} - \langle n_{\mathbf{k}}n^1_{-\mathbf{k}}\rangle)$ or to $\mathbf{p}_{\mathbf{k}}n^1_{-\mathbf{k}}$. The sum over b means b takes on the values of $Q_2(n_{\mathbf{k}}n^1_{-\mathbf{k}} - \langle n_{\mathbf{k}}n^1_{-\mathbf{k}}\rangle)$ and each component of $\mathbf{p}_{\mathbf{k}}n^1_{-\mathbf{k}}$. The projection operator Q projects a variable orthogonal to $\mathbf{v}_1$, $\mathbf{F}_1$, and the bilinear variables.

It now remains to evaluate the correlation functions in Eq. (3.33). The most difficult one is the third term on the RHS, when b and c refer to $\mathbf{p}_{\mathbf{k}}n^1_{-\mathbf{k}}$. Let us first note that

$$QiL\mathbf{p}_{\mathbf{k}}n^1_{-\mathbf{k}} = Q\left(i\mathbf{k}\cdot\boldsymbol{\sigma}_{\mathbf{k}} e^{i\mathbf{k}\cdot\mathbf{r}_1}\right) \tag{3.34}$$

where

$$\boldsymbol{\sigma}_{\mathbf{k}} = \sum_{i>1} \left\{ \frac{\mathbf{p}_i \cdot \mathbf{p}_i}{m} + \frac{1}{2} \sum_{j>i} \mathbf{f}_{ij}\mathbf{r}_{ij} \frac{1 - e^{-i\mathbf{k}\cdot\mathbf{r}_{ij}}}{i\mathbf{k}\cdot\mathbf{r}_{ij}} \right\} e^{i\mathbf{k}\cdot\mathbf{r}_i} \tag{3.35}$$

Here $\mathbf{f}_{ij}$ is the force exerted by fluid particle j upon fluid particle i, and $\mathbf{r}_{ij}$ their relative displacement. $\boldsymbol{\sigma}_{\mathbf{k}}$ is just the microscopic stress tensor. There are no terms present describing the force exerted on a fluid molecule by the tagged particle, because any such term is a linear combination of $\mathbf{F}_1$ and $Q_2(n_{\mathbf{k}}n^1_{-\mathbf{k}})$, and so is projected out by Q. The approximation then made was, for small $\mathbf{k}_1$, $\mathbf{k}_2$,

$$\left\langle [z - QiL]^{-1} QiLp^{\alpha}_{\mathbf{k}_1} n^1_{-\mathbf{k}_1} \left(QiLp^{\beta}_{\mathbf{k}_2} n^1_{-\mathbf{k}_2} \right) \right\rangle$$

$$\cong -\frac{k_B T}{\rho} \left\{ \left(\eta_B - \tfrac{2}{3}\eta \right) k_1^{\alpha} k_2^{\beta} + \eta \left(k_1^{\alpha} k_2^{\beta} + \mathbf{k}_1 \cdot \mathbf{k}_2 \delta_{\alpha\beta} \right\} \times \left[S_1(\mathbf{k}_1 + \mathbf{k}_2) - 1 + N\delta_{\mathbf{k}_1, -\mathbf{k}_2} \right] \tag{3.36}$$

where

$$S_1(\mathbf{k}) - 1 = \rho \int d\mathbf{r}\, e^{i\mathbf{k}\cdot\mathbf{r}} [G(r) - 1] \tag{3.37}$$

and $G(r)$ is the radial distribution function for fluid particles around the sphere. Plausibility arguments for this approximation are given by Masters and Madden[56]—essentially it depends upon the stress–stress correlation function being spatially short-range, and upon its decaying away on a time scale rapid compared to that of correlation functions involving $e^{i\mathbf{k}\cdot\mathbf{r}_1}$.

The remaining static correlation functions were evaluated exactly in nearly all the cases. The only approximations used were to write

$$\left\langle (Q_2 iLF_1^{\alpha}) p^{\beta}_{\mathbf{k}} n^1_{-\mathbf{k}} \right\rangle = k_B T \left[W_l(kR_1) \hat{\mathbf{k}}_{\alpha} \hat{\mathbf{k}}_{\beta} + W_t(kR_1)(\mathbf{1} - \hat{\mathbf{k}}\hat{\mathbf{k}})_{\alpha\beta} \right] \tag{3.38}$$

where

$$W_t(kR_1) = \left[\left(\phi + \frac{\psi}{3} \right) j_0(kR_1) - \frac{2\psi}{3} j_2(kR_1) \right] \tag{3.39}$$

$$W_t(kR_1) = \left[\phi j_0(kR_1) + \psi \frac{j_1(kR_1)}{kR_1} \right] \tag{3.40}$$

and

$$\langle \mathbf{F}_1 \cdot \mathbf{F}_1 \rangle = k_B T (3\phi + \psi) \tag{3.41}$$

If one sets $\phi = 0$, this reduces to Eq. (3.27), a result corresponding to a central potential with a harshly repulsive core. The setting of ϕ equal to zero corresponds to a proposed "stick" model discussed in Refs. 56.

It was found possible to solve Eqs. (3.31) and (3.33) after substituting the above into Eq. (3.33) for various choices of $G(r)$. Considering only the $\phi = 0$ case (the central potential of interaction), it was shown that if the unphysical choice of $G(r) = 1$ everywhere was made, then the calculation reduced to the Brownian limit of the calculation of Bosse et al., yielding $= 5\pi\eta R_1$. If the more physical choice of $G(r) = 1$, $r > R_1$, and $G(r) = 0$, $r < R_1$, was made, then for $z = 0$ the slip result $\nu = 4\pi\eta R_1$ was recovered. Furthermore for finite z, upon letting $\psi \to \infty$ (corresponding to letting the potential of interaction become hard-sphere-like), then Eq. (3.22) reduced to the form of Eq. (3.4), with $\nu(z)$ exactly the same as that calculated from hydrodynamics and the slip boundary condition. Lastly, with the same step-function form of $G(r)$, setting $\psi = 0$ and letting $\varphi \to \infty$, we again regain Eq. (3.4), this time with $\nu(z)$ having the hydrodynamics-plus-stick-boundary form.

All this is rather encouraging. It strongly suggests that the culprit in the previous mode-coupling theories was the factorization approximation—an approximation seen to be equivalent to setting $G(r) = 1$ everywhere, which unphysically permits fluid to be present in the interior of the Brownian particle. Upon expelling the fluid from the interior by making $G(r)$ a step function, the correct hydrodynamic result is regained. The price paid for this, though, is a severe increase in the complexity of the equations to be solved: without the factorization there is wavevector mixing (i.e., bilinear modes of differing wavevector can couple), and $C(t)$ can only be obtained by solving coupled integral equations instead of by evaluating a single k integral. Because of this added complexity, no progress was made in trying to extend the theory to deal with diffusion away from the Brownian limit.

There are still outstanding problems. Although it was shown that the Gaussian approximation is very poor and unphysical (at least when applied to the Brownian limit, and probably when applied to smaller particles also), several approximations were still made without much justification. The most important of these is contained in Eq. (3.36). Furthermore, even for a Brownian particle, $G(r)$ is not a step function in dense fluid—that is, $G(R_1) \neq 1$. Also, even though the correct answer came out, the connection of these equations with the equations of hydrodynamics plus boundary conditions is not very transparent, and a clear, physical picture of the approximations made is not forthcoming.

Much of the blame for this can be attached to working in **k** space as opposed to real space. If one wants to go beyond the factorization approximation (which is essential to get the correct Brownian limit), then there remains no computational advantage in working in **k** space, and the physical nature of the approximations and manipulations made is considerably more obscure. Thus in the later sections much of the analysis will be carried out in real space, where the content of the theory may have a more physically intuitive interpretation.

Before passing on to this, it is of interest to consider another limiting case for diffusion of a tagged particle—that is, the Lorentz-gas limit. As discussed the previous section, the Lorentz-gas model is that of a single, point particle moving in a sea of fixed, specularly reflecting, hard spheres, or discs in two dimensions. In the particular version of this model that we shall consider, the scatterers are arranged completely randomly, so overlapping configurations are permitted. At a critical density of scatterers, the tagged particle always finds itself trapped in a cage of scatterers, so the diffusion constant must vanish. Thus this system provides quite a rigorous testing ground for any theory of diffusion, because of this "percolation" threshold property, and also because the system is well characterized[13,14] by computer simulation. The kinetic-theory approach has been discussed previously; here we briefly outline some of the mode-coupling methods that have been used.

Using methods rather similar to those used in their Brownian-particle calculation, Masters and Madden[57] investigated the properties of D and $C(t)$ at densities high enough that D was small—that is, at densities just below the threshold density. They worked in a canonical ensemble, and began by assuming a continuous potential of interaction between the tagged particle and a scatterer, taking the hard-sphere limit only at the end of the calculation. Thus, once again, the starting point of the calculation was Eq. (3.23). $Q_2 iL\mathbf{F}_1$ was analyzed in terms of the bilinear variables $Q_2(n_{\mathbf{k}}n^1_{-\mathbf{k}} - \langle n_{\mathbf{k}}n^1_{-\mathbf{k}}\rangle)$, $\mathbf{v}_1 n_{\mathbf{k}} n^1_{-\mathbf{k}}$, and some components of $Q_2(\mathbf{v}_1\mathbf{v}_1 n_{\mathbf{k}} n^1_{-\mathbf{k}})$ (note that in the Lorentz gas the tagged-particle number density is the only conserved variable). The analog of Eq. (3.31) then allowed $\xi(z)$ to be expressed in terms of D, the true diffusion constant. This result, with Eq. (3.4), gives a self-consistent theory for D. The result was a cubic equation for D, which showed that D vanished at a critical density of $\rho_c^* = 3/2\pi$, where $\rho^* = \rho a^3$, ρ being the number density of the scatterers, and a being their radius. Furthermore, for ρ^* a little less than ρ_c^*, this equation predicted that

$$D \propto \left(\frac{\rho_c^* - \rho^*}{\rho_c^*} \right)^{1/2} \tag{3.42}$$

For $\rho^* > \rho_c^*$, the diffusion constant appeared to go complex instead of remaining equal to zero. However, as we shall discuss later, at $\rho^* = \rho_c^*$ there appears another solution to the equations which does give $D = 0$ for all $\rho^* > \rho_c^*$, and which is the physical solution. This sort of behavior was first discovered by Goetze et al.[36] in their mode-coupling theory, to be described presently.

Solution of the cubic equation for D for all values of ρ^* gave results in reasonable agreement with the computer simulation for $\rho^* > 0.1$, but for lower densities the agreement was poorer. In common with most mode-coupling theories, the low-density limit was not easily attained. Also, it was possible to investigate $C(t)$; the agreement with computer simulation was reasonable, getting better at higher densities, and the theory predicted a $t^{-5/2}$ long-time tail, as is also obtained at low density from kinetic-theory methods.

So, again the theory seemed to work fairly well. This time the approximation that $G(a)=1$ is rigorously true, but the analog of Eq. (3.36) is still a big assumption. Furthermore, it is clear that although this approach seems well suited exploring the behavior of systems in or near the hydrodynamic regime, it is not good in the dilute-gas limit, largely because the free streaming of the tagged particle is not well described.

A much more comprehensive mode-coupling theory was constructed by Goetze, Leutheusser, and Yip.[36] The basic ingredients of this approach are as follows. Firstly they consider a microcanonical ensemble, and unlike the previous mode-coupling approaches, use Mori's generalized Langevin equation in a form appropriate for dealing with hard-sphere interactions. Thus there is no need to begin with continuous interactions and then take the hard-sphere limit—a procedure that is often by no means straightforward. We now give a brief sketch as to how one may derive the equations of Goetze et al., though the method outlined here differs somewhat from the original presentation.

Let us first do Mori theory on the variable

$$A(\mathbf{v}) = \delta(\mathbf{v}-\mathbf{v}_1) \tag{3.43}$$

for all values of $\mathbf{v}$. This leads to the result

$$\langle A(\mathbf{v}, z) A(\mathbf{v}') \rangle = \varphi_0(\mathbf{v}) \varphi_0(\mathbf{v}') R_{AA}(\mathbf{v},\mathbf{v}') \tag{3.44}$$

where $\varphi_0(\mathbf{v}) = \delta(|v| - v_0)/4\pi v_0^2$ in three dimensions and v_0 is the constant speed of the tagged particle. R_{AA} is given by

$$R_{AA}(\mathbf{v},\mathbf{v}') * R_{AA}^{-1}(\mathbf{v}',\mathbf{v}'') = \delta(\mathbf{v}-\mathbf{v}'') \tag{3.45}$$

where

$$\begin{aligned} R_{AA}^{-1}(\mathbf{v},\mathbf{v}') = z\langle A(\mathbf{v}) A(\mathbf{v}') \rangle - \langle (iL_+ A(\mathbf{v})) A(\mathbf{v}') \rangle \\ - \langle \{ iL_+ [z - Q_A iL_+]^{-1} Q_A iL_+ A(\mathbf{v}) \} A(\mathbf{v}') \rangle \end{aligned} \tag{3.46}$$

The asterisk means integration over repeated velocity variables, iL_+ is the pseudo-Liouville operator that propagates a variable forward in time,

$$iL_+ = \sum_i \mathbf{v}_i \cdot \nabla_{\mathbf{r}_i} + \sum_{i<j} T_+(ij) \tag{3.47}$$

and the Mori projection operator projects a variable orthogonal to $A(\mathbf{v})$ for all $\mathbf{v}$. For future convenience we write

$$K(\mathbf{v},\mathbf{v}',z)=\left\langle\left\{iL_{+}[z-Q_{A}iL_{+}]^{-1}Q_{A}iL_{+}A(\mathbf{v})\right\}A(\mathbf{v}')\right\rangle \tag{3.48}$$

At low density we can ignore K, in which case Eqs. (3.46), (3.47), and (3.48) are the equivalent of the LB equation. At higher densities K is not at all negligible. It involves a correlation function of an irreducible two-body term, so it is logical to analyze it in terms of bilinear variables. Thus let us begin again, this time doing Mori theory on the variables $A(\mathbf{v})$ and $B_{\mathbf{k}}(\mathbf{v}')$, for all values of $\mathbf{v}$, $\mathbf{v}'$, and $\mathbf{k}$, where

$$B_{\mathbf{k}}(\mathbf{v})=\delta(\mathbf{v}-\mathbf{v}_{1})\left(n_{\mathbf{k}}^{1}n_{-\mathbf{k}}-\left\langle n_{\mathbf{k}}^{1}n_{-\mathbf{k}}\right\rangle\right) \tag{3.49}$$

Equation (3.44) still holds, but R_{AA} is now given by

$$R_{AA}(\mathbf{v};\mathbf{v}')*R_{AA}^{-1}(\mathbf{v}';\mathbf{v}'')+R_{AB}(\mathbf{v};\mathbf{v}',\mathbf{k}')*R_{BA}^{-1}(\mathbf{v}',\mathbf{k}';\mathbf{v}'')=\delta(\mathbf{v}'-\mathbf{v}'') \tag{3.50a}$$

$$R_{AA}(\mathbf{v};\mathbf{v}')*R_{AB}^{-1}(\mathbf{v}';\mathbf{v}'',\mathbf{k}'')+R_{AB}(\mathbf{v};\mathbf{v}',\mathbf{k}')*R^{-1}(\mathbf{v}',\mathbf{k}';\mathbf{v}'',\mathbf{k}'')=0 \tag{3.50b}$$

where now the asterisk means integration over repeated velocities and summation over repeated wavevectors. The R^{-1} functions are then given by

$$R_{AA}^{-1}(\mathbf{v};\mathbf{v}')=z\langle A(\mathbf{v})A(\mathbf{v}')\rangle-\langle[iL_{+}A(\mathbf{v})]A(\mathbf{v}')\rangle \tag{3.51}$$

$$R_{AB}^{-1}(\mathbf{v};\mathbf{v}',\mathbf{k}')=-\langle[iL_{+}A(\mathbf{v})]B_{\mathbf{k}'}(\mathbf{v}')\rangle \tag{3.52}$$

$$R_{BA}^{-1}(\mathbf{v},\mathbf{k};\mathbf{v}')=-\langle[iL_{+}B_{\mathbf{k}}(\mathbf{v})]A(\mathbf{v}')\rangle \tag{3.53}$$

$$R_{BB}^{-1}(\mathbf{v},\mathbf{k};\mathbf{v}',\mathbf{k}')=z\langle B_{\mathbf{k}}(\mathbf{v})B_{\mathbf{k}'}(\mathbf{v}')\rangle-\langle(iL_{+}B_{\mathbf{k}}(\mathbf{v}))B_{\mathbf{k}'}(\mathbf{v}')\rangle - \left\langle\left\{iL_{+}[z-QiL_{+}]^{-1}QiL_{+}B_{\mathbf{k}}(\mathbf{v})\right\}B_{\mathbf{k}'}(\mathbf{v}')\right\rangle \tag{3.54}$$

Q projects a variable orthogonally to the variables $A(\mathbf{v})$ and $B_{\mathbf{k}}(\mathbf{v}')$, for all values of the field variables. The approximations to be made in order to obtain the equations of Goetze et al. are the following. Firstly the factorization approximation is used to simplify the first two terms in Eq. (3.54)—this means that only $\mathbf{k}=-\mathbf{k}'$ terms survive. It remains to approximate the third term, which we shall denote by $M(\mathbf{v},\mathbf{k};\mathbf{v}',\mathbf{k}')$. To do this it is first assumed that functions of the velocity decay on a time scale fast compared to that of the decay of $n_{\mathbf{k}}$ and $n^1_{\mathbf{k}'}$, so that these quantities may be regarded as staying put at their zero-time values. This approximation is good for small values of $\mathbf{k}$ and $\mathbf{k}'$. By making a further approximation similar to that made in Eq. 3.36, we have

$$\begin{aligned} M(\mathbf{v},\mathbf{k};\mathbf{v}',\mathbf{k}';z) &\cong K(\mathbf{v};\mathbf{v}';z)\Big\langle \left(n^1_{\mathbf{k}}n_{-\mathbf{k}} - \langle n^1_{\mathbf{k}}n_{-\mathbf{k}}\rangle\right) \\ &\qquad \times \left(n^1_{\mathbf{k}'}n_{-\mathbf{k}'} - \langle n^1_{\mathbf{k}'}n_{-\mathbf{k}'}\rangle\right)\Big\rangle \\ &\cong N\delta_{\mathbf{k},-\mathbf{k}'}K(\mathbf{v};\mathbf{v}';z) \end{aligned} \tag{3.55}$$

where, to get the second line, we have factorized the static, bilinear correlation function.

We now have a self-consistent theory for $K(\mathbf{v};\mathbf{v}';z)$ and hence the velocity correlation function (VCF). To make this more explicit, let us introduce the functions $\Phi(\mathbf{v})$ and $\Theta_{\mathbf{k}}(\mathbf{v})$, given by

$$\Phi(\mathbf{v}) = \varphi_0(\mathbf{v}')\mathbf{v}' * R_{AA}(\mathbf{v}',\mathbf{v}) \tag{3.56}$$

$$\Theta_{\mathbf{k}}(\mathbf{v}) = \varphi_0(\mathbf{v}')\mathbf{v}' * R_{AB}(\mathbf{v}';\mathbf{v},\mathbf{k}) \tag{3.57}$$

We then have the following equations:

$$C(z) = \int d\mathbf{v}\,\varphi_0(\mathbf{v})\mathbf{v}\cdot\Phi(\mathbf{v}) \tag{3.58}$$

$$z\Phi(\mathbf{v}) - \rho\Lambda_{\mathrm{LB}}(\mathbf{v})\Phi(\mathbf{v}) - \Phi(\mathbf{v}') * K(\mathbf{v}';\mathbf{v};z) = \mathbf{v} \tag{3.59}$$

$$z\Phi(\mathbf{v}) - \rho\Lambda_{\mathrm{LB}}(\mathbf{v})\Phi(\mathbf{v}) - \frac{\rho}{2\pi^3}\int d\mathbf{k}\,\overline{T}_{\mathbf{k}}\Theta_{\mathbf{k}}(\mathbf{v}) = \mathbf{v} \tag{3.60}$$

and

$$z\Theta_{\mathbf{k}}(\mathbf{v}) - i\mathbf{k}\cdot\mathbf{v}\Theta_{\mathbf{k}}(\mathbf{v}) - \rho\Lambda_{\mathrm{LB}}(\mathbf{v})\Theta_{\mathbf{k}}(\mathbf{v}) - \rho\Theta_{\mathbf{k}}(\mathbf{v}') * K(\mathbf{v}';\mathbf{v};z) = T_{\mathbf{k}}\Phi(\mathbf{v}) \tag{3.61}$$

We note in passing that if the fourth term on the LHS of Eq. (3.61) were to be dropped, Eqs. (3.60) and (3.61) would be exactly the Fourier transform of a ring equation. We shall explore the connection between ring kinetic theory and the factorization approximation in mode-coupling theories in the next section. In principle these self-consistent equations could be solved, but this would be a hard task. In order to make them more tractable, Goetze et al. made a further approximation to the operator $K(\mathbf{v}',\mathbf{v};z)$ along the lines of a BGK approximation. They wrote, for an arbitrary function f,

$$\rho\Lambda_{\text{LB}}(\mathbf{v})f(\mathbf{v})+f(\mathbf{v}')*K(\mathbf{v}';\mathbf{v};z)\cong-\nu(z)\left[f(\mathbf{v})-\int d\mathbf{v}''\varphi_0(\mathbf{v}'')f(\mathbf{v}'')\right] \tag{3.62}$$

where $\nu(z)$ is the frequency-dependent friction coefficient as given in Eq. (3.4). This choice of model operator preserves the particle-conserving property of the true operator and makes Eq. (3.59) lead directly to Eq. (3.4). On using this form in Eqs. (3.60) and (3.61), a self-consistent, ringlike theory for $\nu(z)$ and hence $C(z)$ emerges.

The properties of these equations have been investigated in a number of papers. Some of the results found were as follows. Firstly, at low density the theory gives the exact LB result, and gives a good approximation to the leading density correction to D. It predicts a critical density at which diffusion vanishes, the values being $\rho_c^* = 9/4\pi$ in three-dimensions and $\rho_c^* = 2/\pi$ in two dimensions. Furthermore, as mentioned previously, they showed that D remained equal to zero for $\rho^* > \rho_c^*$. They also showed that as $\rho^* \to \rho_c^*$ from below,

$$D \propto \frac{\rho_c^* - \rho^*}{\rho_c^*}$$

—that is, a linear dependence. The agreement with computer simulation data was reasonably good in three dimensions, but was less good in two dimensions at the higher densities. For instance, computer studies and the percolation theory suggested $\rho_c^* \cong 0.37$ in this dimensionality, whereas the calculated $\rho_c^* = 2/\pi \cong 0.64$ is considerably different. In three dimensions, Monte Carlo simulations suggest $\rho_c^* \cong 0.81$, in somewhat better agreement with the calculated value of $9/4\pi \cong 0.72$.

They also investigated $C(t)$. For $\rho^* < \rho_c$ they found a long-time $t^{-5/2}$ tail in three dimensions and a t^{-2} tail in two, with a preasymptotic tail of $t^{-3/2}$ in three dimensions and $t^{-3/2}\log(1/t)$ in two.

As $\rho^* \to \rho_c^*$ the formerly preasymptotic tail carried on indefinitely. The long-time behavior was found to be in reasonably good agreement with the simulation data[14] of Alder and Alley. Finally, in the localized phase, that found that D vanished, and that $C(t)$ possessed no long-time power-law behavior.

Although we have reported here only some of the results of this theory, we hope that is enough to show something of its wide scope. As we have seen, it is essentially a self-consistent ring theory, even though it was derived from mode-coupling-like arguments and approximations. It thus has no trouble treating the low-density regime, unlike previous mode-coupling theories. As shown in the derivation, though, several approximations had to be made. Firstly there was the normal factorization approximation, secondly there was the "factorization" of velocity from positional variables in Eq. (3.35), and thirdly there was the simplifiying approximation to the operator $K(\mathbf{v};\mathbf{v}';z)$. We shall return to the first factorization approximation later, but it is because of this that the predictions of Goetze. et al. for the values of ρ_c^* and the behavior of D near there differed from the mode-coupling theory of Master and Madden, who did not make this approximation. The second approximation led to a k-independent estimate for $M(\mathbf{k},\mathbf{v};\mathbf{k}',\mathbf{v}')$. Later work by Leutheusser[36] has sought to improve upon this by constructing a self-consistent, **k**-dependent theory. Plausibility arguments for the third approximation are given in Ref. 37. Near the threshold density, though, one can obtain most of the results reported by using the exact operator $K(\mathbf{v};\mathbf{v}';z)$. Leutheusser has carefully analyzed the various approximations made in the various self-consistent theories.

Returning to the diffusion of a Brownian particle, we shall now proceed to analyze the problem in real space. The methods used will closely follow those in Ref. 43. Let us begin with Eq. (3.4), with $\nu(z)$ given by Eq. (3.3). We then introduce the following variables:

$$a_n(\mathbf{r}) = \sum_{i>1} \delta(\mathbf{r}-\mathbf{r}_{1i}) - \rho G(r) \tag{3.63a}$$

$$\mathbf{a}_v(\mathbf{r}) = \sum_{i>1} \mathbf{v}_i \delta(\mathbf{r}-\mathbf{r}_{1i}) \tag{3.63b}$$

and

$$a_e(\mathbf{r}) = \sum_{i>1} \left[\frac{1}{2} m v_i^2 + U(1i) + \frac{1}{2} \sum_{\substack{j>1 \\ j\neq i}} U(ij) \right] \delta(\mathbf{r}-\mathbf{r}_{1i}) \tag{3.63c}$$

If these variables are Fourier-transformed, we end up with the **k**-dependent, bilinear variables used in previous work. Thus working with the above variables for all values of **r** is completely equivalent to working with all Fourier components of the **k**-dependent bilinear variables; here we make the former choice. Incidentally, this brief discussion brings out the point that physically the bilinear variables describe fluctuations of a fluid variable a given distance away from the tagged particle—that is, they describe local fluctuations.

As is often the case in Mori theory, it is more convenient to use variables that are fluctuating and that are orthogonal to one another at zero time, so we replace $a_e(\mathbf{r})$ by $a_t(\mathbf{r})$, which is closely related to local temperature fluctuations in the fluid and is given by the fluctuating part of $a_e(\mathbf{r})$ that is orthogonal to $a_n(\mathbf{r}')$ for all $\mathbf{r}'$. We note that $a_t(\mathbf{r})$ contains no $U(1i)$ term even though $a_e(\mathbf{r})$ does. This is because this term is a linear combination of the $a_n(\mathbf{r}')$, and hence cannot be present in $a_t(\mathbf{r})$, which must be orthogonal to $a_n(\mathbf{r}')$.

Application of Mori theory then gives the following results:

$$\nu(z) = \frac{1}{m_1^2\langle \mathbf{v}_1 \cdot \mathbf{v}_1 \rangle} \int d\mathbf{r}\, \mathbf{f}_n(\mathbf{r}) \langle \mathbf{F}_1 \cdot a_n(\mathbf{r}) \rangle \tag{3.64a}$$

where

$$\mathbf{f}_a(\mathbf{r}) = \langle \mathbf{F}_1 a_n(\mathbf{r}') \rangle * \mathrm{RR}_{na}(\mathbf{r}', \mathbf{r}) \tag{3.64b}$$

and

$$\sum_a \mathbf{f}_a(\mathbf{r}') * \mathrm{RR}_{ab}^{-1}(\mathbf{r}', \mathbf{r}) = \delta_{nb} \langle \mathbf{F}_1 a_n(\mathbf{r}) \rangle \tag{3.64c}$$

where a and b take on the values n, v, and t, and the asterisk means to take a scalar product and integrate repeated variables over all space. RR_{ab}^{-1} is given by

$$\begin{aligned} \mathrm{RR}_{ab}^{-1}(\mathbf{r}, \mathbf{r}') = {} & z\langle a_a(\mathbf{r}) a_b(\mathbf{r}') \rangle - \langle (Q_1 iLa_a(\mathbf{r})) a_b(\mathbf{r}') \rangle \\ & + \langle \{ [z - Q_1 iL]^{-1} Q_1 iLa_a(\mathbf{r}) \} Q_1 iLa_b(\mathbf{r}') \rangle \end{aligned} \tag{3.65}$$

where Q_1 is defined after Eq. (3.4) and Q projects a variable orthogonally to $\mathbf{v}_1$ and the bilinear variables. We also have

$$\langle \mathbf{F}_1 a_n(\mathbf{r}) \rangle = \rho k_B T \nabla G(r). \tag{3.66}$$

These equations are, so far, exact. We now specialize to the case of the Brownian particle, which essentially means treating the tagged particle as stationary in Eq. (3.65), because its speed is, on average, so much lower than that of a fluid particle.

Equation (3.64c) corresponds to three coupled equations for the functions $f_\alpha(r)$. Let us introduce the microscopic length, λ, such that for $r \geq R_1 + \lambda$ the fluid structure is essentially unperturbed by the presence of the tagged particle—for example, $G(R_1+\lambda)=1$. Then for $r \geq R_1+\lambda$, Eq. (3.64) takes on the form

$$zS(0)f_n^\alpha(\mathbf{r})+\frac{\rho k_\mathrm{B}T}{m}\nabla^\beta f_v^{\alpha\beta}(\mathbf{r})=0 \tag{3.67a}$$

$$zf_v^{\alpha\beta}(\mathbf{r})+\nabla^\beta f_n^\alpha(\mathbf{r})+\left[\frac{(\gamma-1)k_\mathrm{B}T^2}{S(0)C_V}\right]^{1/2}\nabla^\beta f_t^\alpha(\mathbf{r})$$
$$-\frac{1}{\rho m}\left\{\eta\nabla^2 f_v^{\alpha\beta}(\mathbf{r})+\left(\eta_B+\frac{\eta}{3}\right)\nabla^\beta\nabla^\gamma f_v^{\alpha\gamma}(\mathbf{r})\right\}=0 \tag{3.67b}$$

and

$$zmC_V f_t^\alpha(\mathbf{r})+\left[\frac{(\gamma-1)k_\mathrm{B}}{S(0)C_V}\right]^{1/2}\nabla^\beta f_v^{\alpha\beta}(\mathbf{r})-\frac{m\kappa}{\rho}\nabla^2 f_t^\alpha(\mathbf{r})=0 \tag{3.67c}$$

Here C_V is the specific heat per particle at constant volume, γ is the ratio of specific heats, $S(0)$ is the $k=0$ limit of the structure factor of the fluid, and κ is the thermal conductivity. These equations may be related to the normal, linearized hydrodynamic variables by noting the results

$$\left(\frac{\partial P}{\partial \rho}\right)_T=\frac{k_\mathrm{B}T}{S(0)} \tag{3.68}$$

$$\left(\frac{\partial P}{\partial T}\right)_\rho=\frac{\rho^2 C_V(\gamma-1)}{T}\left(\frac{\partial P}{\partial \rho}\right)_T \tag{3.69}$$

and then by introducing the variables

$$\delta n(\mathbf{r})=S(0)\mathbf{f}_n(\mathbf{r})\cdot\mathbf{U}_\infty(z) \tag{3.70}$$

$$v^\alpha(\mathbf{r})=\left(\frac{k_\mathrm{B}T}{m}\right)U_\infty^\beta(z)f_v^{\beta\alpha}(\mathbf{r}) \tag{3.71}$$

and

$$\delta T(\mathbf{r})=\left(\frac{k_\mathrm{B}T^2}{\rho C_V}\right)\mathbf{U}_\infty(z)\cdot f_t(\mathbf{r}) \tag{3.72}$$

Thus, upon taking the scalar product of Eqs. (3.67) with $\mathbf{U}_\infty(z)$ and making the above substitutions, we regain the linearized hydrodynamic equations appropriate for fluid flow with frequency-dependent velocity field $\mathbf{U}_\infty(z)$ at infinity. For $r < R_1 + \lambda$, Eqs. (3.67) cannot be correct. The hydrodynamic equations assume that the fields are varying on a large length scale [here of $O(R_1)$] and long time scale. Because for $r < R_1 + \lambda$, the fluid structure is rapidly changing over a molecular length scale, and we cannot expect the hydrodynamic equations to work in this region. However, Eq. (3.64a) tells us that to determine $\nu(z)$ this is exactly where we need to know $f_n(\mathbf{r})$. We now list several possible approaches to solving this problem.

The first approach is simply to assume that Eqs. (3.67) in fact hold for all r—the only change to be made being to add $\langle \mathbf{F}_1 a_n(\mathbf{r}) \rangle$ to the RHS of Eq. (3.67a). The equations may then readily be solved by means of a Fourier transform, and the result for $\nu(z)$ and D is identical to that obtained from Eq. (3.29). As noted there, this procedure fails even to obtain the functional form of the SE relation. The reason is clearly associated with totally ignoring the effects of the Brownian particle on evaluating the LHS of Eq. (3.64c)—in particular, the very sharp changes that occur upon penetrating the harshly repulsive core of the Brownian particle. This is a very serious shortcoming of the factorization approximation.

We next consider an approximate scheme that at least tries to take some account of the rapid variation of the fluid properties near the Brownian particle. To do this we return to the exact expressions given in Eqs. (3.64) and (3.65). Several of the static correlation functions may be evaluated exactly. Others, though [in particular, the third term on the RHS of Eq. (3.65), the memory term], are less easy to evaluate, so we make some approximations. As an illustration consider $\langle [(z - QiL)^{-1} QiL\mathbf{a}_v(\mathbf{r}')] QiL\mathbf{a}_v(\mathbf{r}) \rangle$. As discussed beneath Eq. (3.35), $QiL\mathbf{a}_v(\mathbf{r})$ contains no term involving the force exerted on a fluid particle by the tagged particle, and is thus of the form of the divergence of something—say $\nabla \cdot \boldsymbol{\tau}(\mathbf{r})$, for example. The correlation function will be spatially short-ranged and for large r can be represented as proportional to $\delta(\mathbf{r} - \mathbf{r}')$ and fluid transport coefficients. Furthermore, for $r < R_1$ it is clear that $\boldsymbol{\tau}(\mathbf{r})$ is very small. Thus we make the physically plausible approximation that the correlation function is proportional to $\nabla_r \nabla_r' \delta(\mathbf{r} - \mathbf{r}') G(r)$. This approximation, combined with similar ones for the other nontrivial correlation functions, yields

$$zS(0)G(r)f_n^{\alpha}(\mathbf{r}) + \frac{\rho k_B T}{m} \nabla^{\beta}\left[G(r) f_v^{\alpha\beta}(\mathbf{r})\right] = \rho k_B T \nabla^{\alpha} G(r) \quad (3.73)$$

$$zG(r)f_v^{\alpha\beta}(\mathbf{r}) + \nabla^{\beta}\left[G(r) f_n^{\alpha}(\mathbf{r})\right] + \left[\frac{(\gamma - 1)k_B T}{S(0)C_V}\right]^{1/2} \nabla^{\beta}\left(G(r) f_t^{\alpha}(\mathbf{r})\right)$$
$$- \frac{1}{\rho m}\left\{\eta \nabla^{\epsilon} G(r) \nabla^{\epsilon} f_v^{\alpha\beta}(\mathbf{r}) + \left(\eta_B - \tfrac{2}{3}\eta\right) \nabla^{\beta} G(r) \nabla^{\gamma} f_v^{\alpha\gamma}(\mathbf{r})\right.$$
$$\left. + \eta \nabla^{\gamma} G(r) \nabla^{\beta} f_v^{\alpha\gamma}(\mathbf{r})\right\} = f_n^{\alpha}(r) \nabla^{\beta} G(r) \quad (3.74)$$

and

$$zG(r)mC_Vf_t^{\alpha}(\mathbf{r})+\left[\frac{(\gamma-1)k_B}{S(0)C_V}\right]^{1/2}G(r)\nabla^{\beta}f_v^{\alpha\beta}(\mathbf{r})$$

$$-\frac{m\kappa}{\rho}\nabla^{\epsilon}G(r)\nabla^{\epsilon}f_t^{\alpha}(\mathbf{r})=0 \tag{3.75}$$

Thus it is seen that sometimes $G(r)$ simply multiplies the terms appearing in Eqs. (3.67), but other times it appears after a gradient operator. Those equations may not be trivially solved by taking a Fourier transform—because $G(r)$ multiplies $\mathbf{f}_n(\mathbf{r})$ and the rest, a Fourier transformation would lead to wavevector mixing.

To make progress, let us note that as $G(r)=1$ for $r > R_1+\lambda$, we may obtain $\nu(z)$ by substituting the LHS of Eq. (3.74) for $\mathbf{f}_n(\mathbf{r})\nabla G(r)$ in Eq. (3.64a) and integrating over the volume V inside a sphere of radius $R_1+\lambda$. The result of this is to give an expression for $\nu(z)$ simply involving a surface integral of the far-field, hydrodynamic fields, provided that z is small. In order to solve for these fields, we require that the far-field equations, Eqs. (3.67), be supplemented by boundary conditions. To obtain these, let us assume the model $G(r)=0$, $r < R_1$, and $G(r)=1$, $r \geq R_1$. Then equating the coefficients of the delta functions in Eqs. (3.73)–(3.75) yields the slip boundary conditions in the form

$$\hat{r}^{\beta}f_v^{\alpha\beta}(\mathbf{r}) = m\hat{r}^{\alpha} \tag{3.76}$$

$$(\mathbf{1}-\hat{\mathbf{r}}\hat{\mathbf{r}})^{\mu\beta}\hat{r}^{\epsilon}\nabla^{\epsilon}f_v^{\alpha\beta}(\mathbf{r})+(\mathbf{1}-\hat{\mathbf{r}}\hat{\mathbf{r}})^{\mu\epsilon}\hat{r}^{\gamma}\nabla^{\epsilon}f_v^{\alpha\gamma}(\mathbf{r})=0 \tag{3.77}$$

and

$$\hat{\mathbf{r}}\cdot\nabla\mathbf{f}_t(\mathbf{r})=0 \tag{3.78}$$

all for $r = R_1$. Equation (3.77) was obtained by multiplying Eq. (3.74) through by $(\mathbf{1}-\hat{\mathbf{r}}\hat{\mathbf{r}})^{\mu\beta}$. If one had instead multiplied it through by $r^{\mu}r^{\beta}$, one would have ended up again with the expression for $\nu(z)$ in terms of the far fields.

It is now simply a matter of solving the hydrodynamic equations using the above slip boundary conditions at the sphere's surface in order to obtain the far fields and hence $\nu(z)$. This procedure yields the slip SE result. The approximation scheme is essentially that used by Masters and Madden. Thus a Fourier transformation of the final term on the LHS of Eq. (3.74) yields the equivalent of Eq. (3.36). The reason it gives the right answer is that account is taken of the rapid variation of fluid transport coefficients and susceptibilities in the neighborhood of the Brownian particle—an effect that is completely missed by making the factorization approximation.

At this point, before proceeding with a more exact analysis of Eq. (3.64), we would like to briefly discuss the hydrodynamic approach used by Peralta-Fabi and Zwanzig[58] to calculate the drag on a large, fixed sphere as a gas flows slowly past it. As discussed before, in the Brownian limit the drag coefficient and the friction coefficient are equal. The point of their calculations was to do away with conventional boundary conditions. Instead they regarded the sphere as a center of force and wrote down equations describing how this force would modify the motion of the fluid.

In their first calculation, they examined the equations

$$\frac{\partial}{\partial t}\delta n(\mathbf{r},t)+\rho\nabla\cdot[G(r)\mathbf{v}(\mathbf{r},t)]=0 \tag{3.79}$$

$$\rho m\frac{\partial}{\partial t}\mathbf{v}(\mathbf{r},t)+k_{\mathrm{B}}T\,\nabla\delta n(\mathbf{r},t) - \eta\{\nabla^2\mathbf{v}(\mathbf{r},t)-\tfrac{1}{3}\nabla[\nabla\cdot\mathbf{v}(\mathbf{r},t)]\}=\nabla u(r)\,\delta n(\mathbf{r},t). \tag{3.80}$$

with $\mathbf{V}(\mathbf{r},t)=U_\infty$ as $r\to\infty$, and $u(r)$ being the potential of interaction between the fixed sphere and a fluid molecule. Because the gas was assumed dilute, $G(r)=\exp[-u(r)/k_BT]$, and $\eta_B=0$. Furthermore, the system was taken to be isothermal. The drag on the sphere, F_0, is then given by

$$\mathbf{F}_0=\int d\mathbf{r}\,[\nabla u(r)]\,\delta n(\mathbf{r},t) \tag{3.81}$$

The steady-state calculation was carried out for two choices of $u(r)$. Firstly a step-function was taken: $u(r)=u_0$, $r<R_1$; $u(r)=0$, $r\geq R_1$. It was then found that the drag coefficient was $5\pi\eta R_1$ and was independent of the value of u_0. Secondly a ramp potential was chosen: $u(r)=u_0$, $r<R_1$; $u(r)=[(R_1+\epsilon)-r]u_0/\epsilon$, $R_1<r<R_1+\epsilon$, and $u(r)=0$, $r>R_1+\epsilon$. In this case, on taking the limit of a hard-sphere potential, the drag coefficient turned out to be approximately $4.132\pi\eta R_1$.

In their second calculation they examined Grad's thirteen-moment equation in the presence of the force due to the fixed sphere. This time they obtained the expected slip result on taking the hard-sphere limit for either choice of $u(r)$.

They concluded that the reason that the hydrodynamic approach gave the wrong answer was that Eqs. (3.79)–(3.80) permitted too much penetration of the fluid into the fixed sphere—a conclusion also reached by the later analysis of Keyes, Morita, and Mercer.[59] This excess penetration was attributed to the use of fluid viscosity that did not vary in the vicinity of the fixed sphere. The 13-moment equation, by giving a more detailed description of the fluid, evidently contained this effect to a sufficient extent to allow the correct slip result to be regained.

We believe that the analysis presented here lends further support to this view of the shortcomings of Eqs. (3.79)–(3.80). Equation (3.74) would suggest that a better result would be obtained if a $G(r)$ were to be slipped in between the two gradient operators acting on $\mathbf{v}(\mathbf{r}, t)$ in Eq. (3.80), equivalent to using a viscosity varying as $\eta G(r)$. We stress that this is still only a very crude approximation scheme, but we believe the formally exact equations would suggest this as a physically plausible model.

Thus, so far we have shown what must be done to the formally exact expression, Eq. (3.64), in order to regain the factorization approximation when the bilinear variables are coupled to the force on the tagged particle. Also it has been shown how the Masters–Madden approximation, Eq. (3.36), could be seen as equivalent to using an r-dependent viscosity given by $\eta G(r)$. Further, it was then easy to see how, upon assuming the step-function form for $G(r)$, the slip boundary conditions arose naturally to supplement the far-field hydrodynamic equations. However, the situation is still unsatisfactory. Firstly, as mentioned before, $G(r)$ is not a step function for a dense fluid surrounding a Brownian sphere, and secondly, the approximations used to get Eqs. (3.73)–(3.75), although reflecting the physical situation to a probably reasonable extent, are still rather unjustifiable. To close this section we return to Eq. (3.64) and use our methods[43] to progress as far as possible without approximations.

The basic plan will be to extract boundary conditions that the far, hydrodynamic fields must obey at $R_1 + \lambda$. This approach is then rather similar to that used by Ronis, Bedeaux, and Oppenheim[60] in their investigation into the microscopic origin of the slip boundary condition.

Let us first consider Eq. (3.64c), setting $b = v$. We then integrate this equation over the volume V. Then, because even in the boundary layer $f_v * R_{vv}^{-1}$ and $f_t * R_{tv}^{-1}$ are, for $z = 0$, of the form of the divergence of something (a result that mathematically comes about because of the presence of the Mori projection operators that project out the direct force of the tagged particle upon the fluid), and because $f_n * R_{nv}^{-1}$ is given by the f_n terms in Eq. (3.74), we obtain once again the result that $\nu(z)$ is given by the surface integral over a sphere of radius $R_1 + \lambda$ of the far hydrodynamic fields—which is in fact easily seen to be the radial component of the hydrodynamic stress tensor.

The boundary conditions that these fields must satisfy are obtained in a similar way. Firstly consider Eq. (3.64c), setting $b = n$. On taking the dot product with $\hat{\mathbf{r}}$ and integrating over V, we regain the boundary condition in

Eq. (3.76)—the equivalent of the normal-velocity boundary condition—provided that z is small and provided that $\int_V d\mathbf{r}[(\mathbf{1}-\hat{\mathbf{r}}\hat{\mathbf{r}})^{\alpha\beta}/r]G(r)f_v^{\alpha\beta}(\mathbf{r})$ is of $O(\lambda/R_1)$ smaller than $R_1^2\hat{\mathbf{r}}\cdot\mathbf{f}_v\cdot\hat{\mathbf{r}}$. That is, provided $\mathbf{f}_v$ is not excessively large inside the volume V, we can obtain the normal-velocity boundary condition without having to make any other approximation. Similarly, if in Eq. (3.64c) one set $b=v$, took the scalar product with $\mathbf{1}-\hat{\mathbf{r}}\hat{\mathbf{r}}$, and integrated over V, the zero-targential-stress condition—Eq. (3.76)—would emerge on making similar assumptions. Lastly, taking the dot product with $\hat{\mathbf{r}}$ of Eq. (3.64c) with $b=t$ and integrating over V yields the zero-temperature-gradient condition.

Thus, on making these fairly mild assumptions, we see that we can obtain the SE result by showing how the equations are equivalent to the hydrodynamic equations plus boundary conditions. As can easily be seen, there is no need to invoke such things as a bare transport coefficient—the full fluid transport coefficients, which describe the far-field fluid flow, naturally appear in the final expressions for the friction coefficient. It is also possible to calculate the long-time behavior of $C(t)$ using the exact real-space equations. The result is once again as given in Eq. (3.16)—details are given in Ref. 43.

To briefly summarize this section on mode-coupling theories, it would seem that the earlier mode theories based upon the Gaussian or factorization approximation are doomed to failure in trying to obtain the SE relation. As the last few pages should have made clear, this factorizing approximation completely misses the rapidly varying fluid properties near the surface of the Brownian particle, and it is just this rapid variation that gives use to the hydrodynamic boundary conditions. The advantages of conducting the analysis in real space are also quite apparent, for it is then easier to see physically what is going on and thus make reasonable approximations. It now remains to attempt to generalize mode-coupling theory so that it can deal with molecular diffusion as well as the Brownian limit. So far little has been achieved.

When the tagged particle has a similar mass to that of a fluid particle, then clearly the thermal speeds are similar, so that the neglect of tagged-particle motion that led to Eqs. (3.67) is no longer valid. Thus the time derivative of $a_n(\mathbf{r})$ involves the relative velocity of the tagged particle and a fluid particle, reflecting the fact that the force can change either due to the fluid number-density fluctuations, or else by a particle moving away from the fluctuation. In kinetic-theory terms, it is important to take recoil effects into account. Following the Lorentz-gas approach, it would certainly be necessary, at the very least, to augment the variables given in Eqs. (3.63) by the presence of $\Sigma_i\mathbf{v}_1\delta(\mathbf{r}-\mathbf{r}_{1i})$, and quite possibly other variables reflecting tagged-particle motion. Again, in kinetic-theory language, it would be hoped that these terms would play a similar role to that played by the LB operator in the propagation between recollisions in the ring or repeated-ring approximations. Secondly, if the sizes of the tagged and fluid particles are similar, it is no longer possible to use **k**-independent, or nonlocal, transport coefficients or fluid susceptibilities. Clearly, far from the tagged particle, the local quantities used in Eq. (3.67) are well approximated by their nonlocal forms, but closer to the tagged particle (inside the volume V) it is not clear what approximations would be good. Finally, it would be desirable for the approximate theory to give the proper low-density limit for a small particle; so far no theory based upon generalized hydrodynamics has achieved this.

In these last two sections, we have attempted to trace some of the modern developments in kinetic theory and in mode coupling. Although the approaches at first sight would appear to be completely disconnected, the work of Goetze et al., for instance, shows the two methods actually to be fairly closely related.

In order to compare the approaches, one must first note that kinetic theory is an inherently more complete description of dynamics than mode coupling with hydrodynamic variables, so we must compare mode coupling with kinetic theory projected onto the hydrodynamic modes. Next consider the form of results obtained with the ring approximation. For any correlation function being calculated, assuming the dynamical variable is diagonal on the full kinetic operator, an expression like Eq. (2.29) will result, with $\mathbf{v}_1$ replaced by the variable of interest. Further evaluation of the matrix element of the ring operator R, as in Eq. (2.26), will lead to the *expression of the inverse of the correlation function as a sum of Boltzmann and ring parts*; *the ring part will involve a single wavevector integral.*

This is clearly nonsense. For example, expression of the inverse of $C(t)$ in this form gives $D^{-1} \propto R^2 - R^3/\eta$, as discussed. On the other hand, direct calculation of the force correlation function in this way would lead to an expression of D as $D \propto R^{-2} + (\eta R)^{-1}$, which is well behaved at large R. An approximation that gives different results from equivalent approaches is unacceptable.

From the last section, the reader will recognize that the above state of affairs is precisely that which applies with the factorization approximation in mode coupling. The factorization approximation always leads to the partition of something, depending on the correlation function calculated, into a "bare" part and a single wavevector integral. Thus, a correspondence between the RA and the factorization approximation is suggested.

On the other hand, the RRA gives rise to multiple wavevector integrals, and is physically well behaved. The same can be said about mode coupling without the factorization approximation, when it is carried out in **k** space. In fact, these two approaches are analogs. The best way to see this in detail is to perform a spatial Fourier transformation of Eq. (2.32b). Then project the equation onto its tagged–bath hydrodynamic modes, such as $n^1_{\mathbf{k}} n_{-\mathbf{k}}$. In the absence of $\bar{T}\Theta$—the term that turns the RA into the RRA—there is no coupling between product variables with different wavevectors, and the equation is in fact a factorization mode-coupling equation. Keeping $\bar{T}\Theta$ adds precisely those terms lost in the factorization.

In short, the RRA is related to mode coupling without the factorization; both these methods have the essential feature that, given that a transport coefficient can be expressed in terms of several possible correlation functions ($\mathbf{v}, \mathbf{F}, \dot{\mathbf{F}}, \ldots$), they give the same answer no matter which correlation is directly calculated. The RA and the factorization approximation to mode coupling do not have this property, and any good results obtained by using them must be regarded as fortuitous.

IV. CONCLUSIONS

Our main conclusions are that good theories of dense systems may be constructed based upon the RRA, or mode coupling without factorization; the equations may need to be made self-consistent. At this point, the Enskog RRAs[40,41], the mode-coupling equations of Goetze et al.,[36] and the other theories discussed herein probably provide a very accurate description of tagged-particle motion in the hard-sphere liquid. It should be clear from the preceeding, however, that solution of these equations away from (e.g.) the Brownian limit is still an unsolved problem.

The methods developed for hard spheres have not yet been applied extensively to particles with attractive potentials, and this will have to be tackled in the future. More interestingly, perhaps, one should recall that one point of studying tagged-particle motion among hard spheres was to develop techniques which could eventually be used on more complicated problems. The methods discussed here are applicable to a whole collection of problems, such as the calculation of other correlation functions than $C(t)$, the study of fluid flow through porous media and of sound propagation in rocks, and so forth. Basically, largely motivated by the AGW computer simulation, a new collection of techniques for treating motion in dense systems has been developed, and all the possible applications of these techniques remain to be seen.

Acknowledgments

Our own work reported here was supported by the National Science Foundation, the Petroleum Research Fund, and the Dreyfus Foundation. T.K. would like to thank J. R. Dorfman for years of good advice.

REFERENCES

1. S. Chapman and T. Cowling, *The Mathematical Theory of Non-Uniform Gases*, 3rd ed., Cambridge University Press, Cambridge, 1970.

2. B. Alder, D. Gass, and T. Wainwright, *J. Chem. Phys.* **53**, 3813 (1970).
3. E. G. D. Cohen, in *Fundamental Problems in Equilibrium Statistical Mechanics*, E. G. D. Cohen (ed.), North Holland, Amsterdam, 1968, Vol. II, p. 228.
4. J. R. Dorfman and E. G. D. Cohen, *Phys. Rev.* **A6**, 776 (1972).
5. K. Kawasaki and I. Oppenheim, *Phys. Rev.* **139**, A1763 (1965).
6. R. Zwanzig, *Phys. Rev.* **129**, 486 (1963).
7. M. Ernst, L. Haines, and J. R. Dorfman, *Rev. Mod. Phys.* **41**, 296 (1969).
8. J. R. Dorfman and E. G. D. Cohen, *Phys. Rev. Lett.* **25**, 1257 (1970).
9. J. Dufty, *Phys. Rev.* **A5**, 2247 (1972).
10. T. Keyes and J. Mercer, *Physica* **95A**, 473 (1979).
11. B. Alder and T. Wainwright, *Phys. Rev. Lett.* **18**, 988 (1967); *Phys. Rev. A* **1**, 18 (1970).
12. M. Ernst and A. Weyland, *Phys. Lett.* **34A**, 39 (1971).
13. C. Bruin, *Physica* **72**, 261 (1974).
14. B. Alder and W. Alley, *J. Stat. Phys.* **19**, 341 (1978); W. Alley, Ph.D. Thesis, Univ. of Calif., Davis, 1979.
15. A. Weyland, *J. Math. Phys.* **15**, 1942 (1974).
16. J. R. Dorfman and E. G. D. Cohen, *Phys. Rev. A* **12**, 292 (1975).
17. G. Mazenko, *Phys. Rev. A* **7**, 209, 222 (1973).
18. P. Resibois and J. Lebowitz, *J. Stat. Phys.* **12**, 483 (1975).
19. P. Resibois, *J. Stat. Phys.* **13**, 393 (1975).
20. P. Futardo, S. Yip, and G. Mazenko, *Phys. Rev. A* **14**, 869 (1976).
21. J. Mehaffey, R. Desai, and R. Kapral, *J. Chem. Phys.* **66**, 1665 (1977).
22. G. Mazenko and S. Yip, in *Modern Theoretical Chemistry: Statistical Mechanics, Part B*, B. Berne (ed.), Plenum, New York, 1977.
23. T. Keyes, *Chem. Phys. Lett.* **51**, 30 (1977).
24. M. Ernst and J. R. Dorfman, *Physica* **61**, 157 (1972).
25. C. McClure, Ph.D. Thesis, Univ. of Maryland, (1972).
26. J. R. Dorfman, H. van Beijern, and C. McClure, *Arch. Mech. Stosow.* **28**, 333 (1976); H. van Beijern and J. R. Dorfman, *J. Stat. Phys.* **23**, 443 (1980).
27. J. Mehaffey and R. Cukier, *Phys. Rev. Lett.* **38**, 1039 (1977).
28. J. Mehaffey and R. Cukier, *Phys. Rev. A* **17**, 1181 (1978).
29. R. Cukier and J. Mehaffey, *Phys. Rev. A* **18**, 1202 (1978).
30. R. Cukier, R. Kapral, J. Lebenhaft, and J. Mehaffey, *J. Chem. Phys.* **73**, 5244 (1980).
31. J. Mercer, Ph.D. Thesis, Yale Univ., 1981.
32. J. Mercer and T. Keyes, *J. Stat. Phys.* **32**, 35 (1983).
33. A. Masters and T. Keyes, *Phys. Rev. A* **25**, 1010 (1982).
34. C. Cercignani and C. Pagani, *Phys. Fluids* **11**, 1395 (1968); C. Cercignani, C. Pagani, and P. Bassinini, *ibid.* **11**, 1399 (1968).
35. P. Bhatnagar, E. Gross, and M. Krook, *Phys. Rev.* **94**, 511 (1954).
36. W. Goetze, E. Leutheusser, and S. Yip, *Phys. Rev. A* **23**, 2634 (1981); *Phys. Rev. A* **24**, 1008 (1981); *Phys. Rev. A* **25**, 533 (1982); E. Leutheusser, *Phys. Rev. A* **28**, 1762 (1983).
37. A. Masters and T. Keyes, *Phys. Rev. A* **26**, 2129 (1982).
38. A. Masters and T. Keyes, *Phys. Rev. A* **27**, 2603 (1983).

39. B. Alder, W. Alley, and J. Dymond, *J. Chem. Phys.* **61**, 1415 (1974); B. Alder and W. Alley, *J. Stat. Phys.* **19**, 341 (1978).
40. W. Sung and J. Dahler, *J. Chem. Phys.* **78**, 6264 (1983).
41. A. Masters and T. Keyes, *J. Stat. Phys.*, **33**, 149 (1983).
42. H. Mori, *Prog. Theor. Phys.* **33**, 423 (1965).
43. A. Masters and T. Keyes, *J. Stat. Phys.*, in press.
44. R. Zwanzig and M. Bixon, *Phys. Rev. A* **2**, 2005 (1970).
45. H. Metiu, D. Oxtoby, and K. Freed, *Phys. Rev. A* **15**, 361 (1977).
46. K. Kawasaki, *Ann. Phys.* (*N.Y.*) **61**, 1 (1970).
47. T. Keyes, in *Modern Theoretical Chemistry, Statistical Mechanics, Vol. 6B*, B. Berne (ed.), Plenum, 1977.
48. M. H Ernst, E. H. Hauge, and J. M. J. van Leeuwen, *Phys. Rev. Lett.* **25**, 1254 (1970); *Phys. Rev. A* **4**, 2055 (1971).
49. T. Keyes and I. Oppenheim, *Phys. Rev. A* **8**, 937 (1973).
50. T. Keyes and I. Oppenheim, *Physica* **81A**, 241 (1975).
51. D. Bedeaux and P. Mazur, *Physica* **73**, 431 (1974); P. Mazur and D. Bedeaux, *Physica* **75**, 79 (1974); D. Bedeaux and P. Mazur, *Physica* **80A**, 189 (1975).
52. J. Bosse, W. Gotze, and A. Zippelius, *Phys. Rev. A* **18**, 1214 (1978).
53. J. R. D. Copley and S. W. Lovesey, *Rep. Prog. Phys.* **38**, 461 (1975).
54. I. A. Michaels and I. Oppenheim, *Physica* **81A**, 221 (1975).
55. M. Tokuyama and I. Oppenheim, *Physica* **94A**, 501 (1978).
56. A. J. Masters and P. A. Madden, *J. Chem. Phys.* **74**, 2450 (1981); *ibid.* **75**, 127 (1981).
57. A. J. Masters, Ph.D. Thesis, Cambridge Univ., U.K., 1980.
58. R. Peralta-Fabi, Ph.D. thesis, Univ. Autonoma de Mexico, 1975; R. Peralta-Fabi and R. Zwanzig, *J. Chem. Phys.* **70**, 504 (1979).
59. T. Keyes, T. Morita, and J. Mercer, *J. Chem. Phys.* **74**, 5281 (1981).
60. D. Ronis, D. Bedeaux, and I. Oppenheim, Physica **90A**, 487 (1978).

ON INVARIANCE OF LOCALIZED HAMILTONIANS UNDER FEASIBLE ELEMENTS OF THE NUCLEAR PERMUTATION–INVERSION GROUP

GRIGORY A. NATANSON†

Department of Chemistry and the James Franck Institute
The University of Chicago
Chicago, Illinois 60637

CONTENTS

I. EQUIVALENCE OF IDENTICAL PARTICLES AND DYNAMICAL SYMMETRY CAUSED BY IT

The concept discussed here is based on a trivial supposition that any approximations to an accurate Hamiltonian must be expressible in a form invariant under its symmetry group. Of course, an *additional* symmetry may

† Present Address: JILA, University of Colorado, Boulder, Colorado 80309

appear, but then the accurate symmetry group is necessarily a subgroup of the approximate one. In particular, there must be a way to cast an approximate Hamiltonian under consideration so that identical particles are treated in an equivalent way, even if this identity is not ordinarily invoked.

If all nuclei are assumed frozen in an equilibrium configuration, the rigid-body Hamiltonian is obviously invariant under permutations which just rotate the molecule as a whole (we consider here *proper* rotations). Hund[1] appears to have been the first who recognized this connection between nuclear permutations and point symmetry operations of the equilibrium configuration in question. Moreover, he even attempted to classify normal vibrations by means of irreducible representations of the appropriate symmetric group. Following Brester,[2] Hund[1] used Cartesian coordinates of nuclei to describe vibrations, and to a certain extent his analysis of vibrational symmetry anticipated the well-known work of Wigner.[3] At the time, Hund could only declare intuitively that the symmetrization of normal vibrations he used was nothing but their symmetrization with respect to some nuclear permutations. The change of molecular variables which made a proof of this assertion possible was suggested by Eckart[4] 8 years later.

After Eckart's work the invariance of the Born–Oppenheimer Hamiltonian under the above permutations could be verified explicitly. However, it took about 30 years[5,6] to do so. If one remembers that the Born–Oppenheimer method itself was only accurately applied to *polyatomic* molecules 40 years[7‡] after publication of Born and Oppenheimer's work,[12] the situation does not seem so paradoxical.

The extension of those ideas to nonrigid molecules was also done in the 1930s[13,14] with reference to ethanelike molecules. The arguments supporting this extension were very close to those mentioned above for quasirigid molecules with frozen vibrations. Again, as long as vibrations around a *semirigid* molecular geometry are assumed frozen, the Hamiltonian invariant under selected permutations can be constructed explicitly.

It may seem that permutational symmetry is broken as soon as some internal motions are considered frozen or just localized in a small domain of the configurational space. However, it should be taken into account that all permutations different from the selected ones transform equilibrium configurations in question to others which cannot be obtained from the so-called "feasible" configurations by any translation and overall proper rotation (we[15–17] call two configurations equivalent or nonequivalent according as they can be connected in such a way or not). The fact that sets of rovibrational

‡Unfortunately both Kiselev's work[7] and our simplification[8] of the Born–Oppenheimer method were overlooked by the authors of Refs. 9 and 10. In particular, three pages of cumbersome computations adopted by Lathouwers and Van Leuven[10] from Born and Huang[11] go in one paragraph of our paper.[8]

functions associated with nonequivalent equilibrium configurations of a quasirigid molecule can be treated as linearly independent and that appropriate degenerate energy states span representations of the complete nuclear permutational group *P* was also recognized by Hund[1] for the particular cases of ammonia and methane. In this connection we should also cite Ref. 13, where this statement was extended to ethanelike molecules, and Wilson's work[18,19] representing a systematic method for calculations of the nuclear statistical weights.

Note that both works[13,14] on ethanelike molecules started with known symmetry properties of the torsional equation. In a similar manner Wilson et al.,[20,21] studying the symmetry of Hamiltonians of rotating nitromethane- and propanelike molecules with torsional internal degrees of freedom, and Kasuya[22] (see also Ref. 23), studying the symmetry of Hamiltonian of the rotating hydrazine molecule with one torsional and two wagging internal degrees of freedom, found the nuclear permutations keeping those Hamiltonians invariant. The main defect of those and similar works[24-26] carried out in the early 1960s is that they turned the problem upside down: instead of answering the question of what minimum symmetry the Hamiltonian under consideration *must have*, their authors just introduced a Hamiltonian *a priori* and then analyze its symmetry properties. To do it one must be sure that the symmetry of the model adequately represents the real symmetry of the molecular system in question and, in particular, that the equivalence of identical nuclei is not broken. Again we come back to the question of how to describe internal nuclear motions localized in some domains of configuration space while treating all identical particles equivalently.

Berry[27] was the first who looked at the matter the other way around. First of all, he explicitly emphasized that the Born–Oppenheimer approximation breaks the symmetry of the molecular Hamiltonian under some permutations. It means that one should distinguish between symmetry groups of global (accurate) and localized Hamiltonians. As discussed in the next section, this conclusion forms the basis of the concept of molecular symmetry. "No molecule containing four or more identical nuclei can be entirely described by the small-oscillation model,"[27] *even if sufficiently high-order terms in the Born–Oppenheimer expansion*[12,7,8] *are taken into account*,[28] that is, almost any molecule may be treated as nonrigid and the only question is whether or not an experimental technique allows one to detect the proper splittings. As long as the probability of tunneling between nonequivalent equilibrium configurations is not negligible (see the next section for discussion), the permutations mapping those configurations onto each other become feasible. As the most impressive illustration of these ideas we can cite Berry's[27] discussion of tunneling processes in the PF_5 molecule, for which *new* (compared with nearly rigid molecules of the same structure) feasible permutations are introduced with no relation to a particular Hamiltonian.

We consider this excursus into history of the problem very important for understanding the feasibility concept formulated by Longuet-Higgins.[29] Going beyond Berry's work,[27] Longuet-Higgins included into consideration the inversion of laboratory-fixed Cartesian coordinates of particles with respect to the center of mass and combined all feasible operation into groups. However, his work still does not answer the question of why the molecular Hamiltonian must be invariant only under "feasible" operations and what happens with its symmetry with respect to all other permutations and permutation–inversions.

It was Watson[30,31] who explicitly stressed the fact that localization of wavefunctions in several disconnected "feasible" domains of the nuclear configurational space manifests itself in clustering of energy levels. This approximate degeneracy, referred to by Bunker[32] as "structural degeneracy," becomes accurate if one replaces all finite unfeasible barriers with infinite ones. We are thus led to an approximate global Hamiltonian which is represented as a sum of localized Hamiltonians.[28,33–35] Each localized Hamiltonian is defined in its feasible domain and must be invariant under permutations and permutation–inversions which convert the domain into itself. All other elements of the complete nuclear permutations–inversion (CNPI) group Π map that domain onto another.[35,36] As a result the approximate global Hamiltonian under discussion turns out to be invariant under any element of Π, and "moreover has a hidden symmetry which results in an additional degeneracy of energy levels."[35] The explicit structure of the appropriate symmetry group responsible for the degeneracy was obtained by the author[37] and is discussed in the next section. We do not think that this group may be actually useful for interpretation of molecular spectra, as it attains unmanageable proportions even more rapidly than the complete nuclear permutation–inversion group does (cf. Ref. 38), whereas (and it is more essential) all selection rules can be obtained by analyzing symmetry properties of a localized Hamiltonian by means of the feasible permutation–inversion (FPI) group (the molecular symmetry group in Bunker's terms[32,38]). We just believe that the explicit construction of the invariance group makes the discussion clearer and shows that there is no significant difference between the structural degeneracy and any others caused by usual symmetries (cf. Ref. 39, p. 161).

The localized Hamiltonian *must be* invariant under all nuclear permutation and permutation–inversions which convert into itself the domain where this Hamiltonian is defined. Hence, as followed from Hougen's works,[5,6,40–43] this Hamiltonian *must be* invariant under appropriate transformations of large- and small-amplitude variables. The natural question is why this Hamiltonian *does* turn out to be invariant. In Hougen's works[5,6,40–43] the invariance under discussion is taken for granted. The answer was given in our works.[28,33–35] As discussed in Section V, at least for nonrigid molecules, an accurate analysis of the problem is not trivial. Of course, if one chooses the localized Hamiltonian *a priori*, there is no such question at all. But what we are actually interested in is how parameters in that Hamiltonian (for example, force constants) are connected to *ab initio* ones, and this part of our paper (Sections V and VII) is original.

As we deal with approximate symmetry, "there is no absolute criterion for the feasibility of a given element" of the CNPI group.[30] The only way to choose between different possibilities is the method of trial and error based on the Watson–Dalton[30,44] "correlation rules for how to observe the splittings due to exchanges between potential minima"[45] and on some general ideas about chemical bonds.

It is worthwhile stressing again that there exists no alternative to this general conclusion. As long as we have fixed an accuracy of our calculations (or observations), we have to select feasible operations from the CNPI group in order to interpret real patterns of energy levels. The next step is to construct a localized Hamiltonian invariant under those operations. Both the isometric theory[46–52] and the invariance group of the Eckart frame[53] are introduced in connection with particular ways of constructing such localized Hamiltonians, and before using those Hamiltonians one should verify that their invariance groups satisfy the criterion of feasibility. The feasibility concept seems "intuitive"[46,48,49,53] only as long as those authors pass over the question: to what extent is the model used able to represent a real molecule, and why does the model of *this* symmetry turn out to be applicable?

As stated in Ref. 54, "the spectroscopic community seems to become increasingly divided into schools that adhere to one or the other of various rival schemes," and our aim is to show once again that there exists only single universal approach. This approach can certainly be developed in different directions, but any of those developments must have the feasibility concept as a foundation.

In Section III we shall give a more detailed analysis of the isometric approach. But even putting aside our objections to some assertions of its authors,[46–52,55,56] we should stress that it deals with very specific molecular systems in which almost all internal motions are forzen, and moreover there must exist a semirigid molecular model (SRMM) of *definite symmetry* to describe nonfrozen large-amplitude internal motions. In this connection Slanina's statement[57] that the "so-called feasibility concept has recently been replaced by a rigorous mathematical definition in the isometric approach" sounds at least stretched.

The main defect of Altmann's arguments[58–61] is that he forms groups from some transformations of nuclear displacements without discussing the question of whether or not they are symmetry operations of any particular

Hamiltonian. In addition, those transformations are determined mainly by means of pictures and hence can be easily misinterpreted.

Clearly, using different combinations of point symmetry operations and nuclear permutations, one is able to construct a whole bunch of groups. As noticed by Ezra,[62-64] the device initially suggested by Altmann[58] differs from the one discussed in the monograph[60] (see also Ref. 61). According to Altmann,[59] the point symmetry group is formed by the point symmetry operations of an equilibrium configuration in a body-fixed frame, which are not symmetry operations of the vibronic molecular Hamiltonian even if nuclei perform just small-amplitude vibrations near the above equilibrium configuration. (Remember that Altmann[59] distinguishes those operations from the Wigner symmetry operations,[3] which are products of point symmetry operations and nuclear permutations). If we side with Altmann's statement[58] that isodynamic operations, are not symmetry operations at all, there is no point in discussing them here. It seems verisimilar[62-64] that Altmann[59-61] means some products of pure nuclear permutations and point symmetry operations. However, we do not think that those products necessarily coincide with isodynamic operations as defined by Ezra.[62-64] In fact, in order to commute with the point symmetry group C_3 of "the ordered configuration of orthoformic acid,"[59] Altmann's isodynamic operation V^I must be a pure permutation, whereas according to Ezra's definition of isodynamic operations it must be a product of that permutation and a reflection in a plane. If our interpretation of Altmann's works[58,59] is correct, the supergroup of the ordered configuration of CH_3BF_2 must be isomorphic to $C_6^I \times C_s$ and has no relation (cf. the note on p. 339 in Ref. 60) with the dynamical symmetry group of the rotation–torsional Hamiltonian found in Ref. 20 (see Section III for comments).

Inasmuch as Altmann does not use the term "supergroup" in his monograph[60] and in a recent review,[61] we can conclude that this term has been withdrawn from use. We would not come back to this notion had it not recently been advocated by Flurry[65-69] and to some extent by Smeyers.[70] It is worthwhile stressing that although some of the decompositions discussed by Smeyers[70] were first given by Altmann,[58] it was Woodman[71] who substantiated Altmann's device for molecules with internal rotation (and only for them).

If we make use of Ezra's definition of isodynamic operations we are led to the group isomorphic to the feasible permutation–inversion group. However, we cannot agree with Altmann's statement[61] that it is easier to deal with those operations than with permutation–inversions. In fact, in order to define an isodynamic operation we need to introduce not only feasible permutation or permutation–inversion but also body-fixed axes. As initially discussed in our papers[33-35] and independently by Ezra,[62-64] one can easily do this if there exists a semirigid model of an appropriate symmetry. But it is not obvious that one can construct such a model in the general case. What must really be credited to Altmann is his idea to use semidirect products as a powerful tool for describing a structure of symmetry groups of molecules with internal rotation. The influence of Altmann's works[58-60] on the authors of the isometric approach,[46-51] as well as on Ezra's[62-64] and Woodman's[71] work (and through Refs. 64 and 71 on our recent work),[37] is undisputable.

Some general factorizations of feasible permutation–inversion groups into semidirect products were recently discussed in detail by Ezra.[62-64,72,73] We completely agree with his conclusion that there exists no general device for an arbitrary molecule. Only *for molecules with internal rotation of atomic groups around a nonlinear immobile frame* can one always carry out the factorization $G^I \wedge (G^P \wedge K)$ (in Ezra's terms), for example, combining into the group $(G^P \wedge K)$ all feasible operations which map the first atom of each top into itself or into the first atom of any other identical top.

Note that the point group $\mathbb{G}^P$ of the SRMM as defined by Ezra has generally nothing to do with "the group of the molecular constraints," in contrast with the statement in Ref. 61. In particular, the point group of the three-parameter SRMM of boron trimethyl has no elements except the unit.

There do exist molecules like acetone for which "the group of molecular constraints" leaves the equilibrium positions of all atoms unaltered, and Maruani et al.[74] appear to mean those molecules when giving their definition of "the Altmann group." However, neither boron trimethyl nor CH_3BF_2 (Altmann's favorite examples) fall under this definition. If we look at Fig. 1 in Ref. 74, only molecules i–n satisfy the aforementioned requirement.

As stated by Flurry and Abdulnur,[68] Altmann's "development is very convenient for chemists, since it draws upon familiar concepts." But what are those concepts? Of course, most chemists are used to treating molecules as rigid structures. But what relation do those concepts have to a picture in which nuclei just move about near an equilibrium configuration? The usual response is to appeal to Wigner's work.[3] But remember that Wigner[3] considered nuclei oscillating near a configuration *fixed in space*. It is a model for a molecule in a crystal but not for a free molecule. It was Eckart[4] who gave a particular change of variables which substantiates Wigner's device.[3] As immediately follows from Eckart's paper,[4] permutations and permutation–inversions just rotating the equilibrium configuration in question generate Wigner symmetry operations in the space of frame-fixed displacements. Of course, by that time physicists appear to have recognized this fact intuitively (at least it is surely true for permutations[18,19]), but Eckart did prove that such rotating axes exist. Unfortunately, a detailed analysis of how transformations of the Born–Oppenheimer variables are related to those permutations and permutation–inversions was given only by Hougen[5,6] a quarter of a century later. As first stressed in our work,[33] each term in the expansion of the Wilson–Howard Hamiltonian[75] as a Taylor series in vibrational coordinates is invariant under the mentioned operations, as the latter generate linear homogeneous transformations of vibrational coordinates.

Note that the concept briefly formulated above concerns nearly rigid molecules, and it must be included in any introductory course in molecular spectroscopy. Bunker's book[32] is an excellent example of such an exposition (see also his reviews[38,76]).

As long as only nearly rigid molecules are under discussion, it is still possible to forget about the grounds and to just use recipes given by standard textbooks. But for nonrigid molecules those recipes change. As stressed in Refs. 77 and 78, the Wigner approach does not work in the general case. We refer the reader to our review[79] for the detailed analysis of the problem for molecules with internal rotation.

It is difficult for us to judge to what extent the pictorial device in Flurry's monograph[66] should be credited to Altmann, but it has certainly led to wrong numbers of infrared- and Raman-active fundamentals for the propane molecule.[65] The correct analysis was carried out by Bunker.[38] It was shown that the symmetry classification of normal vibrations is exactly the same at the low- and infinite-barrier limits (see also comments in Section 7). Arguments similar to the ones made by Crawford and Wilson[80] allow one to conclude that the selection rules, at least for infrared spectra, are also the same at the two limits.

Note that propane is a particular representative of molecules with internal rotation of methyl groups, and hence its **G** matrix can be made independent of torsional angles by means of a special change of variables.[32,38,71,79–82] The problem becomes much more complicated for *n*-alkanes with a larger number of carbons if one has to consider internal rotations around all C—C bonds as done by Flurry and Abdulnur.[68]. The difficulties appear because for those molecules, in contrast with propane, one should generally use double groups generated by internal rotations of atomic groups C–CH_2–C around C—C bonds (see Section VII). Flurry and Abdulnur[68] completely ignore this problem for *n*-alkanes.‡

The classification of elements of the FPI group necessary for a symmetry analysis of normal vibrations in molecules with internal rotation was given in our papers[28,35,79,85,86] and briefly reviewed in Section VII.

The organization of this paper is as follows. In Section II we give a more detailed analysis of the feasibility concept. We revise general properties of a global Hamiltonian resulting from the assumption that some barriers be-

‡We cannot understand either the motives which induced those authors to introduce a new factorization of the symmetry group[83] with a nontrivial character table instead of the correct one used by Flurry[65] earlier (see also Refs. 70, 84).

tween domains can be treated as infinite. The direct consequence of the assumption is an additional "hidden" symmetry of the approximate global Hamiltonian in question and resultant structural degeneracy of energy levels discussed in Section II.A. The aim of Section II.B is to show that only tunneling between right- and left-handed forms of optical isomers necessarily requires the use of the nonstationary theory. The nuclear probability density of any optically nonactive molecule or particular optical isomer must be the same near each equilibrium configuration, and one cannot introduce wave packets moving from one equilibrium configuration to another without violating the exclusion principle. In Section III we consider molecular structures with a few internal degrees of freedom. The section is subdivided into four subsections. In Section III.A we trace the relation between the symmetry group $\mathbb{H}$ of an SRMM as defined by Ezra,[62-64] an invariance group of an appropriate Hamiltonian, and the so-called dynamical PI (DPI) group $\mathscr{F}_{\theta,\rho}$ of a free-rotating SRMM. As discussed in that subsection, there always is a homomorphism (or just isomorphism) from $\mathbb{H}$ onto $\mathscr{F}_{\theta,\rho}$, but in contrast with the former the group $\mathscr{F}_{\theta,\rho}$ is a subgroup of the invariance group of the Hamiltonian, as explicitly shown in Section III.C. In Section III.B we briefly revise the symmetry theory of an impurity rigid nonlinear molecule rotating in a crystal field. Section III.D contains a review of some additional "hidden" symmetries of the localized Hamiltonian in question which arise for some SRMMs.

The original part of this work, including consideration of nuclear vibrations, is mainly presented by Sections IV–VI, where we formulate sufficient requirements for a localized nuclear–electronic Hamiltonian to be invariant under elements of the FPI group. In Section VII we briefly discuss some peculiarities of the symmetry classification of normal vibrations in molecules with internal rotations (one can find more details in our review[79]). In appendices we give a critical analysis of different methods to express the nuclear Hamiltonian in terms of small- and large-amplitude variables. In Appendix A we also select the term corresponding to frozen vibrations. If one uses least-squares large-amplitude variables, this term precisely coincides with the localized Hamiltonian discussed in Section III.C.

II. A POLYATOMIC MOLECULE AS A CLUSTER IN THE NUCLEAR CONFIGURATIONAL SPACE

A. Hidden Symmetry of a Global Hamiltonian and Structural Degeneracy of Energy Levels

A very important peculiarity of most polyatomic molecules, which forms a basis of the approach under consideration here, is the fact that wavefunctions are localized in several symmetrically equivalent disconnected domains of nuclear configuration space, and elements of the CNPI group Π map those domains into each other.[30,31,36] To keep the discussion simple we assume the present adiabatic approach is a good zeroth-order approximation; this assumption then makes it possible to delimit different *feasible* domains of the nuclear configuration space by means of values of the energy barrier between them[30,31] (see also Refs. 28, 33–35, 39, 42). We shall show in the next subsection how this restriction can be removed.

Let us consider different nonequivalent stable equilibrium configurations‡ which are mapped onto each other by elements of Π. All those configurations have the same value of the potential energy. The smallest maximum value of the adiabatic potential along all the paths between two equilibrium configurations is called the *barrier* between the configurations in question. All motions within a connected domain of the nuclear configuration space are referred to as *feasible* if the potential energy within that domain is less than a fixed value W_0. This value is determined eventually by the accuracy (energy resolution) to which the Schrödinger equation is solved.[28,35] As illustrated in Ref. 31 by means of a one-dimensional system, the cluster behavior of wavefunctions is to a considerable extent a mathematical problem[36] (see also Ref. 87, p. 52). In order to stress this fact we prefer to refer to the accuracy of calculations instead of the energy resolution of a given experiment, in contrast with Refs. 30, 31, 39, 42, 44. Of course, if one is interested in calculations of molecular spectra, the choice of accuracy is determined by the spectral resolution of observations.

To achieve the idealized limit in which the hidden approximate symmetry becomes exact, we suppose that all finite barriers exceeding W_0 are treated as *infinite*.[31] As a result, some levels become degenerate. This degeneracy corresponds to unresolved spectral levels if the accuracy of the calculations is assigned to correspond correctly to the resolving power of the given experiment.

Of course the above definition of feasibility is difficult to apply to a concrete molecular system, as the barriers are generally unknown, but it allows a precise definition of *feasible operations* as the set of those converting the domain under consideration into itself,[30] with the corresponding splittings resolved. It is easily verified that operations feasible in the domain under consideration form a group, which is denoted by $G^{(i)}$ for the ith domain. All groups $G^{(i)}$ are isomorphic to each other.

One of the ways to define a feasible domain directly is to label all nonequivalent equilibrium configurations which are included in it[44] (we call a domain *elementary* if it contains only equivalent equilibrium configurations).

‡The term "nonequivalent configurations" adopted from our earlier papers[15–17] seems misleading in the present context as we refer to equilibrium configurations which are symmetrically equivalent but inconvertible by a motion.

It should be stressed that any choice of a semirigid molecular model implies that the model when deforming goes through all of the configurations in the domain. Besides this implication, one also has to choose intermediate configurations, and hence any approach based on study of the symmetry of the model needs some assumptions in addition to the concept of feasibility.

Once feasible domains have been selected, the next step is to construct a localized Hamiltonian $\hat{h}^{(i)}$ $(i=1,2,\ldots,n)$ in each of them. Eigenfunctions of each of these Hamiltonians are assumed to vanish beyond the domain in which the Hamiltonian is appropriate, and hence the proper Schrödinger equation may be solved independently for each domain. It should be noted that the term "localized Hamiltonian" was adopted by us[33] from the work[88] of Gilles and Phillipot but as explained below for *nonrigid* molecules we use this term in a different sense.

The global approximate Hamiltonian is represented as a sum

$$\hat{H}=\sum_{i=1}^{n}\hat{h}^{(i)} \tag{2.1}$$

[cf. (22) in Ref. 88]. To retain its invariance under the Π group we require that each localized Hamiltonian $\hat{h}^{(i)}$ be invariant under the subgroup $G^{(i)}$. Of course the molecular Hamiltonian may be only *approximately* invariant if splittings caused by symmetry breaking are less than the maximum resolving power of the given experiment (see Ref. 39, pp. 148, 149). It is however easier on the whole to verify the exact invariance than to estimate errors resulting from its breaking.

To prove that the invariance of $\hat{h}^{(i)}$ under $G^{(i)}$ leads to the invariance of the global Hamiltonian (1.1) under Π, we should study[35] how different localized Hamiltonians $\hat{h}^{(i)}$ are connected with each other. Let us fix a domain (say 1), expand the Π group into left cosets of $G^{(1)}$, and select a single representative $\hat{\pi}_i$ from each coset. Each selected element $\hat{\pi}_i$ generates a mapping of the first domain into the ith, and two different elements $\hat{\pi}_i$ and $\hat{\pi}_j$ $(i \neq j)$ map it into two different domains. We define $\hat{h}^{(i)}$ as

$$\hat{h}^{(i)}=\hat{\pi}_i\hat{h}^{(1)}\hat{\pi}_i^{-1} \tag{2.2}$$

As any element of Π is represented as $\hat{\pi}_i\hat{g}\hat{\pi}_j^{-1}$, where $\hat{g}\in G^{(1)}$, one can easily verify that the Hamiltonian (1.1) does commute with all elements of Π. A very similar analysis of the problem in slightly different terms was given by Pedersen.[36] Gilles and Philippot[88] used an analogous argument to prove that a *localized* Hamiltonian of a nonrigid molecule (the delocalized Hamiltonian in terms of Ref. 88) represented as a sum of the Born–Oppenheimer Hamiltonians (i.e., of ones having elementary domains as their support) is invariant under operations permuting elementary domains, but those authors confused the issue by using new operations, the so called "perrotations," instead of feasible permutations and permutation–inversions.

As stressed in Ref. 28, expanding a feasible domain complicates the solution of the Schrödinger equation, that is (paraphrasing Ref. 39, p. 148) one should only do "as much as necessary but not as much as possible." Note that combining some domains into one leads to a higher symmetry of the localized Hamiltonians. We are led to the apparent paradox: "an increase in symmetry is associated with a splitting of degeneracies"[42] (see also Ref. 89, p. 232; and Ref. 90). But one should remember that $\hat{h}^{(i)}$ is just an auxiliary Hamiltonian, with the property that only its eigenvalues but not their degeneracies can be used for interpretation of real spectra. To find the required degeneracies one must consider the *global* Hamiltonian having exactly the same energy spectrum. Its symmetry does become lower as the combined domain is considered. The structure of the proper hidden symmetry group responsible for degeneracies of levels of the global Hamiltonian was found in Ref. 37. The group contains all possible mappings of domains onto each other such that for any mapping and any pair of domains (say i and j) there exists either a permutation or a permutation–inversion converting the ith domain into the jth in exactly the same way as the mapping in question does. Those mappings form the group T^{glob}, which can be represented as semidirect product

$$T^{\text{glob}} = (M^{(1)} \times M^{(2)} \times \cdots \times M^{(n)}) \wedge \mathrm{S}_n, \tag{2.3}$$

where S_n is the symmetric group and any element of the subgroup $M^{(i)}$ converts the ith domain into itself in exactly the same way as the proper element of $G^{(i)}$ does but, in contrast with the latter, keeps all other domains unchanged. If we choose operations $\hat{\pi}_i$ as described above and define the *standard projection* of the ith domain into the jth as the mapping generated by $\hat{\pi}_j \hat{\pi}_i^{-1}$, the group S_n is formed by automorphisms which map any domain into one of its standard projections.

In order to explain structural degeneracy of energy levels for states totally symmetric with respect to any subgroup $M^{(i)}$, we[37] had to introduce additional symmetry operations $\hat{C}_i$ which just multiply functions localized in the ith domain by -1. The degenerate states in question span *irreducible* representations of the group

$$D^{\text{glob}} = \Delta \wedge T^{\text{glob}}, \tag{2.4}$$

where the invariant subgroup Δ is generated by operations $\hat{C}_i$. One has[37]

$$G_{\text{rig}}^{(i,\alpha)} \subset G_{\text{nonrig}}^{(i)} \subset \Pi \subset D_{\text{nonrig}}^{\text{glob}} \subset D_{\text{rig}}^{\text{glob}} \tag{2.5}$$

where the index α numbers nonequivalent equilibrium configurations included in the ith feasible domain and $G_{\text{rig}}^{(i,\alpha)}$ is the symmetry group of the Born–Oppenheimer Hamiltonian (for nonlinear molecules it is isomorphic to the point symmetry group of the equilibrium configurations). If the whole space is treated as feasible, both groups $G_{\text{nonrig}}^{(i)}$ and $D_{\text{nonrig}}^{(i)}$ coincide with Π. Of course the chain in Eq. (1.4) can be extended by introducing some intermediate nonrigid symmetry groups G_{nonrig}.

Let us consider the PF_5 molecule and choose the alternant subgroup A_5 of S_5 as the feasible permutation group.[91] In this case any feasible permutation–inversion must be the product of an *odd* permutation and the inversion (the group Q in Table 2 of Ref. 91). Then one finds

$$G_{\text{rig}}^{(i,\alpha)} \overset{\text{iso}}{=} D_{3h}, \qquad i=1,2; \quad \alpha=1,2,\ldots,10 \tag{2.6a}$$

$$G_{\text{nonrig}}^{(1)} = G_{\text{nonrig}}^{(2)} = G_{120} \tag{2.6b}$$

$$\Pi = S_5 \times \{\hat{e}, \hat{\imath}\} \tag{2.6c}$$

where $\hat{\imath}$ is the inversion,

$$D_{\text{nonrig}}^{\text{glob}} = \left[\prod_{i=1}^{2} \left(M^{(i)} \times C_2^{(i)}\right)\right] \wedge S_2 \tag{2.6d}$$

$$D_{\text{rig}}^{\text{glob}} = \left[\prod_{i=1}^{2} \prod_{\alpha=1}^{10} \left(M^{(i,\alpha)} \times C_2^{(i,\alpha)}\right)\right] \wedge S_{20} \tag{2.6e}$$

Splittings of energy levels corresponding to the chain $D_{\text{rig}}^{\text{glob}} \supset \Pi$ are discussed by Dalton[92] (see also Refs. 90, 93).

Therefore we[37] gave the explicit expression for the "group whose irreducible representations correspond to the degenerate states, and from which the splittings, such as those derived by Dalton, can be derived as consequences of *symmetry breaking*, rather than by inducing a higher symmetry group G from a lower symmetry group G^0 and generator G'"[39]. It is worthwhile noting that Berry (Ref. 39, pp. 148, 149), in defining the molecular symmetry group (the FPI group in our terms), follows Dalton's interpretation[44] (see also Refs. 90, 94) of Longuet-Higgins's work.[29] The crucial difference between the induction procedure suggested by Dalton[44] and the account developed here is that we, like Watson[30,31] (see also Ref. 36), answer the question *why one can exclude unfeasible operations from consideration*, whereas Dalton takes "the viewpoint of asking *how to interpret newly-resolvable splittings in terms of new feasible operations*." Remember that permutations of *identical* particles are symmetry operations for any molecular system (we do not consider here chemical isomers like the isomers of the butyl ion[90,94]), and the fact that we can limit ourselves to consideration of only some of them does need explanation.

More than 15 years ago we asked this question ourselves, running across the assertion of Petrashen and Trifonov (see p. 232 in their monograph[89]) that "transition to a more approximate model reduces the degree of symmetry rather than otherwise." We have cited the similar statement of Hougen,[42] but at that time neither the authors of Ref. 89 nor we knew the works of Hougen[5,6,42] and Watson.[30] So, just studying Wilson's works[18,19] carefully, we came to the idea of a "hidden" symmetry which arises when some barriers are assumed infinite. The answer to the first question immediately led to a second one: why molecular Hamiltonians used by spectroscopists *turn out* to be invariant under permutations and permutation–inversions rotating a rigid molecular model (of course at that time we were interested only in nearly rigid molecules). An accurate analysis of Eckart's paper[4] gave the answer. A few months later we found similar ideas in Watson[30] and Hougen's[5,6] papers and considered the question closed. We were recently surprised to find that the term "hidden" or "approximate" symmetry in the context explained above has been *explicitly* used only by us,[35] and moreover that we did not apply this term in Refs. 28, 33, 34 ourselves when discussing this problem.

Of course, all the discussion here concerns only definitions, not the induction procedure itself. We must define this group as suggested by Watson,[30,31] but in practice it would be very difficult to apply this definition for such molecules as PF_5[91,92,95,96] or XeF_6.[97–99] For those purposes Dalton's induction procedure[44,92] especially as developed by Brocas and Fastenakel[91,95,100] (see also Refs. 96, 98, 99, 101 and Brocas's review[102]), is extremely useful.

Although the Pauli principle does not generally permit one to limit oneself to consideration of a single domain (even if it is separated by infinite barrier from the rest of space), this can be done as long as only energy levels and matrix elements used in different applications of the *stationary* theory are of interest. We are interested only in this part of the problem below, and so select a single domain and omit the superscript i in the notion $\hat{h}^{(i)}$, $G^{(i)}$.

As initially stressed by Berry,[27] the localized Hamiltonian $\hat{h}$ is invariant only under feasible operations $\hat{g} \in G$, not under other elements of Π. So as long as the localized Hamiltonian is applied, one has to use its symmetry group G. It is not a matter of convenience as declared in Ref. 103. The statement in Ref. 103 that the Π group is too large to be useful has no sense. For example, for PF_5 we have to use this group[92,93] or ones[91,95,96] which are just slightly smaller (see also Ref. 98 in this connection). One need consider only the subgroup G of this group because of the "cluster" structure

[Eq. (2.1)] of the global Hamiltonian. As a result, the "inconveniently large"[30,32,104] CNPI group produces "a symmetry labeling in which accidental degeneracies occur in a systematic fashion."[32] So we actually need even a larger group D^{glob},[37] but fortunately the wavefunctions under consideration span only such irreducible representations of this supergroup as are induced from different irreducible representations of the G group (see also Ref. 105, p. 46), and hence the latter allows us "to label the energy levels as fully as necessary."[38]

All that was said above about the hidden symmetry group D^{glob} can be extended to the nonrelativistic Hamiltonian dependent explicitly on spin variables of nuclei and electrons. But as long as we consider total wavefunctions including nuclear spins, we should take into account that the Pauli principle allows only states spanning one-dimensional representations of the FPI group,[18] and hence a statistical weight of any energy level coincides with a multiplicity of degeneracy of eigenvalues of the localized Hamiltonian in question. It was Ezra[83] who called our attention to the fact that structural degeneracy can be broken even if all barriers are still assumed infinite, and to experiments of Borde et al.[106,107] where such splittings were observed. Only if the localized Hamiltonian has additional *nontrivial* symmetry does the hidden symmetry group [Eq. (2.4)] manifest itself in structural degeneracy of energy levels in addition to degenerate doublets characteristic of optically active molecules.

Note that the analysis given above is applicable to any symmetry operations which interconvert domains separated by nearly infinite barriers. In particular, if we place a molecule in a crystal field, the global Hamiltonian of the molecule is invariant only under a discrete number of proper or improper overall rotations. Following Gilles and Philippot[88] and Ezra,[63,64] we use the term "perrotations" for products of those rotations and nuclear permutations, but of course their rotations differ from ours (see Section III). Those products form the complete nuclear perrotation (CNPR) group. We have not included in our consideration proper overall rotations for isolated molecules, as in that case any of those rotations is feasible by definition. In an external field even a proper overall rotation may bring a molecule from one minimum to another. We shall briefly discuss a structure of the feasible perrotation (FPR) groups in the next section.

It should be stressed again that we introduced the group D^{glob} just to explain the structural degeneracy of energy levels. We by no means suggest using the global Hamiltonian instead of a localized one. Certainly it is much more difficult to deal with its symmetry operations than with the FPI group. That is why we oppose the idea, due to Louck and Galbraith,[53] of considering both proper and improper rotations even if the barrier between right and left forms is treated as infinite. The Hamiltonian corresponding to such a description[64,108] has the hidden symmetry, which manifests itself not only in the structural degeneracy of energy levels but also in the fact that any matrix elements between wavefunctions localized in different domains must vanish. We have to deal with a number of wavefunctions, which is twice larger than we actually need, and to employ artificial symmetry operations (see Appendix C in Ref. 109 for example) in order to explain degeneracies of energy levels and selection rules dictated by the hidden symmetry. When we limit ourselves to consideration of only a localized Hamiltonian, this kind of symmetry is taken into account in a natural way. We refer the reader to our paper[110] for a more detailed discussion.

Of course we must combine the two feasible domains corresponding to right- and left-handed forms of a molecule if the inversion is assumed feasible. However, the question of how the formalism treating improper matrices as dynamical variables can be applied to the interpretation of inversion splittings caused by tunneling between the two forms has been never discussed.

To be correct we should mention that it was Landau and Lifshitz[111] who implicitly suggested the idea of combining two feasible domains converted into each other by the inversion in order to include in consideration a larger number of permutations for calculating nuclear statistical weights. We also advocated this device in our work,[28,33] because in contrast with Wilson's recipe[18] it allows making use of the well-known representations of the symmetric group when decomposing the space of nuclear spin functions into irreducible subspaces for such molecules as ammonia and methane. However, we did not recognize the fact that one can get the same advantage employing only feasible operations, as suggested by Longuet-Higgins [29].

Let us call the subgroup of nuclear permutations converting the combined domain into itself the *semifeasible permutation* (SFP) *group* S^{P}, as it contains an equal number of feasible and unfeasible permutations. By analogy we define the *semifeasible permutation–inversion* (SFPI) group" S^{PI} by adding the inversion to the SFP group. As the inversion $\hat{\imath}$ keeps spin variables unchanged, feasible operations act on spin variables in exactly the same way as elements of the SFP group do, and hence we obtain the same symmetry species whether we decompose the space of spin variables into irreducible subspaces with respect to the FPI or the SFP group.

Note that there exists a crucial difference between the labeling schemes discussed by Landau and Lifshitz[111] and by Oka.[112] The latter uses symmetry types of the FPI group as main symbols, whereas the symmetry classification introduced by Landau and Lifshitz and recently advocated by several authors[64,105–107,113–115] implies relabeling energy levels according to symmetry species of the SFP group S^{P}. Clearly there exists no reason to use either scheme of labeling energy levels by irreducible representations of the SFPI group S^{PI} if hyperfine patterns are unresolved. However, in order to study those patterns Oka needs just to introduce the parity as an additional superscript, not to relabel all the levels. Nevertheless, we believe that labeling energy levels by the parity even within a hyperfine pattern is at least methodologically incorrect unless the inversion is feasible. In contrast to the statement on p. 318 in Ref. 115, we are sure that the labeling scheme dealing with irreducible representations of the FPI group can never lead to erroneous conclusions,—presuming, of course, it is correctly applied. What we should do is to specify the symmetry of both a total wavefunction (including spins) and its spatial part with respect to *feasible* operations. For instance, energy levels of the *para*-, *ortho*-, and *meta*-modifications of methane are labeled by symbols ${}^1A_i(E)$, ${}^3A_1(F_2)$ or ${}^3A_2(F_1)$, and ${}^5A_i(A_i)$, respectively ($i = 1,2$).[110] Of course, total wave functions of A_1 type, for example, are localized components of odd global wave functions of the methane molecule; but, as stressed by Oka,[112] "the knowledge of the parity of molecular levels does not introduce any new symmetry distinction between energy levels other than those which are already given by using the permutation–inversion group of Longuet-Higgins." We refer the reader to Ref. 110 for a more detailed discussion.

B. Molecular Structure as a Superposition of Born–Oppenheimer Wavefunctions

Keeping in mind some of Woolley's remarks[116–118] let us first show that the adiabatic approximation plays no role in our arguments (see also Ref. 36). Suppose we could find eigenvalues and eigenfunctions of the accurate nonrelativistic spinless Hamiltonian [Eq. (2.1)] discussed by Woolley.[118] We expect, in most cases, that eigenstates corresponding to a lower part of the spectrum would exhibit two essential features:

1. Energy levels cluster in quasi-degenerate multiplets.
2. Probability densities are localized in some domains of the nuclear configurational space.

If we then impose *new boundary conditions* which require wavefunctions to vanish outside some boundaries between domains (for nearly rigid molecules those boundaries can be defined, e.g., by means of the least-squares form[119,120] as discussed in Refs. 16, 121), we shall obtain a new Hamiltonian which has the form of Eq. (1.1). As a result the additional symmetry [Eq. (2.4)] arises, which manifests itself in degenerate multiplets of energy levels. These levels become only quasidegenerate when we return to the accurate Hamiltonian. Like any symmetry, such a reduction comes about not in a molecule but in our description of it. In contrast with Woolley's statement in Ref. 117, it is perfectly inconceivable (at least for us) "that nonadiabatic, nonrelativistic calculations on, say, NH_3 (or any other small 'inversion' system) would predict a pattern of degenerate states in place of the 'inversion splitting' that dominates the electronic spectrum." From our point of view, this degeneracy will appear only if we impose the mentioned boundary conditions obviously opposed by Woolley.[116]

We can agree with Woolley[116] that nobody has solved the problem with such boundary conditions and we do not advocate them either, because the new problem may be even more complicated than the accurate one. We just want to illustrate the real nature of the symmetry to be faced. In the usual approach, one uses wavefunctions *localized* in a particular domain as a basis set in a version of the perturbation theory. Examples are expansions of molecular Hamiltonians[7,8,12,122] in the Born–Oppenheimer parameter κ. The question of the extent to which such calculations are adequate to the auxiliary problem with the additional boundaries certainly needs more accurate mathematical studies for each particular numerical method in question. Each Born–Oppenheimer wavefunction approximates an accurate solution only in the appropriate elementary domain. The approximate global wavefunction is represented as a superposition of Born–Oppenheimer functions localized in different elementary domains. As initially stressed by Berry[27] (see also Ref. 28), only planar molecules with no more than three identical atoms can exist in pure stationary Born–Oppenheimer states. We thus cannot agree with the statement in Ref. 123 that "usually one encounters only localized states." The conclusion reached by Bixon[123] that "the exact eigenstates, which are generally delocalized, are rarely found in nature as such" and that there are only a few exceptions to this rule (such as ammonia) sounds at least strange to us. Wavefunctions of NH_3 and, for example, PH_3 differ from each other only by a probability for nuclei to be in saddle D_{3h} configurations. All of them are localized in several elementary domains, and this by no means contradicts the fact that *an isolated molecule does have a shape*,[87,124–126] in contrast with statements in Refs. 116–118, 123, 127–129.

It is exactly true "that the derivation of the aforementioned approximate molecular Hamiltonians cannot be achieved by continuous mathematical approximations"[116] and we may be "thus dealing with a *qualitative* change in the theory."[117] So each time we doubt the accuracy of our calculations, we *must* combine some feasible domains into a new one and repeat all calculations for a new localized Hamiltonian defined in that domain. Of course, it may happen that the accuracy of our calculations will require considering the whole nuclear configurational space (see Refs. 92, 93 for an example), and in this case any difference between localized and global Hamiltonians certainly disappears.

As discussed above, it is true that localized Hamiltonians do not transform as invariants under the CNPI group, and their eigenfunctions

(Pedersen's functions on a truncated support[36]) do not diagonalize the accurate Hamiltonian,[116] but they do diagonalize the approximate global Hamiltonian [Eq. (2.1)], which is invariant under this group (cf. Ref. 36, p. 4030). There is no sense in expressing those functions "as a superposition of the exact eigenstates with time-dependent coefficients,"[116] because if we knew the latter we would not have used the approximate ones. The assertion is that a required accuracy of our calculations allows us to employ the approximate Hamiltonian of Eq. (2.1) instead of the accurate one, and *each particular case* in which this assumption does not work must be considered separately. Analyzing the concept of molecular structure, Woolley[116–118,127–129] did not recognize the fact that this concept is based just on localization of *stationary accurate* wavefunctions in different domains of the nuclear configurational space. The role of the localization in the concept of the molecular structure has been discussed by several authors[87,124–126] and we do not think that it is worthwhile coming back to this question again. But the assertion that those states are *time-independent* and do have full symmetry of the eigenstates,[35,36,87] in contrast with Woolley's statement on p. 31 of Ref. 118 or with similar assertions in Refs. 27, 39, 128, 130, immediately leads to the conclusion that the effect of the environment may be essential for the existence of optical[117,118,123,129,131–138] or (probably) chemical (Ref. 118, p. 32) isomers, but not of molecular structure in any case (cf. Refs. 117, 118, 123, 125, 127–129). To summarize this discussion, we cite a closely related remark of Harris[133]:

> Everything we describe is exact. Surely a system being optically active has nothing to do with the Born–Oppenheimer approximation. However, for convenience and also because the description is so adequate, let us assume the Born–Oppenheimer approximation.

As stressed in our paper,[28] a molecule may be observed in right- and left-handed forms if only *pure* nuclear permutations are feasible or, in other words, if any nuclear permutation–inversion is assumed unfeasible. At the same time Frei and Günthard[55] published their interpretation of the chirality of nonrigid molecules based on the isometric approach. But it should be stressed that this question (in contrast with interpretation of microwave spectra discussed in Section III) is solved solely by means of the feasibility concept, with no relation to whether or not one can construct SRMMs describing the tunneling process between the two forms. By analogy it is not precisely correct to state[56] that "a satisfactory determination of the symmetry number requires a detailed definition of SRMM and its finite internal coordinates." As explained in Refs. 13, 28, 139, 140, the feasibility concept allows one to conclude that the symmetry number is equal to the number of feasible permutations if high excited states span so-called[140] "multiregular" representations of the FPI group. Of course, the latter requirement can be proved only for each particular model in question. In particular, for rigid molecules the symmetry number is accurately *derived*[13,141] but not "defined," in contrast with the statement in Ref. 56. An analysis of a concrete model allows one to conclude whether or not it is possible to estimate the partition function correctly just by dividing the Boltzman partition function (calculated without the exclusion principle) by the symmetry number predicted by the feasibility concept. If the answer is no, it means that one has to calculate nuclear statistical weights in the appropriate sum, but not to look for another symmetry number. For example, the symmetry number of the phenol molecule is equal to 2 whether we assume internal rotation[142] or inversion[143] of the hydroxyl. However, the possibility of using this number for calculating the partition function may depend on a particular supposition about the nature of the large-amplitude degree of freedom.

But let us come back to our main discussion. Treating the problem of optically active isomers from the same point of view as we did in Ref. 28 (see the class (c) in Ref. 144), Philippot and Senders[144] subdivide optically inactive molecules into two classes according to whether or not the inversion belongs to the FPI group (the classes (a) and (b) of [144] respectively). It should be noticed that molecules studied by Mislow[145] (see Fig. 1 in Ref. 144) are examples of classes (b) and (c) whether *ABC* tops are improperly or properly congruent. The above authors mixed the inversion of laboratory axes with the permutation–inversion generating the inversion of molecule-fixed axes. As an example of the class (a) we can cite an umbrella motion in ammonia[146,147] through a planar configuration.

As a matter of fact, Philippot and Senders[144] did not use the mentioned subdivision of optically inactive molecules at all. However, it does seem useful to subdivide those molecules into two classes, though in a different way. We prefer to use the term "chiral" in the same sense as Lord Kelvin did,[148] that is, for molecules in which equilibrium configurations are improperly congruent with their mirror images. Clearly only chiral molecules may be optically active. This happens if there exists no feasible permutation–inversion.

We assert that for nonchiral molecules a classical picture of identical nuclei moving from one equivalent minimum to another has no quantum-mechanical analog, because of the exclusion principle. Only an interconversion of right- and left-handed forms can be described in such a way.

For example, if we consider the NHDT molecule as Harris does,[133] we can limit ourselves to consideration of the ground state, construct nonstationary wavefunctions Ψ_R and Ψ_L describing right- and left-handed forms, and then study the one-dimensional tunneling from one form to another. But for NH_3 we can never localize a wave packet *only* near one pyramid, whether the molecule is in a stationary or a nonstationary state. For example, the ground states of NH_3 and ND_3 are described by wavefunctions $\Psi_R - \Psi_L$ and $\Psi_R + \Psi_L$ respectively, whereas the second level of each inversion doublet is forbidden by the Pauli principle.‡ Therefore even in a nonstationary state there always exist two wave packets which meet each other in the D_{3h} configuration. As long as the barrier is high, we can follow only one of those packets, which will move back and forth somewhere between C_{3v} and D_{3h} configurations. To observe a molecular structure our experiment must be fast enough compared with those oscillations.

It is also unnecessary to have recourse to the nonstationary theory in order to explain the molecular polarizability (cf. Refs. 27, 87, 125). The electric field, which is not invariant under the inversion, mixes quasidegenerate *stationary* rotational states of opposite parity.[149]

All that was said above about the ammonia molecule is easily extended to an interpretation of the NMR spectra. From our point of view, the paradox[27,39] that "chemists continually do experiments that distinguish one hydrogen atom from another in the same molecule, because they occupy chemically inequivalent sites"[39] is connected with not precisely correct interpretations of those experiments and can be resolved without assuming that the exclusion principle does not apply to nonstationary states. First of all, we do not necessarily need to employ a nonstationary theory to interpret NMR spectra. The usual procedure is to introduce an effective spin Hamiltonian (see, for example, Ref. 71), supposed to be obtained from the accurate one by averaging the latter with a spatial probability density. We obtain one or another Hamiltonian depending on whether the nuclear wavefunctions under consideration are localized near equilibrium configurations or delocalized in space between them. In experiments that are fast enough we may be able to catch molecules near their equilibrium configuration, but, for example, no experiment can catch a *particular* fluorine at the base or apex of the trigonal pyramid PF_5. Again we can consider nonstationary states, but we cannot construct a wave packet localized at a particular minimum. We always have 20 wave packets which meet each other from time to time in C_{4v} configurations when performing the Berry pseudorotation.[150] As a matter of fact, it was Dalton (Ref. 44, p. 270) who first came to the assertion that tunneling is a nonphysical process because the symmetrization principle has been ignored. But in contrast with his discussion in Ref. 44, we believe that only the probability amplitude of tunneling between C_{3v} and D_{3h} configura-

‡One can observe the inversion doublets only in states of E type, but not in the ground state (cf. Refs. 125, 126). The two wavefunctions describing such states are linear combinations of *four* Born–Oppenheimer functions.

tions of ammonia or between D_{3h} and C_{4v} of phosphorus pentafluoride should be considered. Of course, there is no difficulty in reformulating the induction procedure using just barriers between the equilibrium configurations in question instead of probability amplitudes of tunneling.

For a similar reason we cannot agree with describing the photodissociation of H_3^+ as a two-dimensional process[151,152] with one *particular* hydrogen at the top of an isosceles configuration. This is a correct picture for the ozone molecule,[153] which has three isosceles equilibrium configurations separated by high barriers. There we have three weakly interfering wave packets, and hence we can study only one of them (of course, we must be careful when calculating statistical weights of states for probability amplitudes). But the equilibrium geometry of H_3^+ is an equilateral triangle.[151] This means that all three wave packets strongly interfere near that geometry and we cannot limit ourselves to consideration of only a single channel. In order to treat photodissociation of H_3^+ as a two-dimensional process we should consider vibrations of E type. We have a complete analogy with the description of the Berry pseudorotations in PF_5 in Refs. 154–157. The only difference is that one (but note: *any one*) apex fluorine in PF_5 moves towards the top of a C_{4v} pyramid, whereas one of the protons in H_3^+ goes away to infinity.

By analogy, one can never tell which local mode in benzene is excited (cf. Ref. 158). Correct wavefunctions are linear combinations of pure local modes spanning irreducible representations of the FPI group

$$G_{24} \stackrel{\text{iso}}{=} D_{6h}$$

(see Table VI in Ref. 158). We can limit ourselves to consideration of a single local mode only so long as we can neglect splittings between the quasidegenerate states in question. Note that the quasidegenerate multiplets of energy levels typical for the local mode picture appear as a result of the approximate separation of variables. One should not confuse this problem and the clustering of energy levels caused by high barriers between different domains of the nuclear configurational space (cf. Refs. 158, 159). Remember that localized Hamiltonians are approximations of a global Hamiltonian in different domains, and hence all they depend on exactly the same variables.

III. MOLECULAR SYSTEMS WITH A FEW INTERNAL DEGREES OF FREEDOM

A. Symmetry Groups Generated by a Molecular Model

In order to separate small- and large-amplitude motions one usually introduces a molecular model, that is, a molecular geometry having only large-amplitude degrees of freedom; the small-amplitude vibrations are considered frozen. Suppose ρ designates a set of model parameters $\rho_1, \rho_2, \ldots, \rho_p$ describing large-amplitude internal motions. After model coordinates $\bar{\mathbf{a}}(\rho)$ have been chosen, one can consider different symmetry groups according to whichever problem is of interest. Let us start with the *dynamical symmetry group* $\mathbb{D}$ *of a free-rotating model.* (Here and below the term "rotation" implies only a *proper* rotation; otherwise we use the term "improper rotation.") The Hamiltonian $\hat{h}_{\theta,\rho}$ describing the motions in question can be expressed explicitly in terms of three Euler angles $\theta \equiv \{\theta_1, \theta_2, \theta_3\}$ and large-amplitude internal variables ρ. The *total* dynamical symmetry group $\mathbb{D}_{\text{total}}$ is formed by all transformations of θ and ρ keeping the Hamiltonian $\hat{h}_{\theta,\rho}$ invariant. Except for some exotic cases of hidden symmetry, this group factors as a direct product:

$$\mathbb{D}_{\text{total}} = \mathrm{SO}(3) \times \mathbb{D}_{\text{super}} \tag{3.1}$$

where the group SO(3) is formed by overall rotations of the model. If an orientation of the model in space is defined by 3×3 orthogonal matrix $\mathbf{S}(\theta)$ with $\det(\mathbf{S}) = 1$, any element of $\mathbb{D}_{\text{total}}$ can be represented as a triad $(\mathbf{R}^{(l)}, \mathbf{R}^{(r)}, \rho \to \tilde{\rho})$, where $\mathbf{R}^{(l)}$ and $\mathbf{R}^{(r)}$ induce respectively left and right shifts on the group SO(3) formed by matrices $\mathbf{S}(\theta)$:

$$\mathbf{S}(\tilde{\theta}) = \mathbf{R}^{(l)}\mathbf{S}(\theta)\mathbf{R}^{(r)} \tag{3.2}$$

[elements of the dynamical supersymmetry group $\mathbb{D}_{\text{super}}$ are represented as diads $(\mathbf{R}^{(r)}, \rho \to \tilde{\rho})$].

Let us give some examples. If a model is rigid, the $\mathbb{D}_{\text{super}}$ is SO(3), D_∞, and D_2 respectively for spherical, symmetric, and asymmetric tops. For ethanelike molecules, overall and internal rotations can be separated exactly and $\mathbb{D}_{\text{super}}$ is isomorphic to $D_\infty \times C_{6v}$. For nitromethane $\mathbb{D}_{\text{super}}$ is isomorphic to D_{6h}.[20] A very interesting family of SRMMs with hidden dynamical symmetry is D_{2d} puckering in four-membered ring molecules, for which the $\mathbb{D}_{\text{super}}$ is isomorphic to $D_{\infty h}$.[160,161] As discussed in Section III.D below, similar extra symmetry takes place for any $2k$-membered n-cycloalkane simulated by a D_{kd} SRMM.

Here we are actually interested in the subgroup $\mathscr{F}_{\theta,\rho}$ of $\mathbb{D}_{\text{super}}$ which is generated by permutations and permutation–inversions; we call it the dynamical PI (DPI) group and represent its elements by means of diads $(R_g^{(r)}, \rho \to \tilde{\rho}^g)$ such that[28,33–35,85]

$$\overline{\Pi}_g \bar{\mathbf{a}}(\rho) = \overline{\mathbf{R}_g^{(r)} \mathbf{a}(\tilde{\rho}^g)} \tag{3.3}$$

where $\overline{\Pi}_g$ is a $3N \times 3N$ matrix representing the gth element of G. (We use the bar to denote $3N$-dimensional columns or $3N \times 3N$ matrices, where N is the number of atoms in the molecule in question; see comments on p. 481 in Ref. 17.) Transformations $\rho \to \tilde{\rho}^g$ form the so-called internal isometric group[46–52] $\mathscr{F}_\rho$ (we do not use the notation $\mathscr{F}(\rho)$ suggested in Ref. 46, because it may lead to the incorrect inference that the group itself depends on ρ). The subgroup of $\mathscr{F}_{\theta,\rho}$ generated by pure permutations is called the dynamical permutation symmetry group and denoted by $\mathbb{H}^+$. As mentioned above, this group has been used in Refs. 14, 18–23 to find feasible permutations and, using them, to calculate statistical weights. The $\mathbb{H}^+$ group is the common subgroup of the dynamical PI symmetry group $\mathscr{F}_{\theta,\rho}$ and the symmetry group $\mathbb{H}$ of the semirigid molecular model defined by Ezra[62–64,72,73] by means of diads $(\mathbf{C}_g^{(r)}, \rho \to \tilde{\rho}^g)$ such that

$$\overline{\mathbf{P}}_g \bar{\mathbf{a}}(\rho) = \overline{\mathbf{C}_g^{(r)} \mathbf{a}(\tilde{\rho}^g)} \tag{3.4}$$

Here $\overline{\mathbf{P}}_g$ is a pure permutation and the 3×3 orthogonal matrix $\mathbf{C}_g^{(r)}$ is allowed to have determinant $+1$ or -1. The group $\mathbb{H}$, which is *defined* as a set of ordered pairs, may be thought of as an extension of the point symmetry group to nonrigid molecules; it is by far the most convenient for investigations of group structure.

The relations in Eqs. (3.3), (3.4) can be also represented as

$$\overline{\mathbf{W}}_g \bar{\mathbf{a}}(\rho) \equiv \overline{\mathbf{P}}_g \overline{\left(\mathbf{C}_g^{(r)}\right)^{-1} \mathbf{a}(\rho)} = \bar{\mathbf{a}}(\tilde{\rho}^g) \tag{3.5}$$

The operations $\overline{\mathbf{W}}_g$ (Ezra[63,64] calls them "perrotations") form the "full isometric group" as defined by Günthard et al.[46–52] As follows from Eq. (3.5), the "isometric" configurations $\bar{\mathbf{a}}(\rho)$ and $\bar{\mathbf{a}}(\tilde{\rho}^g)$ must have the same potential energy, that is, the potential energy is invariant under the internal isometric group $\mathscr{F}_\rho$. However, Eq. (3.5) says nothing of symmetry properties of the kinetic energy (see Section III.C for a more detailed discussion).

Note that the authors of Refs. 74, 162–164 limit themselves to consideration of large-amplitude *internal* motions, and hence they need the internal but not the full isometric group. The difference appears for semirigid molecular models having nontrivial point symmetry groups.[62–64]

If the electronic function $\Psi(\eta; \rho)$ in question, where η is a set of body-fixed Cartesian coordinates of electrons, is single-valued, the group $\mathbb{H}$ coincides with the dynamical PI symmetry group $\mathscr{F}_{\eta,\rho}$ in the space of variables η, ρ. The fact that feasible *nuclear* permutations and permutation–inversions generate point-symmetry operations in the space of the body-fixed Cartesian coordinates of electrons was initially recognized by Hougen.[6] However, for periodic nuclear internal motions the function $\Psi(\eta; \rho)$ may be double-valued if nuclei move around a conical intersection between potential-energy surfaces,[165-168] and this may happen whether the symmetry group $\mathbb{H}$ contains a primitive period (isometric) transformation (see Refs. 47–52, 62, 64 for definition) or not. In the latter case there exists an element $(\mathbf{1}_3, \rho \to \tilde{\rho})$ of $\mathscr{F}_{\eta,\rho}$ (*trivial periodic transformation*) which is not included in $\mathbb{H}$,and hence $\mathbb{H}$ is just a subgroup of $\mathscr{F}_{\eta,\rho}$. (Alternatively, we consider double-valued or projective representations of $\mathbb{H}$.)

Operations $\overline{\Pi}_g$ form a subgroup $G^{(\mathrm{a})}$ of Π which is uniquely determined by a choice of the model, and this fact may lead to the conclusion that one can do without any concept of feasibility just by studying symmetry properties of the model; but as mentioned above, the choice of the model is limited by symmetry requirements itself.

Let us give some examples. As stated by Brand et al.,[169] the FPI group of aniline ($C_6H_5NH_2$) is isomorphic to D_{2h} if both wagging and torsional motions are feasible. However, there are two other possibilities when only one of those internal motions is feasible, and the analysis carried out by Brand et al.[169] showed that it is sufficient to use the group G_4 including the permutation which generates the wagging but not a torsional motion. On the contrary, an assignment of microwave transitions in methylamine (CH_3NH_2) may be done by means of the SRMM with internal rotation ($\mathscr{F}_{\theta,\rho} \overset{\text{iso}}{=} C_{3v}$), whereas the inversion splitting can be estimated by means of some corrections.[170] Thus Ito[170] did use the classification of energy levels according to the FPI group[171-173]

$$G_{12} \overset{\text{iso}}{=} D_{3h}$$

in spite of employing the model with the lower symmetry. When many years later Tsuboi et al.[174] and recently Kreglewski[171] made use of two-parameter SRMMs having the DPI group $\mathscr{F}_{\theta,\rho}$ isomorphic to G_{12}, they already knew the real patterns of the energy levels they were studying.

By analogy a general analysis[175,176] shows that accurate calculations of energy spectra of methylaniline must be done by means of three-parameter models in which the DPI group $\mathscr{F}_{\theta,\rho}$ is isomorphic to D_{3h}.[175] If we compare this analysis with the one made by Brocas and Fastenakel[91,95] for PF_5 or by Brocas and Rusu[98] for XeF_6, we see that the only difference is that for molecules with wagging and torsion motions we can construct SRMMs, whereas we cannot do so for PF_5 and XeF_6. But if we consider possible FPI groups of the RPF_4 molecule,[101,102] a one-parameter SRMM has been constructed by Russeger and Brickmann[177] for[64,73]

$$\mathscr{F}_{\theta,\rho} \overset{\text{iso}}{=} \mathbb{H} \overset{\text{iso}}{=} C_{4v}$$

In this connection we should also cite Hougen's papers[178,179] (see also Ref. 180), in which he even managed to introduce the double groups of the ethane and hydrazine molecules without constructing appropriate SRMMs.

It should be stressed again that the choice of the model is itself correct only if the subgroup $G^{(a)}$ coincides with the FPI invariance group G or contains it as a subgroup.

As an example, let us consider the calculation of the partition function for a molecular system whose equilibrium configuration has a slightly distorted symmetry.[141] If this configuration is used as the rigid model of the molecule in question, the group $G^{(a)}$ induced by this model is just a subgroup of G.[28] With this model, neither energy levels nor their statistical weights are calculated correctly. This is the *real* reason why the "jump in the value of the rotational partition function" occurs[141] when the symmetry of the model changes slightly. We should also keep this fact in mind when analyzing the Jahn–Teller coupling (see Ref. 181, p. 633). If the point symmetry of the equilibrium configuration is slightly broken by the Jahn–Teller coupling, the FPI group is *always* isomorphic to the point-symmetry group of the crossing point of the electronic surfaces in question (cf. Ref. 182, p. 92; the authors of Ref. 182 confuse the CNPI and FPI groups: the term "the Longuet-Higgins group" is certainly used for the latter).

Below, the semirigid model of an isolated molecule is assumed chosen in such a way that the PI subgroup $G^{(a)}$ induced by it coincides with the FPI invariance group G, that is, $G^{(a)} = G$.

The procedure for calculating nuclear statistical weights is a good argument against replacing the concept of feasibility with the analysis of symmetry properties of the model as done by Moret-Bailly[183] (see also Ref. 114). As stated in Ref. 114 and cited in Ref. 64,[‡] one can make use of the fact that elements of the point symmetry group of the model transform this model into its physical equivalents. But the cardinal question is whether or not any transformations of the generalized coordinates in question are induced in such a way. For example, similar arguments treating proper and improper rotations in an equivalent fashion led the authors of Ref. 89 to the wrong (but at least consistent) conclusion that *continuous* rotational functions have a definite symmetry with respect to any permutation in Eq. (3.4) whether or not it generates a proper or improper rotation of the rigid model under study. Using permutations instead of feasible permutation–inversions, those authors gave the wrong recipe for calculating nuclear statistical weights.

We do not think that any study based only on the point symmetry of the model can answer the question why the total dynamical symmetry groups

[‡]As explained in Ref. 110, the symmetry classification suggested by Moret-Bailly[183] for spherical tops has nothing to do with the approach advocated by Ezra.[64]

$\mathbb{D}_{\text{total}}$ of free-rotating linear and planar triatomic rigid models ABC are isomorphic to each other:

$$\mathbb{D}_{\text{total}} \overset{\text{iso}}{=} O(3)$$

(Rotations of the linear model are described by only two angles; for the planar molecules in question the matrices $\mathbf{R}_g^{(r)}$ in Eq. (3.2) form the point group C_2). The isomorphism occurs because no transformation of generalized coordinates corresponds to elements of the subgroup C_∞ of the point symmetry $C_{\infty v}$ of the linear model under consideration (see Ref. 6 for a more detailed discussion).

B. An Impurity Molecule Rotating in a Crystal Field

The isomorphism between $\mathbb{H}$ and the FPI group for *isolated nonlinear nearly rigid* molecules usually disguises the real nature of the symmetry. However, removing one of the mentioned constraints immediately forces us to distinguish unambiguously between the two groups.

As a nontrivial example let us consider a rigid impurity molecule rotating in a crystal field. An excellent study of the dynamical symmetry of such a system has been carried out by Miller and Decius,[184] so we are able just to review their results briefly, clarifying some points. Note that Kiselev and Luders[103,185,186] appear to have missed the work of Miller and Decius[184] and did not recognize the real nature of the symmetry operations in question when developing their own approach.

According to the general recipe we should study the law of transformation of dynamical variables under feasible perrotations (see previous section for the definition). It is worth stressing that we consider rotations of an impurity molecule in an external field of the point symmetry $\mathbb{F}$ which coincides with the point subgroup of a host crystal (see Ref. 60, p. 105). We do not think that it is reasonable to represent elements of this group in the form $S(P)$ and $S(P^*)$ as done by Miller and Decius[184] (see also Refs. 187, 188; cf. Refs. 103, 185, 186), because P and P^* here imply permutations of all atoms of a host crystal.

Any proper rotation $\mathbf{R}_f^{(l)}$ belonging to the group $\mathbb{F}$, as well as any permutation from the group $G^{(\mathbf{a})}$ is obviously a feasible operation. However, the same is not generally true for an improper rotation $\mathbf{C}_f^{(l)} \in \mathbb{F}$ or permutation–inversion $\overline{\Pi}_g \in G^{(\mathbf{a})}$, but for different reasons. The former leads to the mirror-image configuration separated from the initial one by an "infinitely" high barrier; the latter is not a symmetry operation of the global Hamiltonian if the point symmetry group of the crystal does not contain the inversion $\mathbf{I} = -\mathbf{1}_3$. Therefore the FPR group G of this system cannot be represented as the direct product of the groups $\mathbb{F}$ and $G^{(\bar{a})}$ in the general case[184] (see also Ref. 189 for an example), in contrast with the results of Ref. 190. Nevertheless, any product of an improper rotation $\mathbf{C}_f^{(l)} \in \mathbb{F}$ and an arbitrary nuclear permutation $\overline{\mathbf{P}}$ keeps the global Hamiltonian invariant. If the permutation $\overline{\mathbf{P}}$ generates an improper rotation of the model, the mentioned product of two operations induces a transformation of rotational coordinates: the matrix $\mathbf{S}(\theta)$ transforms under this operation as

$$\mathbf{S}(\tilde{\theta}^{f,g}) = \mathbf{R}_f^{(l)}\mathbf{S}(\theta)\mathbf{R}_g^{(r)} \tag{3.6}$$

where $\mathbf{R}_f^{(l)} = -\mathbf{C}_f^{(l)} \in \mathrm{SO}(3)$ and the matrix $\mathbf{R}_g^{(r)}$ is determined by (3.3) with $\overline{\Pi}_g = -\overline{\mathbf{P}}$. This operation is therefore necessarily included in the FPR group G.[184] We see that no elements of G correspond to the model symmetry elements conforming to reflections if the group $\mathbb{F}$ contains just proper rotations. Moreover, if just *small-amplitude* nutations are allowed, the group

$$\mathbb{D}_{\text{total}} \overset{\text{iso}}{=} G_0$$

contains only such diads $(\mathbf{R}_g^{(l)}, \mathbf{R}_g^{(r)})$ that

$$\mathbf{R}_g^{(l)}\mathbf{R}_g^{(r)} = \mathbf{1}_3 \tag{3.7}$$

Here $\mathbf{R}_g^{(l)} = \pm\mathbf{C}_g^{(l)}$, where $\mathbf{C}_g^{(l)} \in \mathbb{F}$, according as $\overline{\Pi}_g$ in (3.3) is a pure permutation (the upper sign) or a permutation–inversion.

Let us consider now the librator limit when an impurity molecule can rotate only around a single axis (say, X). We should consider only perrotations generating such diads $(\mathbf{R}_f^{(l)}, \mathbf{R}_g^{(r)})$ that

$$\mathbf{R}_f^{(l)}\mathbf{T}_X(\gamma)\mathbf{R}_g^{(r)} = \mathbf{T}_X(\tilde{\gamma}) \tag{3.8}$$

(we shall use similar notation $\mathbf{T}_Y(\gamma)$, $\mathbf{T}_Z(\gamma)$ for rotations around the two other axes Y and Z). If vibrations are frozen, the angle γ is the sole dynamical variable, and the dynamical perrotation (DPR) group $\mathscr{F}_\gamma$ is composed of transformations $\gamma \to \tilde{\gamma}$ which are similar to internal isometric substitutions $\rho \to \tilde{\rho} \in \mathscr{F}_\rho$.

Let us review in the above terms the discussion of the librator limit given by Kiselev and Luders[103,185,186] for a C_{2v} triatomic ABA in an O_h external field. The molecule is assumed to rotate over its symmetry axis, which is chosen to coincide with a C_ν axis of the crystal (the C_ν model according to Ref. 103, where $\nu = 2, 3, 4$; following Ref. 103, we refer to the C_3 axis, instead of S_6 as in Ref. 191). To label elements of the FPR group $G^{(\nu)}$ we make use of the isomorphism between this group and the group $\mathbb{C}^{(\nu)}$ formed by diads $(R_f^{(l)}, R_g^{(r)})$ satisfying Eq. (3.8). The group $\mathbb{C}^{(\nu)}$ is generated by the diads $\hat{C} \equiv (\mathbf{T}_X(2\pi/\nu), \mathbf{1}_3)$, $\hat{P} \equiv (\mathbf{1}_3, \mathbf{T}_X(\pi))$, $\hat{A} \equiv (\mathbf{T}_Y(\pi), \mathbf{T}_Z(\pi))$ and isomorphic to $D_{\nu h}$. Here the XZ plane is chosen as one of the symmetry planes of the crystal, and the operation $\hat{P}$ is generated by the permutation of the identical atoms in the triatomic molecule. The operation $\hat{A}\hat{P} = (\mathbf{T}_Y(\pi), \mathbf{T}_Y(\pi))$ changes sign of the librational angle γ. For $\nu = 3$ the dynamical perrotation group $\mathscr{F}_\gamma^{(3)}$ also contains the invariant cyclic subgroup C^0 generated by the operation $\hat{P}\hat{C}$ which transforms γ into $\gamma - \pi/3$, that is,

$$C^0 \stackrel{\text{iso}}{=} C_6 \quad \text{and} \quad \mathscr{F}_\gamma^{(3)} \stackrel{\text{iso}}{=} C_{6v} \stackrel{\text{iso}}{=} \mathbb{C}^{(3)} \stackrel{\text{iso}}{=} D_{3h}$$

(see Table 4 in Ref. 103). For even ν there exists an element of the FPR group which keeps γ unchanged. This perrotation generates the invariant subgroup

$$G^P \stackrel{\text{iso}}{=} C_i$$

which is the kernel of the homorphism $\mathbb{C}^{(\nu)} \stackrel{\text{ho}}{\rightarrow} \mathscr{F}_\gamma^{(\nu)}$ and is analogous to the point group of a SRMM as defined by Ezra.[62-64,72,73] For $\nu = 2, 4$ we thus find $\mathbb{C}^{(\nu)} = G^P \wedge K$, where the group

$$K \stackrel{\text{iso}}{=} \mathscr{F}_\gamma^{(\nu)}$$

is generated by the operations $\hat{C}$ and $\hat{A}$. Therefore

$$\mathscr{F}_\gamma^{(\nu)} \stackrel{\text{iso}}{=} C_{\nu v}$$

It should be pointed out that although Kiselev and Luders[103,185,186] obtained symmetry groups also isomorphic to $D_{\nu h}$ for $\nu = 2, 3, 4$ and the same law of transformation of the librational angle γ under elements of these groups, these authors deal with different symmetry operations. For example, for $\nu = 4$ their operation $\hat{B}$ which keeps the angle γ unchanged is a permutation–inversion, but not a pure permutation of atoms of the impurity molecule and crystal in the language of Miller and Decius.[184] Summarizing, we conclude that n in the n-fold potential hindering the libration,

$$V = V_0(1 - \cos n\gamma)$$

is equal to 6 for the C_3 model and to 2 and 4 for the C_2 and C_4 models respectively.[191]

The above arguments concerning the librator limit can be easily extended to a molecule absorbed by a surface *provided the molecule rotates around the axis perpendicular to the surface*. In contrast with Nichols and Hexter,[187,188] we do not think that all other rotations as well as out-of-plane vibrations can be treated in a similar way. We should also mention that the FPR group at the librator limit does not generally factor into a direct product as stated in Ref. 188. We refer the reader to our work[189] for counterexamples.

There is also no difficulty in including in consideration nonrigid molecules rotating in a crystal, as done by Fredin and Nelander[192] for the ammonia molecule trapped in a site of C_{3i} symmetry.

C. The Dynamical PI Group as an Invariance Group of the Hamiltonian

For many SRMMs, G-invariance of the Hamiltonian corresponding to the particular case of frozen vibrations follows from the fact that it is a term of definite structure in the expansion of the global Hamiltonian as a Taylor series in vibrational coordinates (see Section IV and Appendix A). However, a more careful examination of our previous arguments,[35] made in Section IV below, shows that an extension of the above conclusion to arbitrary semirigid models cannot be carried out easily, and moreover, as stressed by Frei et al.,[51] it does not seem reasonable to complicate the problem by introducing vibrational coordinates if we can do without them.

Bauder et al.[46] were the first who recognized the fact that for all known semirigid models internal isometric substitutions $\rho \to \tilde{\rho}$ are nothing but linear transformations

$$\tilde{\rho}_{\nu}^{g} = c_{\nu}^{g} + \sum_{\mu} B_{g}^{\mu\nu} \rho_{\mu} \tag{3.9}$$

In our previous work[28,33-35] we certainly underestimated the significance of this result. Only due to Eq. (3.9) can we interchange a matrix element $\partial\rho_{\mu}/\partial\tilde{\rho}_{\nu}^{g}$ and the operator $\partial/\partial\rho_{\mu}$ in the quantum-mechanical Hamiltonian, when proving its invariance under operations $\overline{\Pi}_g$ in Eq. (3.3). The previously mentioned invariance of the Hamiltonian

$$\hat{h} \equiv \tfrac{1}{2}\mu_0^{1/2} \sum_{s,t=1}^{p+3} \hat{N}_s\{\mu_c\}_{st}\mu_0^{-1/2}\hat{N}_t + V(\rho) \tag{3.10}$$

where $\mu_0 = \det \boldsymbol{\mu}_0$, $\{\boldsymbol{\mu}_0^{-1}\}_{st} \equiv \sum_A m_A \mathbf{g}_{0,A}^s \cdot \mathbf{g}_{0,A}^t$,

$$\mathbf{g}_{0,A}^s(\rho) \equiv \begin{cases} \mathbf{A}_s \mathbf{a}_A, & s = 1,2,3 \\ \dfrac{\partial \mathbf{a}_A}{\partial \rho_s}, & s = 4,\ldots, p+3 \end{cases} \tag{3.11}$$

$\{\mathbf{A}_s\}_{tu} \equiv e_{stu}$, and m_A are nuclear masses, is directly verified after one takes into account that the components $\hat{K}_j' \equiv \hat{N}_j$ $(j = 1,2,3)$ of the total angular momentum along molecule-fixed axes [see Eq. (A6) in Appendix A], the operators $\hat{N}_{3+\nu} \equiv i\,\partial/\partial\rho_\nu$, and the vectors in Eq. (3.11) transform according to the law

$$\hat{K}_j' \overset{\hat{g}}{\rightarrow} \sum_{l=1}^{3} \hat{K}_l' \{\mathbf{R}_g^{(r)}\}_{lj} \tag{3.12a}$$

$$\hat{N}_{3+\nu} \overset{\hat{g}}{\rightarrow} \sum_{\mu=1}^{p} \{\mathbf{B}_g^{-1}\}_{\nu\mu} \hat{N}_{3+\mu} \tag{3.12b}$$

$$\bar{\mathbf{g}}_0^s(\tilde{\rho}^g) = \begin{cases} \overline{\Pi_g \mathbf{A}_s \left[\mathbf{R}_g^{(r)}\right]^{-1} \mathbf{a}(\rho)}, & s = 1,2,3 \\ \overline{\Pi}_g \displaystyle\sum_{\mu=1}^{p} \overline{\left[\mathbf{R}_g^{(r)}\right]^{-1} \mathbf{g}_0^{\mu+3}(\rho)} \{\mathbf{B}_g^{-1}\}_{\nu\mu}, & s = 3+\nu \end{cases} \tag{3.12c}$$

Of course, all credit for this result must go to Bauder et al.[46] However, it is worth stressing that the concept of isometry of nuclear configurations plays no role in the proof: the configurations $\bar{\mathbf{a}}(\rho)$ and $\bar{\mathbf{a}}(\tilde{\rho}^g)$ remain isometric even if the coordinates ρ and $\tilde{\rho}^g$ are connected by a nonlinear transformation, in which case the above arguments do not work.

We find a close relationship between the concept of isometry[46-52] and the device developed by Moret-Bailly[183] for spherical tops. Like Moret-Bailly, the authors of the isometric approach have to conclude that the molecular Hamiltonian under discussion is invariant under the improper rotations (see p. 452 in Ref. 47 and p. 60 in Ref. 49), but it is not true in the general case. Frei et al.[47] apparently misunderstood a very clear recipe given by Hougen[5] and used since then by spectroscopists (see, for example, Ref. 193): transitions may occur only between states of the same permutational symmetry, and for electric dipole transitions those states must in addition have opposite parity. If we look at Table 2 in Ref. 194, we immediately conclude that the operation V_2 corresponds to the permutation in the FPI group of the glyoxal molecule, and hence according to Table 3 in the same work the only possible electric dipole transitions are $\Gamma_1 \rightarrow \Gamma_2$ and $\Gamma_3 \rightarrow \Gamma_4$. Of course, those selection rules are independent of whether the domain of the internal rotation angle τ is chosen as $-\pi/2 \leq \tau < \pi/2$ or $0 \leq \tau < \pi$. By analogy the analysis of Table 2 in Ref. 195 leads to the same selection rules for nitroethylene. We would have been extremely surprised if observed transitions in the microwave spectrum of nitroethylene[196] did not satisfy those rules (cf. Ref. 47, p. 457).

D. Hidden Symmetry of the Localized Hamiltonian

Although the DPI group $\mathscr{F}_{\theta,\rho}$ and dynamical supersymmetry group $\mathbb{D}_{\text{super}}$ usually coincide, some examples have been listed above in which these two differ. Let us start our discussion with the rigid dichloromethane molecule. Its DPI group $\mathscr{F}_{\theta}$ is D_2 and coincides with the $\mathbb{D}_{\text{super}}$. But we know that any other asymmetric top has exactly the same dynamical supersymmetry group. The difference is that any *effective* rotational Hamiltonian obtained by averaging with vibrational wavefunctions *must* still be invariant under this group for dichloromethane, but the symmetry under discussion generally disappears for arbitrary asymmetric tops.

It was Altmann[60] who first stressed the fact that there exists the *homomorphism* from $\mathbb{D}_{\text{super}}$ on $\mathscr{F}_{\theta,\rho}$ for nitromethanelike molecules. But it is erroneous[197] to believe that Altmann gave any explanation to this "hidden" symmetry. The only way to find the dynamical supersymmetry group $\mathbb{D}_{\text{super}}$ is to analyze symmetry properties of the particular Hamiltonian $\hat{h}_{\theta,\rho}$ as, for example, was done in Refs. 20, 161. (In this connection the technique suggested by Renkes[198] may be useful). We do not need to "save" Longuet-Higgins's analysis by adding extra operations to the FPI group as done in Ref. 197. The feasibility concept predicts only *minimum* dynamical symmetry groups and does not exclude additional "hidden" symmetries.

It is worth mentioning that the supersymmetry group of nitromethane is isomorphic to D_{6h} only if we consider internal rotation of a *rigid* top around a *rigid* frame. The *additional* symmetry breaks if we consider a relaxation model.[199,200] However, it should be stressed that the DPI group

$$\mathscr{F}_{\theta,\rho} \overset{\text{iso}}{=} D_6$$

remains; that is, in contrast with Eq. (10) of Ref. 199, the potential function of the torsional angle τ contains no term proportional to $\cos 3\tau$, because the SRMM *must* satisfy Eq. (3.3) in any case. By analogy the invariance group of any effective rotation–torsion Hamiltonian $\hat{h}_{\theta,\tau}$, obtained by averaging with vibrational wavefunctions,[201,202] generally coincides with $\mathscr{F}_{\theta,\tau}$. There is however a slight difference between the two cases. As long as we include in the effective Hamiltonian quantum-mechanical corrections obtained by averaging the kinetic or potential energy with vibrational wavefunctions, it is not sufficient just to satisfy the condition in Eq. (3.3) when choosing the SRMM: we must be sure that *our vibrational coordinates* are chosen in an appropriate way (see Section IV); otherwise we may break permutation–inversion symmetry and as a result a term proportional to $\cos 3\tau$ will appear.

As another example of a hidden symmetry let us consider ring puckering of cyclohexane.[203] This is a natural extension of the work of Baltagi et al. on four-membered rings.[160,161] But before coming to the discussion of *models* we should emphasize, following Stone and Mills,[204] that even if the equilibrium configuration of N-membered cycloalkane is puckered, the FPI group is isomorphic to the point symmetry group D_{Nh}, "so long as we recognize splitting due to tunneling through the inversion barrier at the planar configuration of the ring." The fact that there exist planar configurations having the same point symmetry group is inessential and has no relation to Longuet-Higgins's arguments[29] (see also Refs. 204–206), in contrast with their interpretation given in Ref. 207.

Nevertheless those configurations may be useful as a starting point for constructing semirigid models linearly dependent on their parameters such that the matrices $\mathbf{C}_g^{(r)}$ in Eq. (3.4) form the point symmetry group D_{Nh}[208] (see also Refs. 209–211 for the particular cases $N = 6$, 7, and 8).

Let us simulate ring puckering of cyclohexane between two D_{3d} equilibrium configurations by a one-parameter SRMM having D_{3d} point symmetry for any value of the model parameter ρ_1 chosen in such a way that the reflection in the plane perpendiclar to the symmetry axis of the model changes its sign. Equation (3.4) is obviously fulfilled for any element of the FPI group. One has

$$\mathscr{F}_{\theta,\rho} \overset{\text{iso}}{=} D_6$$

whereas $\mathbb{H} = D_{6h}$, that is,

$$\mathbb{H} \overset{\text{ho}}{\rightarrow} \mathscr{F}_{\theta,\rho}$$

However, elements of $\mathscr{F}_{\theta,\rho}$ are not the only symmetry operations which keep the Hamiltonian invariant. In fact, the model in question is the symmetric top for any values of its parameters. Because the large-amplitude internal motions span the totally symmetric representation of D_{3d} in contrast with infinitesimal rotations, they do not interact with the latter. We have mentioned above that the dynamical supersymmetry group of rigid symmetric top is D_∞. Adding to this group the "intrinsic" operation $(\mathbf{1}_3, \rho_1 \to -\rho_1)$ of the DPI group $\mathscr{F}_{\theta,\rho}$, we immediately come to the group $\mathbb{D}_{\text{super}}$ isomorphic to $D_{\infty h}$, in agreement with the results of Baltagi et al.[160,161] for four-membered ring molecules. Clearly the above arguments are applicable to any $2k$-membered n-cycloalkane simulated by a SRMM of D_{kd} point symmetry.

Note that for D_{3d} SRMMs the factorization

$$D_{6h} = D_{3d} \times C_i \tag{3.13}$$

can be treated as a particular case of the Ezra factorization[62-64] using the *covering point symmetry group* G^P as an invariant subgroup (cf. Ref. 67). But in contrast with Flurry's arguments[67] (see also Section I), this decomposition results from the symmetry analysis of the SRMMs in question. In particular, Pickett and Strauss[209,212] consider a three-parameter SRMM in order to describe tunneling between the chair and boat forms of cyclohexane. Although the FPI group does not certainly change, the SRMM no longer has the D_{3d} point symmetry group, whereas, whatever model is employed, there always exists the invariant subgroup

$$G^I \overset{\text{iso}}{=} C_2$$

formed by the intrinsic operation of the DPI group $\mathscr{F}_{\theta,\rho}$ $(\mathbf{R}_g^{(r)} = \mathbf{1}_3)$ if we apply Ezra's definition of intrinsic operations[62-64] to diads $(\mathbf{R}_g^{(r)}, \rho \to \tilde{\rho}^g)$ forming this group. Therefore we generally have the factorization $G^I \times K$ (cf. Eq. (3.45) in Ref. 64).

IV. LOCALIZED HAMILTONIAN AS A TRUNCATED TAYLOR SERIES IN VIBRATIONAL COORDINATES

As long as vibrations are assumed frozen, one can construct the localized Hamiltonian and study its symmetry properties explicitly; hence it might seem natural[46-51] to connect a dynamical symmetry of the molecular system in question with the symmetry of its model. However, there is no reason for using the symmetry group of the model to classify vibrational levels, because nuclear vibrations break this symmetry. As stated above, any localized Hamiltonian (including its vibrational part) must nevertheless be invariant under all feasible elements of the CNPI group, and hence its energy levels can be classified by means of the symmetry species of the group G. Contrary to the statement in Ref. 53, (p. 104), we can conclude that it is point symmetry operations, not feasible permutations and permutation–inversions, that have little to do with nuclear motions (see also Ref. 64, p. 39). Only because the FPI group of a *free nonlinear quasirigid* molecule is isomorphic to the point symmetry group of its equilibrium configuration can one label rovibrational levels by symmetry species of the point symmetry group. As soon as there is no isomorphism (linear molecules, nonrigid molecules, parity nonconservation, impurity molecules in crystal), the point symmetry group of the equilibrium configuration becomes completely useless for classification of *nuclear* energy levels.

It is also not correct that the FPI group G "cannot be used on its own to determine the internal motions" as stated in Ref. 53. In order to illustrate how one can do without any point symmetry group in explaining the symmetry properties of a set of nuclei moving on an adiabatic potential surface near an equilibrium configuration, let us consider the NH_2Cl molecule. From each set of equivalent configurations we select the single standard configuration $\bar{\mathbf{y}}$ (see Ref. 17 for the proper references) such that the principal axes of its inertial tensor coincide with the coordinate axes (in the usual approach based on the Eckart frame[4,53,119,120] only the *equilibrium* standard configuration satisfies this condition). We make the convention that any standard configuration has zero values of rotational and translational coordinates, that is, an arbitrary nuclear configuration $\bar{\mathbf{r}}$ can be obtained from one and only one (equivalent) standard configuration by translation and *proper* rotation

$$\mathbf{r}_A = \mathbf{z} + \mathbf{S}(\theta)\mathbf{y}_A \tag{4.1}$$

($A = 1,2$ for the hydrogen atoms, $A = 3$ for chlorine, and $A = 4$ for nitrogen). We place the origin of coordinates at the center of mass of any standard configuration, so the vector $\mathbf{z}$ in Eq. (4.1) is the center of mass of the nuclear configuration $\bar{\mathbf{r}}$ in question. There must exist a one-to-one correspondence between a set of standard configurations and the space of internal variables Я, and no two standard configurations can be equivalent. We must still select which direction of a principal axis (from two possible ones) is to be positive. The axes are chosen so that projection of the vectors $\mathbf{y}_2 - \mathbf{y}_1$, $\mathbf{y}_3$ and $(\mathbf{y}_1 - \mathbf{y}_3) \times (\mathbf{y}_2 - \mathbf{y}_3)$ onto the axes X, Y, Z respectively are positive. The matrix $\mathbf{S}(\theta)$ in Eq. (4.1) can be found uniquely almost everywhere.

Because the molecule is treated as quasirigid, we are interested only in the law of transformation of $\mathbf{S}(\theta)$ under the feasible permutation–inversion $(12)^*$. When acting on a configuration $\bar{\mathbf{y}}$, this operation does not change the direction of the principal axis coincident with the X axis, whereas the directions of the other two are reversed. Therefore

$$\bar{\mathbf{P}}^*\bar{\mathbf{y}}(r_1, r_2, r_3, \alpha_{12}, \alpha_{13}, \alpha_{23}) = \overline{\mathbf{T}_X(\pi)\mathbf{y}(r_2, r_1, r_3, \alpha_{12}, \alpha_{23}, \alpha_{13})} \tag{4.2}$$

where r_A ($A=1,2,3$) is the interatomic distance between atoms A and 4, and $\alpha_{AA'}$ ($A<A'=1,2,3$) is the bond angle $A4A'$. It is easily shown that the matrix $\mathbf{S}(\theta)$ is transformed under $\overline{\mathbf{P}}^*$ in Eq. (4.2) as

$$\mathbf{S}(\tilde{\theta}^g)=\mathbf{S}(\theta)\mathbf{R}_g^{(r)} \tag{4.3}$$

where $\mathbf{R}_g^{(r)}=\mathbf{T}_X(\pi)$. It is crucial that the law of transformation in Eq. (4.3) is independent of internal variables. Note that Eq. (4.3) is equivalent to the requirement that the set of standard configurations $\bar{\mathbf{y}}(Я)$ in question is invariant under operations $\overline{\mathbf{W}}_g$ defined by Eq. (3.5), that is,

$$\overline{\mathbf{W}}_g\bar{\mathbf{y}}(Я)=\bar{\mathbf{y}}(\tilde{Я}^g) \tag{4.4}$$

(cf. Refs. 53, 62–64, 73, 108, 213). A second very important property of this set of external and internal coordinates lies in the fact that deviations q of internal coordinates from their equilibrium values undergo a homogeneous linear transformation:

$$\tilde{q}_\gamma^g=\sum_\beta q_\beta Q_{\beta\gamma}^g \tag{4.5}$$

[the order of the coordinates q is the same as in Eq. (4.2), i.e., the nonzero matrix elements are $Q_{12}=Q_{21}=Q_{33}=Q_{44}=Q_{56}=Q_{65}=1$].

Let us expand the global Hamiltonian as a Taylor series of coordinates q, that is, we represent it as an infinite sum

$$\hat{H}=\sum_{m,n}\hat{C}_{m,n}\left(q,\frac{\partial}{\partial q}\right), \qquad n=0,1,2 \tag{4.6}$$

where $\hat{C}_{m,n}$ are homogeneous polynomials whose degrees of homogeneity are m for the coordinates q, and n for the derivatives. The localized Hamiltonian is obtained by truncating Eq. (4.6). It is easily seen that the aforementioned properties of the change of variables under consideration give rise to an invariance of each polynomial under the feasible operation $\hat{g}\in G$.[28,33–35] Note that we constructed the localized Hamiltonian to be invariant under the FPI group G without any model at all. We made no *formal* reference to the point symmetry C_{2v} of the equilibrium configuration. For this reason we believe that PI and not point symmetry is actually fundamental (contrast Ref. 213, p. 58); the latter appears in the symmetry analysis of nuclear vibrations only if a specific change of variables is used (see Section VI).

If the matrix $\mathbf{S}(\theta)$ transforms under a permutation–inversion $\overline{\mathbf{P}}^*$, according to Eq. (4.3), electronic body-fixed coordinates undergo the point symmetry operation $-\mathbf{R}_g^{(r)}$ (the minus sign appears because only the inversion of both nuclear and electronic Cartesian coordinates is a symmetry operation commuting with the global Hamiltonian[6]). In particular, those coordinates for the NH_2Cl molecule in the body-fixed axes under consideration undergo reflection in the *YZ* plane under the permutation–inversion (1 2)* for *any* nuclear configuration. Note that the law of transformation of η is the same whether or not this configuration has a symmetry plane.

Analyzing Eqs. (42a)–(42e) in Ref. 62 (see also the similar relations [Eq. (3.30′)] in Ref. 64), one can conclude that we have managed to obtain exactly the same transformations as Ezra's without using any model (the only difference is that we have standard configurations $\bar{\mathbf{y}}$ instead of displacements $\bar{\mathbf{d}} = \bar{\mathbf{y}} - \bar{\mathbf{a}}$). That is one of the reasons why we cannot agree with Ezra's statement[62–64,73] that those transformations are induced by elements of the symmetry group $\mathbb{H}$ of the molecular semirigid model. The main reason why Ezra's statement cannot be taken without qualification is that, in our point of view, no symmetry operation is able to induce transformations in a space of variables in which this operation is not defined. All transformations discussed by Ezra are induced only by elements of the PI group. Of course it is necessary that any feasible operation $\bar{\Pi}_g$ transform the model according to Eq. (3.3), that is, the symmetry group $\mathbb{H}$ of the molecular model must contain the operation generated by the permutation $\bar{\mathbf{P}}_g = \bar{\Pi}_g \det \mathbf{C}_g$ according to Eq. (3.4). A correct choice of the model and hence of its symmetry group $\mathbb{H}$ is possible only after all feasible operations have been selected.

Our remark remains true even if multivalued variables are used as a result of considering *periodic* large-amplitude internal motions. In this case there exists a homomorphism from the DPI group $\mathscr{F}_{\theta,\rho,q,\eta}$ onto the *G* group. All operations forming the kernel of the homomorphism coincide with the identity transformation in the space of Cartesian coordinates. As a rule, the group $\mathscr{F}_{\theta,\rho,q,\eta}$ is isomorphic to $\mathbb{H}$, and hence, as shown by Ezra[62–64,72,73] the $\mathbb{H}$ group can be used as a very powerful tool to study the structure of the *abstract* group in question. However, some comments are appropriate.

First of all, note that one can study the structure of $\mathscr{F}_{\theta,\rho}$ using exactly the same formalism as developed by Ezra for the $\mathbb{H}$ group. The only defect of $\mathscr{F}_{\theta,\rho}$ compared with $\mathbb{H}$ is that there exists a homomorphism from $\mathbb{H}$ onto $\mathscr{F}_{\theta,\rho}$ if the model has a center of symmetry for any values of its parameters ρ, as happened for the D_{3d} SRMM of cyclohexane discussed in the previous section.

There may also exist a homomorphism of $\mathscr{F}_{\theta,\rho,q,\eta}$ onto $\mathbb{H}$. This happens for single-valued semirigid molecular models if the electronic function is double-valued (as mentioned in Section III) or if double-valued normal modes appear[35,86] (see also Section VII). In those cases one has to use projective representations regardless of whether we are looking at $\mathbb{H}$ or $\mathscr{F}_{\theta,\rho}$.

For nonrigid molecules one should distinguish between large- and small-amplitude internal variables, denoted by ρ and q respectively (we also call q "vibrational coordinates"). First of all, it is necessary to require that the law of transformation of the large-amplitude variables θ and ρ under feasible operations $\hat{g} \in G$ be independent of q, that is,

$$\tilde{\theta}^g(\theta, \rho, q) = \tilde{\theta}^g(\theta, \rho, 0) \tag{4.7a}$$

$$\tilde{\rho}^g(\rho, q) = \tilde{\rho}^g(\rho, 0) \tag{4.7b}$$

This usually means that the matrix $\mathbf{S}(\theta)$ transforms according to Eq. (4.3), where $\mathbf{R}_g^{(r)}$ may depend only on ρ, and the coordinates ρ undergo a linear transformation according to Eq. (3.9).[46] The vibrational coordinates q must undergo a linear homogeneous transformation according to Eq. (4.5), where the matrix $\mathbf{Q}^g$ may depend[28,33,35] on ρ (see also Section VI). If those matrices are independent of ρ, each term in Eq. (4.6) is invariant under G. Otherwise this group has as invariant any polynomial

$$\hat{C}'_k = \sum_{n=0}^{2} \hat{C}_{n+k,n} \tag{4.8}$$

(This can be proved by scaling all vibrational coordinates q by $\varkappa$ and considering different terms in the expansion of the global Hamiltonian H as a Taylor series in $\varkappa$.[35])

In particular, one has to use similar arguments to show that different terms in the expansion of the Hamiltonian of a linear molecule as a Taylor series in q are invariant under overall rotations because (in contrast to nonlinear molecules) those rotations do act on transverse vibrational coordinates. In fact, one has

$$\mathbf{R}^{(l)}\mathbf{S}(\varphi, \theta)\mathbf{y}_A(q) = \mathbf{S}(\tilde{\varphi}, \tilde{\theta})\mathbf{y}_A(\tilde{q}) \tag{4.9}$$

where $\mathbf{S}\mathbf{y}_A$ is the radius vector of the Ath nucleus drawn from the center of mass,

$$\mathbf{S}(\varphi, \theta) = \mathbf{T}_Z(\varphi)\mathbf{T}_Y(\theta) \tag{4.10}$$

and

$$\mathbf{y}_A(\tilde{q}) = \mathbf{T}_Z(\chi(\varphi, \theta))\mathbf{y}_A(q) \tag{4.11}$$

and hence overall rotations induce linear transformations $q \to \tilde{q}$ parametrically dependent on rotational variables φ and θ.

Note that the above analysis requires one to distinguish only between large- and small-amplitude motions, not between large-amplitude *external* and *internal* degrees of freedom. Let us consider van der Waals dimers as an example. We start our discussion from an $(A_2)_2$ dimer[214-217] (the extension of our arguments to $(AB)_2$ dimers[214,215,218-220] is obvious). Let $\mathbf{r}_{D\alpha}$ be the radius-vector of the αth nucleus ($\alpha = 1, 2$) in the Dth diatomic monomer (D = A, B). Following Refs. 214–217, we express the vectors $\mathbf{r}_D = \mathbf{r}_{D2} - \mathbf{r}_{D1}$ and the vector $\mathbf{R}$ connecting the centers of mass of the diatomic monomer in terms of spherical coordinates: r_D, φ_D, θ_D and R, φ, θ respectively, which undergo linear transformations[214-216] under the action of feasible operations. It is even more essential that the law of transformation is independent of particular values of the vibrational coordinates $q_D = r_D - r_D^e$, where r_D^e is an equilibrium value of r_D. As feasible operations may just change sign and permute the coordinates q_D, we conclude that each term in the expansion of the accurate Hamiltonian as a Taylor series in q is invariant under the G group discussed in Refs. 214–217. It is worth emphasizing that we could do without any SRMM at all.

Note that the third *external* Euler angle χ must be chosen, if at all, only in a symmetric way with respect to both diatomic monomers (cf. Ref. 217, p. 5566). In particular, if ϕ_A and ϕ_B are body-fixed longitude angles (see Fig. 1 in Ref. 217), one should put $\chi = \frac{1}{2}(\phi_A + \phi_B)$, and hence this angle is a double-valued function of Cartesian coordinates. If we also put $\gamma = \frac{1}{2}(\phi_A - \phi_B)$, we obtain that the angles φ, θ, χ, γ transform under the action of elements of the double group $G_{16}^{(2)}$[221,222] according to Table II in Ref. 222. The procedure applied to the dimer $(HF)_2$ leads us to the double group $G_4^{(2)}$ (see Table I in Ref. 223).

By analogy we can attach Eckart frames[4] to the centers of mass of identical nonlinear molecules forming a dimer.[224] (See also Refs. 225, 226; we shall come back to the discussion of these papers below.) We then construct localized Hamiltonians invariant under appropriate FPI groups, expanding the global Hamiltonian as a Taylor series of vibrational coordinates. Some of these groups have been discussed by Dyke et al.[227,228]

The problem becomes much more complicated if we try to select overall rotations from large-amplitude internal motions for dimers formed by identical molecules having a symmetry axis, such as $(H_2O)_2$,[227] $(CH_4)_2$,[228] $(C_6H_6)_2$,[228] $(NH_3)_2$,[229] and so on. The crucial point is how to introduce torsional angles χ_A and χ_B in a symmetric way with respect to both molecules. We shall see below that exactly the same problem arises for molecules with internal rotation. The torsional angles which we are introducing just now for $(H_2O)_2$ and $(NH_3)_2$ dimers can be used, for example, respectively for R—NO_2 and R—CH_3 molecules with internal rotation if R is an *asymmetric* frame. Let us start with the $(H_2O)_2$ dimer. It is the simplest case: we need only require the Dth torsional axis to cross the line $H \cdots H$ (for example, this axis can coincide with the height of the triangle H_2O perpendicular to that line). As a result, the torsional angle χ_D transforms under action of the permutation of the hydrogens in the Dth molecule as $\chi_D \to \chi_D + \pi$ for an *arbitrary* nuclear configuration. Then we can use the longitude angles ϕ_A and ϕ_B of the torsional axes to define the third Euler angle χ in exactly the same way as has been done for diatomic monomers.

The choice of the torsional angle for NH_3 must be more specific. Let us place all the hydrogens of the Dth molecule in the plane XY in such a way that the angle between any two vectors $\mathbf{r}_{D\alpha}$ $(\alpha = 1, 2, 3)$ is equal to 120°. One finds

$$|\mathbf{r}_{D\alpha}|^2 + |\mathbf{r}_{D\beta}|^2 + |\mathbf{r}_{D\alpha}|\,|\mathbf{r}_{D\beta}| = \rho_{D,\alpha\beta}^2, \qquad 1 \le \alpha < \beta \le 3 \qquad (4.12)$$

where $\rho_{D,\alpha\beta} \equiv |\mathbf{r}_{D\alpha} - \mathbf{r}_{D\beta}|$. The solution exists, at least near a C_{3v} equilibrium configuration, because the determinant of the system

$$\sqrt{3}\,(dr_\alpha + dr_\beta) = 2\,d\rho_{\alpha\beta}, \qquad 1 \le \alpha < \beta \le 3 \qquad (4.13)$$

is not zero. The Dth torsional axis is chosen coincident with the Z axis. Note that it does not go through the nitrogen in the general case. It is essential that the torsional angle χ_D transforms under action of the permutation (123) of the hydrogens in the Dth molecule as $\chi_D \to \chi_D + 2\pi/3$, again for an *arbitrary* nuclear configuration NH_3. All other angles are introduced as described above. We are thus led to the double group $G_{36}^{(2)}$ (see Ref. 40, pp. 1923, 1924) for the law of transformation of the angles φ, θ, χ, γ under the action of its elements).

As mentioned above, we can use exactly the same torsional angles to describe internal rotation of such atomic groups as NO_2 or CH_3 around an *asymmetric* frame. The latter requirement is very important, because, by the proper choice of the torsional axis, we can satisfy the symmetry requirements only for a single top. For example, we cannot use those torsional angles for CH_3–NO_2 or BH_2–BD_2 molecules. Nor can the device be extended to more complicated tops of the same symmetry (for example, to the phenyl group, which has exactly the same point symmetry as the nitro group).

For molecules with internal rotation of one or more atomic groups, the simplest way to make matrix elements $Q_{\beta\gamma}^{g}$ independent of torsional angles is to include atoms connecting two molecular fragments in turn in each of them.[79] In such a way one can reduce the problem to a choice of vibrational coordinates for nearly rigid fragments. A choice of coordinates satisfying the condition of Eq. (4.5) for nearly rigid molecules is discussed in standard textbooks (see Ref. 230, for example). In contrast with van der Waals molecules, it does not seem reasonable to select *three-dimensional* rotations of each top as suggested in Refs. 225, 226, 231–233. Lyaptsev and Kiselev's[231] idea of treating some of the Euler angles describing rotations of tops around the axis of the molecule as vibrational coordinates cannot be easily realized if at least one of the tops has a symmetry axis. Remember that in this case those angles must undergo linear *homogeneous* transformations under the action of feasible operations, whereas, for example, the angle θ_A of nutation of the nitro group defined as explained above transforms under the permutation of the oxygens as $\theta_A \to \pi - \theta_A$ (cf. Table I in Ref. 217).

V. ELIMINATING MOMENTUM COUPLINGS BETWEEN LARGE- AND SMALL-AMPLITUDE MOTIONS

In the previous section we discussed only *symmetry* limitations imposed on molecular variables. One should also make momentum couplings as small as possible. This usually means eliminating the couplings between large- and small-amplitude motions in the kinetic energy at $q = 0$. This requirement is obviously fulfilled if we use Eckart coordinates[4] for each molecular fragment as discussed in the previous section. In the general case, when choosing internal variables one should satisfy the so-called Hofacker–Marcus conditions[234,235]

$$\sum_{A=1}^{N} \frac{1}{m_A} \nabla_A \rho_\nu \cdot \nabla_A q_\beta \bigg|_{q=0} = 0, \qquad \nu = 1,2,\ldots, p, \quad \beta = 1,2,\ldots, \Gamma \quad (5.1)$$

A very elegant method to satisfy these conditions for an arbitrary set of large- and small-amplitude internal variables (τ, q) was suggested by Quade[236] (see also Refs. 237, 238). For this purpose one makes use of new large-amplitude internal variables ρ connected with the initial ones by the transformation[236]

$$\rho_\nu = \tau_\nu + \sum_{\beta=1}^{\Gamma} \mathscr{R}_\nu^\beta(\tau) q_\beta, \qquad \nu = 1,2,\ldots, p \quad (5.2)$$

where the coefficients $\mathscr{R}_\nu^\beta(\tau)$ are determined by the relation

$$\mathbb{G}_{\beta\,\Gamma+\nu}(\tau,0) + \sum_{\gamma=1}^{\Gamma} \mathbb{G}_{\beta\gamma}(\tau,0)\mathscr{R}_\nu^\gamma(\tau) = 0, \qquad \nu = 1,2,\ldots, p, \quad \beta = 1,2,\ldots, \Gamma \quad (5.3)$$

where the $\mathbb{G}$ matrix is calculated in internal variables τ and q ($p+\Gamma=3N-6$). Compared with the least-squares large-amplitude internal variables[17] discussed in the next section, Quade's variables can be more easily expressed in terms of Cartesian coordinates. In addition, the initial variables τ as well as q are chosen independently of the nuclear masses, and hence we can study isotopic substitution, starting from the same potential represented as a function of τ and q and then reexpressing this potential in terms of mass-dependent variables ρ by means of Eq. (5.2).

The question which was not discussed by Quade is how to satisfy the symmetry requirements for molecules such as ammonia[237] or for molecules with internal rotation of a C_{3v} top.[238] The crucial point is how to find large-amplitude internal variables τ that undergo linear transformations as in Eq. (3.9) under the action of feasible operations. Taking into account that the $\mathbb{G}$ matrix transforms as a covariant tensor of rank 2 (we use covariant coordinates τ_ν and q_β), one concludes that the new large-amplitude internal variables in Eq. (5.2) have exactly the same law of transformation [Eq. (3.9)] as τ.

We have shown in the previous section how one can choose the torsional axis for C_{3v} tops. In order to define an inversion coordinate for ammonia let us distinguish between the upper and lower surfaces of the triangle formed by the hydrogens by means of the vector

$$\boldsymbol{\eta} \equiv (\mathbf{r}_1-\mathbf{r}_3)\times(\mathbf{r}_2-\mathbf{r}_3)=\mathbf{r}_1\times\mathbf{r}_2+\mathbf{r}_2\times\mathbf{r}_3+\mathbf{r}_3\times\mathbf{r}_1 \tag{5.4}$$

perpendicular to the plane. The vector (5.4) is invariant under both the permutation (123) of the hydrogens and the inversion; the permutation (12) multiplies it by -1. Let an *arbitrary* configuration NH_3 (no point symmetry is assumed) be oriented in space in such a way that the hydrogens and nitrogen are in the plane XY and on the Z axis, respectively, and the directions of the latter and the vector $\boldsymbol{\eta}$ coincide. We can choose as an inversion coordinate τ, for example, the Z projection of the nitrogen ($-\infty<\tau<\infty$) or the angle $\tau=\frac{1}{3}(\theta_1+\theta_2+\theta_3)$, where θ_i is the colatitude angle of the ith hydrogen (cf. Ref. 237, p. 2108); thus, in the latter case, $0<\tau<\pi$ ($\tau=\pi/2$ for the planar configuration[237]).

The next question is how to choose the body-fixed axes, that is, how to define a set of standard configurations satisfying Eq. (4.4). The principal axes of the inertial tensor of nuclear configurations $\bar{\mathbf{r}}$ are not suitable for the definition of standard configurations in the cases of the G groups of higher order, as for an *arbitrary* configuration $\bar{\mathbf{y}}$ the matrix $R_g^{(r)}$ in Eq. (4.3) may differ from a 3×3 permutational matrix only by the signs of some elements. By contrast, for example, the permutation (123) of the hydrogens in the ammonia molecule must rotate any C_{3v} configuration $\bar{\mathbf{y}}$ through 120°. That is the reason why we considered the NH_2Cl molecule above but not the ammonia molecule.

Let us show that one can satisfy the condition of Eq. (4.7a) if there exists an SRMM transformed according to Eq. (3.3) under the action of feasible operations. First of all, note that a choice of vibrational coordinates uniquely determines the geometrical structure of such a model at $q = 0$. For example, the use of the vibrational coordinates suggested for ammonia by Cress and Quade (see Fig. 1 in Ref. 237) implies that those authors use the umbrella model with fixed N—H bonds.[239] Then the large-amplitude internal variables ρ are assumed to undergo transformations as in Eq. (3.9) for any values of q, that is, internal isometric substitutions in Eq. (3.3) are known. The problem is whether or not we can orient the model in space for different values of ρ in such a way that Eq. (3.3) is fulfilled. We can do it, for example, for the mentioned model of ammonia and for SRMMs with internal rigid-top rotation.

Let us define the set of standard configurations $\bar{\mathbf{y}}$ minimizing the form[119] (see also Refs. 33, 120)

$$L = \sum_A m_A \left| \mathbf{r}_A - \mathbf{S}(\theta')\mathbf{a}_A(\rho') - \mathbf{z}_A \right|^2 \tag{5.5}$$

with respect to θ' only; the internal variables ρ' are assumed to be known functions of $\bar{\mathbf{r}}$. That is, in contrast with Ref. 33, we do not minimize this form with respect to ρ', and hence the global Sayvetz conditions are not fulfilled. But in order to eliminate zeroth-order velocity couplings between large- and small-amplitude internal motions it is sufficient to satisfy only the local Sayvetz conditions[17,33]

$$\sum_A m_A \left. \frac{\partial \mathbf{a}_A}{\partial \rho_\nu} \cdot \frac{\partial \mathbf{r}_A}{\partial q_\beta} \right|_{\bar{\mathbf{r}} = \bar{\mathbf{a}}(\rho)} = 0 \tag{5.6}$$

as long as the Eckart conditions are fulfilled. (Remember that the "global Eckart conditions"[4] are nothing but the requirement for the form in Eq. (5.5) to have an extremum at $\theta' = \theta$.[119,120]) In order to verify that the local Sayvetz conditions [Eq. (5.6)] are actually fulfilled, one need just prove the relation

$$\sum_A m_A^{-1} \nabla_A \theta_i \cdot \nabla_A q_\beta \Big|_{\bar{\mathbf{r}} = \bar{\mathbf{a}}(\rho)} = 0, \qquad i = 1,2,3, \quad \beta = 1,2,\ldots,\Gamma \tag{5.7}$$

as the internal variables have been chosen to satisfy the local Hofacker–Marcus conditions (5.1) (see Ref. 17, p. 485). The derivatives of θ_i with respect to $\mathbf{r}_A$ can easily be calculated (see Appendix B) by differentiating the global Eckart conditions

$$\sum_A m_A \mathbf{r}_A \times \mathbf{S}(\theta)\mathbf{a}_A(\rho') = \mathbf{0} \tag{5.8}$$

with respect to $\bar{\mathbf{r}}$. (Note that both θ and ρ' are functions of $\bar{\mathbf{r}}$. We have already used this idea in Ref. 33 to derive the quantum-mechanical Hamiltonian; the only difference now is that we do not have the global Sayvetz conditions.) Using the local Hofacker–Marcus conditions [Eq. (5.1)] and taking into account the invariance of vibrational coordinates under overall rotations, one can verify that the condition in Eq. (5.7) is actually fulfilled. The model $\bar{\mathbf{a}}(\rho)$ is assumed chosen to satisfy Eq. (3.3), and hence the set of standard configurations used is invariant under operations $\overline{\mathbf{W}}_g$ according to Eq. (4.4).

As internal variables τ, q, and hence ρ' are assumed known as functions of $\bar{\mathbf{r}}$, and derivatives of θ with respect to $\mathbf{r}_A$ can be calculated in the way mentioned above, there is no difficulty in constructing the quantum-mechanical Hamiltonian.

As mentioned above, the Quade method[236] does seem preferable if one can find large-amplitude internal variables τ undergoing linear transformations as in Eq. (3.9) under the action of feasible operations. Unfortunately, in the general case this is difficult to do by means of conventional coordinates. In the next section we shall discuss the so-called "least-squares large-amplitude internal variables,"[17] which always transform according to Eq. (4.7b) as long as there exists an SRRM satisfying Eq. (3.3).

VI. LEAST-SQUARES SETS OF EXTERNAL AND INTERNAL VARIABLES

Let us come back to the form in Eq. (5.5), minimizing it now with respect to both θ' and ρ'. Besides the Eckart conditions [Eq. (5.8)] one obtains the global Sayvetz conditions

$$\sum_A m_A \frac{\partial \mathbf{a}_A}{\partial \rho_\nu} \cdot [\mathbf{S}^{-1}(\theta)(\mathbf{r}_A - \mathbf{z}) - \mathbf{a}_A(\rho)] = 0 \tag{6.1}$$

as necessary conditions for this form to have an extremum at $\rho' = \rho$, $\theta' = \theta$. As a result, we find that the least-squares set of standard configurations[17,33] $\bar{\mathbf{y}}^{(\mathrm{ls})}$ satisfies the conditions

$$\sum_A m_A \mathbf{d}_A \cdot \frac{\partial \mathbf{a}_A}{\partial \rho_\nu} = 0 \tag{6.2a}$$

$$\sum_A m_A \mathbf{d}_A \times \mathbf{a}_A(\rho) = \mathbf{0} \tag{6.2b}$$

where $\bar{\mathbf{d}} = \bar{\mathbf{y}} - \bar{\mathbf{a}}$. Note that this set is obviously invariant under feasible operations if the model transforms according to Eq. (3.3); the law of transformation of least-squares large-amplitude variables is uniquely determined by the symmetry properties of the model[33] (see also Refs. 28, 34, 35, 62–64, 73). It should be stressed that the least-squares large-amplitude internal variable differs from the one considered in the previous section even if we use exactly the same semirigid model of the ammonia molecule, choosing the height of the pyramid as the model parameter.

It is important that the least-squares set of standard configurations $\bar{\mathbf{y}}^{(\mathrm{ls})}$ satisfies the *linear* conditions [Eqs. (6.2a), (6.2b)] and hence one can define internal variables q by means of simple relations

$$\bar{\mathbf{d}} = \sum_{\beta} \bar{\mathbf{c}}^{\beta}(\rho) q_{\beta} \tag{6.3}$$

As the displacements $\overline{\mathbf{W}}_g \bar{\mathbf{d}}$ satisfy the global Eckart–Sayvetz conditions with respect to the vectors $\mathbf{a}_A(\tilde{\rho}^g)$, one has

$$\overline{\mathbf{W}}_g \tilde{\mathbf{c}}^{\gamma}(\rho) = \sum_{\beta} \bar{\mathbf{c}}^{\beta}(\tilde{\rho}^g) Q^g_{\beta\gamma}(\rho) \tag{6.4}$$

that is, rectilinear vibrational coordinates q undergo linear homogeneous transformations $\mathbf{Q}^g$ generally dependent on ρ. Therefore each term in the expansion in Eq. (4.8) is actually invariant under feasible operations $\hat{g} \in G$. As shown in Ref. 79 (see Appendices A and B therein), for molecules with internal rotation one can always find a set of rectilinear vibrational coordinates such that the law of their transformation under feasible operations is independent of ρ. One can easily construct rectilinear vibrational coordinates with a *ρ-independent* law of transformation [Eq. (4.5)] for SRMMs of cyclohexane linearly dependent on ρ,[208] making use of the point symmetry of the supporting configuration. (Note that we have a complete analogy with the SRMM of the ammonia molecule suggested by Newton and Thomas.[240])

SRMMs with internal rotation of *rigid* tops have the obvious defect: with the exception of those few molecules for which the symmetry of frame and top are the same, like nitrobenzole or methylsilane, the point symmetry of atomic groups in an equilibrium configuration is lower than the symmetry of the proper top in SRMM. For example, in order to simulate internal rotation of the methyl group one has to use a C_{3v} pyramid, whereas equilibrium configurations generally have no symmetry axis. In order to estimate geometrical parameters of equilibrium configurations one had to apply *rigid* tops instead of SRMMs (see, e.g., Refs. 241–244). The only way to take this effect into account without breaking the equivalence of the hydrogens is to allow bond lengths and angles to change during internal rotation. Another advantage of such relaxation is the possibility of interpreting microwave transitions in torsionally excited and ground states by means of the same SRMM.[245,246] Following Meyer,[247] we call such SRMMs *flexible models* with internal rotation. *Ab initio* calculations of the models of H_2O_2[248,249] and CH_3ONO[250] have been reported. The only questions essential to answer for this discussion are whether or not flexible models satisfy the condition in Eq. (3.3) and, if the answer is yes, whether one can choose coefficients $\bar{\mathbf{c}}^{\beta}$ in Eq. (6.3) in such a way that matrices $\mathbf{Q}^{g}$ in Eq. (6.4) are independent of ρ.

It should be stressed that the role played by the Eckart conditions in the theory of molecular symmetry was to a considerable extent overstated by Louck and Galbraith;[53] it is important[28] but by no means fundamental, in contrast with Ezra's conclusion (Ref. 64, p. 39). The fact that symmetry operations $\overline{\mathbf{W}}_g$ generate internal motions of the particles compatible with molecule-fixed axes does not mean that "the Hamiltonian which describes these motions must accordingly be invariant" (Ref. 53, p. 103) under $\overline{\mathbf{W}}_g$. First of all, for an accurate Hamiltonian this is true only if that Hamiltonian is invariant under the appropriate operation $\overline{\Pi}_g$. We can use exactly the same change of variables when considering parity nonconservation,[251] and hence the invariance group of the Eckart frame is still isomophic to the point symmetry group of an equilibrium configuration, whereas the invariance group of the localized Hamiltonian contains only feasible permutations, i.e.,

$$\mathbb{H} \overset{\text{ho}}{\rightarrow} \mathscr{F}_{\eta,q,\theta}.$$

Even if we do not consider such an exotic case as parity nonconservation, Eq. (4.3), which is equivalent to the requirement for $\overline{\mathbf{W}}_g$ to belong to "the invariance group of the frame," does not ensure on its own the invariance of an approximate Hamiltonian. It is also necessary that symmetry operations $\overline{\mathbf{W}}_g$ generate linear homogeneous transformations *in the space of vibrational coordinates*. For example, one can use the Eckart frame to define rotational coordinates but choose an arbitrary set of internal variables as vibrational coordinates. (The only limitation is that those variables have zero values at the equilibrium configuration.) Then the condition in Eq. (4.3) is obviously fulfilled, but in the general case only infinitesimal variations of vibrational coordinates undergo linear homogeneous transformations under the action of feasible operations. As a result one can conclude that those operations usually keep invariant only the first nonconstant term in the expansion of any operator in q. As the expansion is carried out near the equilibrium configuration, the linear term in the expansion of the adiabatic potential vanishes, and hence the approximate Hamiltonian turns out to be invariant under feasible operations in the *harmonic* approximation, whatever internal coordinates are considered. Taking into account anharmonic terms breaks this invariance unless special symmetrized internal variables are used.

It is useful to note that for *any* molecule one can introduce additional terms in the form of Eq. (5.5), for example, minimizing the form

$$L^{(4)} = L(\theta', \rho') + \gamma \sum_{A} m_A \left| \mathbf{r}_A - \mathbf{z} - \mathbf{S}(\theta')\mathbf{a}_A(\rho') \right|^4 \qquad (6.5)$$

with respect to θ' and ρ'. As long as one can define a complete set of conventional vibrational coordinates, all conditions for different terms in Taylor series in q to be invariant under the group G are fulfilled. Derivatives of the vibrational coordinates in question with respect to Cartesian coordinates $\mathbf{r}_A$ of nuclei are well known. As for derivatives of the large-amplitude variables θ and ρ, they can be found by differentiating the proper extremum conditions with respect to the Cartesian coordinates in exactly the same way as we have done previously[33,34] for changes of variables satisfying the *global* Eckart–Sayvetz conditions. After all derivatives of the new variables with respect to Cartesian coordinates have been obtained, the explicit expression for the Hamiltonian in terms of the new variables is easily found. Of course the fact that we have the nonlinear extremum conditions instead of the linear ones makes the operator for the kinetic energy more complicated, but this is not relevant for symmetry analysis. It should be stressed that the changes of variables under consideration satisfy the local Eckart–Sayvetz conditions (see Ref. 17 for details) and hence the form of Eq. (6.5) ensures separation of large- and small-amplitude motions no less efficient than that obtained with the form $L(\theta', \rho')$. In particular, for quasirigid molecules one can make use of the Born–Oppenheimer method,[12] applying the same mathematical formalism as has been applied[7,8] for the Wilson–Howard Hamiltonian.[75] Again we can see that there is no reason for the Eckart vectors and Eckart frame to "be considered to be basic conceptual constructions on which the theory of the vibration–rotation spectra of molecules is erected."[53] They are just a very useful auxiliary tool, nothing more.

Of course, we do not think that the introduction of new terms into Eq. (5.5) is reasonable, and we consider the form in Eq. (6.5) only as a counterexample to arguments of Louck and Galbraith.[53] But the idea of using the Eckart frame together with any conventional vibrational coordinates may turn out to be very fruitful. First, the use of rectilinear vibrational coordinates makes calculations of Franck–Condon factors much more complicated, as those coordinates in different electronic states are connected by *nonlinear* relations.[16,121,252] Also, different assumptions about a force field are usually formulated in conventional coordinates[253–255] which are connected with rectilinear vibrational coordinates of a polyatomic molecule by complicated transcendental equations (only in the simplest case of a triatomic molecule ABA does this equation take a compact form[121]), and one has to use their expansion as a Taylor series.[256] The abovementioned method allows one to construct a Hamiltonian whose rotational part coincides with the rotational part of the Howard–Wilson Hamiltonian, whereas the Hamiltonian used by Gribov et al. (see Ref. 257 and references therein) to calculate force constants coincides with the pure vibrational part of the Hamiltonian under discussion after omitting extra potential terms in both Hamiltonians. (This term depends on the Jacobian of the transformation in question; the authors[257] took into account only the metric in the space of internal coordinates; i.e., when calculating the Jacobian they omitted the additional factor of the square root of the determinant of the inertial tensor—see Eq. (2.34) in Ref. 258 or Eq. (D5) in Ref. 17.) It is worth noting that one does not need Wigner symmetry operations $\overline{\mathbf{W}}_g$ to study symmetry properties of the Hamiltonian constructed in the way discussed. Clearly it is much easier to construct this Hamiltonian than to apply Quade's device,[255‡] which is to choose an arbitrary set of standard configurations and then switch them by rotations $e^{\mathbf{A}(q)}$, where

$$\mathbf{A}(q) = \sum_{i=1}^{3} \mathbf{A}_i \sum_{\gamma} \rho_\gamma^i q_\gamma \tag{6.6}$$

and the coefficients ρ_γ^i are chosen to satisfy the local Eckart conditions.

VII. NORMAL VIBRATIONAL COORDINATES

Up to now we have considered vibrational coordinates introduced *a priori*, and hence we have been able to study their symmetry directly. However, in order to separate vibrations in the zeroth-order perturbation limit one has to introduce normal vibrational coordinates, and their symmetry species may differ from the initial ones[40,42,81] (see also Refs. 28, 35, 71, 77–79, 82, 85, 86, 222, 223, 259, 260).

Let $\mathbf{G}(\rho)$ and $\mathbf{F}(\rho)$ be matrices with elements

$$G_{\beta\gamma}(\rho) \equiv \sum_A \frac{1}{m_A} \mathbf{s}_A^\beta(\rho) \cdot \mathbf{s}_A^\gamma(\rho) \tag{7.1}$$

where

$$\mathbf{s}_A^\beta(\rho) \equiv \nabla_A q_\beta \big|_{\bar{\mathbf{r}} = \bar{\mathbf{a}}(\rho)} \tag{7.2}$$

‡We doubt that one can find a reasonable alternative to the Eckart frame for symmetric- and spherical-top molecules.

and

$$F_{\beta\gamma}(\rho) = \left.\frac{\partial^2 W}{\partial q_\beta\, \partial q_\gamma}\right|_{q=0} \tag{7.3}$$

where $W(q, \rho)$ is the adiabatic potential. We use the term "normal vibrational coordinates" for linear combinations of q,

$$q_\beta^{(n)} = \sum_{\gamma=1}^{\Gamma} l_\gamma^\beta(\rho) q_\gamma \tag{7.4}$$

diagonalizing the **GF** matrix:

$$\left[\mathbf{G}(\rho)\mathbf{F}(\rho) - \omega_\beta^2(\rho)\mathbf{1}_\Gamma\right]\mathbf{l}^\beta(\rho) = 0 \tag{7.5}$$

The vibrational frequencies $\omega_\beta(\rho)$ as functions of ρ have been calculated for pseudorotation in CH_3PF_4[177] and for two molecules with internal rotation: dimethylmercury[261] and 1,4-dichlorobutyne.[260] Similar calculations have been reported for some isomerization processes and chemical reactions.[262-264]

Even if we employ exactly the same SRMM, the vibrational frequencies $\omega_\beta(\rho)$ will generally depend on the particular choice of large-amplitude internal variables ρ.[17] They are uniquely defined only in equilibrium configurations. This means that interpretation of microwave spectra by means of any harmonic model allowing relaxation of force-field parameters[265] (see also Ref. 200) can give only the difference $\Delta\omega = \omega(\rho_e) - \omega(\rho_e')$, where ρ_e and ρ_e' are the values of the parameters ρ respectively in a stable equilibrium configuration and at a saddle point. It also implies that one cannot carry out *ab initio* calculations of relaxation of force-field parameters without specifying the large-amplitude internal variables ρ as functions of Cartesian coordinates. For example, the choice of ρ was not specified by Russegger and Brickmann.[177] The relation given by Eq. (1) in Ref. 177 may give the impression that they use the least-squares large-amplitude internal variable ξ, but as explained in the last section, it is extremely difficult to express the potential energy in terms of such a variable and it is unclear whether Russegger and Brickmann[177] did choose this way. As an alternative one can place the center of mass of the four fluorines of CH_3PF_4 at the origin, draw the Z axis through the P atom, and put $\tau = 2\pi - \theta_1 - \theta_2 - \theta_3 - \theta_4$, where θ_i is the colatitude angle of the ith fluorine. Compared to using the set (ξ, S), it is much easier to calculate the force field in the variables τ, S, where the vibrational coordinates S are defined by Eq. (48) in Ref. 177. As the Quade transformation [Eq. (5.2)] contains no quadratic terms in q, we conclude that it does not change the force field (see Eq. (12) in Ref. 79).

The important point is that the **GF** matrix undergoes a similarity transformation under action of feasible operations as long as the large- and small-amplitude internal variables transform according to Eqs. (4.7b) and (4.5). This result was initially proved by us[33] for the simplest case of rectilinear vibrational coordinates in which the **G** matrix differs from the unit $\mathbf{1}_\Gamma$ by just a constant. The extension to arbitrary vibrational coordinates was discussed in Ref. 79. The additional condition which must be proved in the general case is that the **G** matrix transforms as a tensor of rank 2 under the action of feasible operations. (Remember that we consider only a *subspace* of internal coordinates.) As for the **F** matrix, its law of transformation immediately follows from the fact that the quadratic form obtained by expanding the adiabatic potential as a Taylor series in q *must* be invariant under feasible operations; the analysis is very simple because the expansion contains no derivatives. It should be emphasized that the **GF** matrix undergoes the similarity transformation even if matrix elements in Eq. (4.5) depend on ρ, and hence it is true for any set of rectilinear vibrational coordinates (see previous section).

Here we consider only the simplest case when surfaces do not have common points (see Refs. 35, 79 for an analysis of other possibilities). As a trivial example we can cite molecules with internal rotation of one or more methyl groups: as long as we neglect relaxation of both the SRMM and the force field, the eigenvalues ω_β^2 of the **GF** matrices of such molecules are independent of the torsional angles γ.[71] Groner's calculations[259] showed that nitroethane and its *deutero*-derivatives give us another example of molecules with no accidental crossing of curves $\omega_\beta(\gamma)$, but in contrast with the first example, we could not be sure of this fact before the calculations[259] were done. For example, accidental crossings have been found in calculations carried out by Lichene et al.[260] for 1,4-dichlorobutyne. (Accidental crossings have also been reported by Russegger and Brickmann,[177] but, as explained above, we cannot be completely sure how they obtained their results.)

If the **GF** matrix has only nondegenerate eigenvalues for any value of ρ, the normal vibrational coordinates span one-dimensional representations of the FPI group. The general proof is given in Ref. 79. Earlier this result was proven by Hougen[40] for nondegenerate vibrations in dimethylacetylene.

It should be stressed that the internal isometric group $\mathscr{F}_\rho$ of any molecule with internal rotation always contains one or more periodic transformations. If vibrations are assumed frozen, we can limit ourselves to consideration of primitive period transformations.[48-52,62,64] However, we must take into account all periodic transformations[‡] when classifying normal

[‡]Remember that periodic transformations $(\mathbf{1}_3, \rho \to \tilde{\rho})$ do not fall under the definition of primitive period transformations as given in Refs. 48–52, 62, and 64.

vibrations with respect to irreducible representations of G, because the vector $\mathbf{l}^{(\beta)}(\rho)$ in Eq. (7.5) may change its sign after rotation of a top through 2π.

Another peculiarity of molecules with internal rotation is an existence of so-called "nontrivial elements"[28,35,79,85] in the G group. We call an element *nontrivial*[28,35,79,85] if there is no fixed point $\tilde{\rho}^g = \rho$ and if this element is represented by -1 in at least one unidimensional real representation of the group G.

Let us consider the phenol molecule. Its FPI group contains four elements: the unit E, permutation P, inversion E^*, and permutation–inversion P^* (see Table I in Ref. 266). If we interpret the large-amplitude motion of the hydroxyl as internal rotation, then the permutation P is a nontrivial element, and as a result we cannot predict the symmetry of normal vibrations with respect to this permutation. In fact, although the conventional vibrational coordinates of the frame C_6H_5O span representations $11a_1 + 10b_2 + 3a_2 + 6b_1$ (there is a misprint in Ref. 266; cf. Ref. 169, p. 363) and another two independent internal coordinates describing vibrations in the top

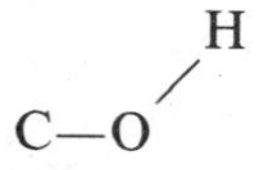

(the stretching mode C—O has been listed) are of a_1 type, one can only state that symmetry species of normal vibrations may be

$$(10+n)a_1 + (13-n)b_2 + na_2 + (9-n)b_1 \tag{7.6}$$

In order to clarify this statement let us introduce another term: *essentially trivial element*,[35,79,86] which is used for elements having fixed points. Symmetry classification of normal modes with respect to those elements can be carried out at the fixed points. Therefore the total number of vibrations symmetric or antisymmetric with respect to essentially trivial elements is uniquely determined by the SRMM in question. Both P^* and E^* are essentially trivial elements of G_4, that is, the total number of vibrations a_1, b_1 and that of vibrations a_1, b_2 are always equal to 19 and 23 respectively.

However, the same permutation P is an essentially trivial element with a fixed point at the C_{2v} configuration if, following Attanasio et al.,[143] we interpret the large-amplitude motion of the hydroxyl as inversion. As a result we can classify normal vibrations uniquely with respect to irreducible representations of G_4 by putting $n = 3$ in Eq. (7.6).

According to the definition, a nontrivial element cannot be the square of any other element. Therefore, internal rotation of $\hat{d}$ of a top through 2π is a trivial element if there exists an even permutation rotating the top around the torsional axis. That is why we considered above only single-valued representations of G. Double-valued normal modes may appear, for example, for *ortho-* or *meta-* (*but not para-*) chlorophenol.

Groner[259] was the first who illustrated this ambiguity with a particular indisputable example. Diagonalizing the $\mathbf{G}(\gamma)\mathbf{F}$ matrix of the nitroethane molecule and its deutero-derivatives $CH_3CD_2NO_2$, $CD_3CH_2NO_2$, and $CD_3CD_2NO_2$, he found that in the expression

$$(14-n)\Gamma_1+(8-n)\Gamma_2+(n+1)\Gamma_3+n\Gamma_4 \tag{7.7}$$

$n=1$ (conventional vibrational coordinates) only for $CD_3CH_2NO_2$; for $CH_3CH_2NO_2$ and $CH_3CD_2NO_2$ $n=2$, and for $CD_3CD_2NO_2$ $n=3$. A similar example (biphenyl and its 4,4′-dihalogeno-derivatives) was found by us[86] (see Ref. 79 for corrections), but it concerns curves $\omega_\beta(\gamma)$ having crossing points.

A much more serious problem is connected with the fact that the symmetry classification of normal modes may depend on a particular choice of a torsional angle as a function of Cartesian coordinates even if exactly the same SRMM is used. As shown in Ref. 79, any change of symmetry of normal modes with respect to a nontrivial element *factored into the product of two essentially trivial elements* is accompanied by the appearance of the crossing points of surfaces $\omega_\beta(\rho)$. Of course, large-amplitude internal variables which do not result in crossing points yield the best separation of motions. However the symmetry classification does depend on the change of variables used if a nontrivial element does not factor in the mentioned way. For example, this may happen for the $CHDClNO_2$ molecule. It means that the analysis carried out by us in Ref. 17 is incomplete and we should use some additional criteria to find a more preferable set of coordinates.

As pointed out by Woodman,[71] the $\mathbf{G}$ matrix of molecules with internal rotation of C_{3v} tops can be made independent of torsional angles γ by means of a special γ-dependent transformation of vibrational coordinates initially used by Howard[267] for ethane and then applied by Crawford and Wilson[80] to other molecules with internal rotation of C_{3v} tops. Of course, we must also cite in this connection Bunker's work[81] (where it was shown that the symmetry classification of normal vibrations in dimethylacetylene is unambiguous at the low-barrier limit) and the paper of Fleming and Banwell.[82] The latter first stressed that the symmetry classification of normal modes with respect to irreducible representations of the FPI group of the nitromethane molecule coincides at the low-barrier limit with the more intuitive classification suggested by Crawford and Wilson.[80]

The γ-dependent transformation making the $\mathbf{G}$ matrix independent of γ can be found for any molecule with internal rotation of one or more symmetric tops around their symmetry axes.[79] As for the potential field, it can be represented in the approximation of free internal rotation as

$$W(q) = \sum_{\tau} W_{\tau}(\{q_{\tau}\}) \tag{7.8}$$

where the coordinates q_{τ} describe vibrations in the τth nearly rigid fragment. (Remember that atoms connecting two fragments are included in both.) As a result of Eq. (7.8) the **F** matrix turns out to be independent of the torsional angles γ. (Note that our interpretation of this fact slightly differs from the one given by Woodman.[71]) Therefore we obtain the **GF** matrix independent of γ, and normal vibrations can be uniquely classified with respect to irreducible representations of G.

However, we cannot agree with Woodman's statement that his discussion justifies the intuitive account which implies neglecting all interactions between atoms of a frame and a top. The latter has been applied, for example, by Harris and Thorley[268] to the phenyl group in *n*-alkylbenzenes although for those molecules the γ dependence of the **G** matrix cannot be eliminated by means of the aforementioned change of variables.

Another important property of C_{3v} tops which is characteristic only of atomic groups rotating around its symmetry axis of *odd* order is that permutations forming the torsional subgroup in Woodman's terms[71] are "conditionally trivial elements,"[35,79,86] that is, the ones represented by $+1$ in any real one-dimensional representation of G. This means that the symmetry classification of normal modes can be carried out by irreducible representations of the frame group.[71] If all elements of this group leave equilibrium positions of all atoms unaltered (remember the definition of "the Altmann group" given in Ref. 74), the symmetry classification of normal vibrations is the same at the low- and infinite-barrier limits. The propane molecule briefly discussed in Section I is a typical example.[38]

It is worth stressing that the **G** matrix of an alkane can be made independent only of the torsional angles γ' of methyl groups; it still depends on the torsional angles γ'' describing internal rotation of fragments CCX_2, where $X = C, H$. This means that normal vibrational coordinates may change their sign after rotation of the ith fragment over 2π and hence the *trivial periodic transformations* $\gamma_i'' \rightarrow \gamma_i'' + 2\pi$ must be included in the internal isometric group.

The symmetry classification of normal vibrations in molecules with the internal rotation of the methyl group becomes ambiguous if relaxation of the SRMM[199,200,246] or force field[200,265] is taken into account. Those effects may change the symmetry of a pair of quasidegenerate normal modes, and (as stressed by us[85]) this immediately leads to an anomalous microwave spectrum in the first excited vibrational state. For example, *in the approximation of free internal rotation* we have three pairs of quasidegenerate normal modes *antisymmetric* with respect to the permutation of the oxygens of CH_3NO_2.[32] This means that the transitions observed in the ground vibrational state are forbidden if any of those vibrations is excited. However, one can observe the same transitions in the ground and all excited vibrational states if the excited vibration turns out to be symmetric with respect to the permutation of the oxygens.

By analogy, the mentioned effects may result in double-valued normal modes in molecules with internal rotation of the methyl group about a C_s frame,[85] and in this case torsion–rotational wavefunctions must be also double-valued. In Ref. 85 we calculated an exact spectrum of the torsion–rotation Hamiltonian with these boundary conditions, using the numerical algorithm developed by us earlier[269] for usual single-valued functions. We obtained that $\nu_A - \nu_E$ splittings mainly have different signs in the ground and first excited *double-valued* vibrational states. Recently Hougen[270] gave a very simple explanation of this result at the high-barrier limit. In fact, we have exactly the same minima as considered by Hougen for the ethane molecule,[178] and Table 4 in Ref. 178 immediately predicts different signs of splittings $\nu_A - \nu_E$ in single- and double-valued torsion–rotational states (we obtained some exclusions, as our calculations were carried out for finite barriers).

Similar problems arise when one classifies normal vibrations of an impurity molecule in a crystal at the librator limit. First of all, a rigid model of the molecule must satisfy the condition

$$\overline{\Pi}_g \bar{\mathbf{a}} = \overline{\mathbf{R}_g^{(r)} \mathbf{a}} \tag{7.9}$$

for any feasible perrotation $(\mathbf{R}^{(l)}, \overline{\Pi}_g)$. Let us describe vibrations by means of displacements $\mathbf{d}_A(q)$ connected with radius vectors of nuclei by the relation

$$\mathbf{r}_A = T_X(\gamma)[\mathbf{a}_A + \mathbf{d}_A(q)] \tag{7.10}$$

Only the one Eckart condition

$$\sum_A m_A (\mathbf{a}_A \times \mathbf{d}_A)_X = 0 \tag{7.11}$$

is assumed fulfilled. It is easily shown that feasible perrotations $(\mathbf{R}_g^{(l)}, \overline{\Pi}_g)$ generate Wigner symmetry operations $\overline{\mathbf{W}}_g$ in the space of displacements $\bar{\mathbf{d}}$:

$$\overline{\mathbf{W}}_g \bar{\mathbf{d}} = \overline{\Pi}_g \overline{\left(\mathbf{R}_g^{(r)}\right)^{-1} \mathbf{d}} \tag{7.12}$$

and the symmetry classification of rectilinear vibrational coordinates is carried out according to the usual rules.[3,111,230]

Let us consider again the C_{2v} triatomic molecule in the O_h field.[103,186] We limit ourselves to the consideration of the most complicated case: the symmetry classification of normal vibrations in the C_4 model. Compared with the C_3 model it has additional nontrivial elements: the pure proper rotation $\hat{C} \equiv (\mathbf{T}_X(\pi/2), \hat{e})$ and its product with the single pure permutation $\hat{p}$ [the perrotation $\hat{P} \equiv (\mathbf{1}_3, \hat{p})$]. The feasible perrotation group G of the C_4 model factors as

$$G \overset{\text{iso}}{=} D_{4h} = C_4 \wedge K^{\text{vib}} \tag{7.13}$$

where the invariant subgroup C_4 is formed by pure overall rotations and the subgroup

$$K^{\text{vib}} \overset{\text{iso}}{=} C_{2h}$$

is generated by the permutation $\hat{p}$ and perrotation $\hat{A} \equiv (T_Y(\pi), \hat{p}^*)$. As pure proper rotations do not act upon rectilinear coordinates, the latter can be only of A type. We use the subscripts g, u and $1,2$ to specify the symmetry of vibrations with respect to $\hat{P}$ and $\hat{A}$ respectively. We find two internal vibrations of A_{1g} type and one of A_{2u} type, in agreement with Ref. 186. The symmetry species of nutations (A_{1u}, A_{2u}) also coincide with the ones given by Kiselev and Luders. However, the symmetry species of translations are A_{1g}, A_{1u}, and A_{2u}, but not A_{1g}, A_{2g}, A_{2u} as stated in Ref. 186. Note that there exists no zero-frequency vibration in the space of Cartesian coordinates as long as the potential depends on γ (cf. Ref. 186, p. 33).

As explained above, to classify rectilinear vibrational coordinates does not mean to classify normal vibrations. Fortunately, all three internal motions have the same symmetry with respect to the essentially trivial element $(\mathbf{T}_Y(\pi), \hat{\imath})$, and hence they cannot change their symmetry with respect to nontrivial elements $\hat{C}$ and $\hat{P}$‡ *if their interactions with external motions are negligible*. This statement does not apply to external motions.

Summarizing, we would like to stress that the problem of the symmetry classification of normal modes with respect to nontrivial elements is apparently the least studied point of the approach developed here. We hope

‡The perrotations $\hat{A}\hat{P} = (\mathbf{T}_Y(\pi), \hat{\imath})$ and $\hat{C}(\hat{A}\hat{P})$ change sign of the librational angle γ; the perrotation $\hat{A}$ generates the transformation $\gamma \to \pi - \gamma$ with the fixed point $\gamma = \pi/2$. Therefore both $\hat{C}$ and $\hat{P}$ can be represented as a product of two essentially trivial elements with $\hat{A}\hat{P}$ as one of them.

that increasing interest in microwave transitions in excited vibrational states will draw more attention to this very peculiar question.

Acknowledgments

This research was supported by a Grant from the National Science Foundation. The author is very grateful to Professor M. N. Adamov for his enthusiastic encouragement and fruitful discussions of many years. The author is also indebted to Professor R. S. Berry and Professor G. S. Ezra for many stimulating communications and invaluable critical comments on the original manuscript of this review. It was Dr. Ph. Bunker and Professor S. Rice who stimulated a complete revision of the first version of this work, whereas numerous discussions with Dr. O. Mullins helped to perfect the arguments.

APPENDIX A

There exists a widespread opinion[32,271-275] that Wilson and Howard[75] derived the quantum-mechanical Hamiltonian for a polyatomic molecule by quantization of the classical kinetic energy expressed in terms of curvilinear coordinates. From our point of view, this assertion results from a not precisely correct interpretation of their work.[75] It has been stressed by Sutcliffe[276] (cf. Ref. 273, 275, 277) that the only difference between the Wilson–Howard[75] and Darling–Dennison[278] derivations is that Wilson and Howard start from the covariant Laplacian

$$\Delta_{3N} = g^{-1/2} \sum_{\nu,\mu} \hat{p}_\nu g^{\nu\mu} g^{1/2} \hat{p}_\mu \tag{A.1}$$

(note that here $\det\|g^{\nu\mu}\| = g^{-1}$), whereas Darling and Dennison just work with the Podolsky transformation[279] of the same Laplacian,

$$\Delta_{3N} \to \tilde{g}^{1/4} \Delta_{3N} \tilde{g}^{-1/4} \tag{A.2}$$

where

$$\tilde{g}^{1/2}(\mathbf{Я}) \equiv \frac{g^{1/2}(\mathbf{Я}, \theta)}{\det[\mathbf{b}(\theta)]} \tag{A.3}$$

[In anticipation of applying this formula to arbitrary internal variables $\mathbf{Я}$, we use here $\mathbf{Я}$ instead of q; the matrix $\mathbf{b}(\theta)$ will be defined below.] Note that Wilson et al. in their monograph[230] follow the derivation of Wilson and Howard[75] after representing the Laplacian in the Podolsky form. Let us reexamine here what Wilson and Howard actually do.

The important peculiarity of rectilinear vibrational coordinates is connected with the fact that one cannot represent them as explicit functions of

Cartesian coordinates. On the other hand, the derivatives of Cartesian coordinates with respect to the aforementioned internal as well as external variables are known if one uses the Eckart frame[4] to describe overall rotations of a molecule. This means that one can easily find the matrix $\|g_{\mu\nu}\|$ from the classical kinetic energy expressed in terms of velocities, and the only problem is to invert this matrix. But to do that one merely needs to reexpress the classical kinetic energy in terms of momenta instead of velocities, and Wilson and Howard did exactly this. They just made use of velocities and momenta as abstract symbols to obtain the Frobenius–Schur relation,[280] which was later suggested by Kiselev[7‡] for the same purpose.

After the Hamiltonian had been transformed to the new variables, the next step was to express derivatives $\partial/\partial\theta_i$ in terms of the components of the total angular momentum relative to the Eckart frame:

$$\hat{K}_j' = i \sum_{k=1}^{3} \{\mathbf{b}^{-1}\}_{kj} \frac{\partial}{\partial\theta_k} \tag{A.4}$$

Using a particular parametrization of matrices $\mathbf{S}(\theta)$ in terms of the Eulerian angles θ, Wilson and Howard proved the relation

$$\det\mathbf{b}\,\hat{K}_j' = i \sum_{k=1}^{3} \frac{\partial}{\partial\theta_k} \det\mathbf{b}\,\{\mathbf{b}^{-1}\}_{kj} \tag{A.5}$$

which allowed them to exclude dependence on θ from matrix elements coupling different momenta.

In such a way that quantum-mechanical operator of the nuclear kinetic energy was transformed to the well-known form just slightly modified by Darling and Dennison.[278] The only comment which should be made concerns Eq. (A.5). As initially stressed in Ref. 282, this relation is nothing but the requirement for $\hat{K}_j'$ to be Hermitian. The latter results from the fact that according to their definition the operators

$$\hat{K}_j' \equiv \sum_{k=1}^{3} S_{kj} \hat{L}_k' \tag{A.6}$$

where

$$\hat{L}_k' \equiv \sum_{A=1}^{N} (\mathbf{r}_A - \mathbf{z}) \times (-i\nabla_A) \tag{A.7}$$

‡A similar suggestion to invert the matrix $\|g_{\mu\nu}\|$ directly, including both nuclear and electronic motions, has been recently made by Webster et al.[281]

are generators of right shifts on the SO(3) group.[15] In fact, expressing the operators $\hat{L}'_k$ in terms of $\hat{K}'_j$ and substituting them into the obvious relation

$$(\hat{L}'_k \mathbf{S}) = i\mathbf{A}_k \mathbf{S} \tag{A.8}$$

one finds[15]

$$(\hat{K}'_j \mathbf{S}) = i\mathbf{S}\mathbf{A}_j \tag{A.9}$$

Equation (A.4) reflects the fact that the components $\hat{L}'_k$ of the total angular momentum relative to the laboratory-fixed frame, and hence their projections [Eq. (A.6)] on molecule-fixed axes, do not act on internal variables. To verify that the relation (A.5) does allow one to make the aforementioned matrix elements independent of θ one should take into account that

$$\frac{\partial \mathbf{r}_A}{\partial \theta_j} = \sum_{k=1}^{3} \mathbf{S}\mathbf{A}_k \mathbf{r}_A b_{jk} \tag{A.10}$$

because of Eqs. (A.4) and (A.9).

Note that the commutation relations for the operators $\hat{\mathbf{K}}'_j$ [Eq. (A.6)] directly follow from Eqs. (A.8), (A.9), and the well-known commutation relations between the operators $\hat{L}'_k$.[15] We need no particular parametrization of matrices **S**. A similar account was independently developed by Louck,[272] who introduced analytical continuation of the operators $\hat{\mathbf{K}}'_j$ from the space of proper rotational matrices **S** to the full real linear group. This continuation has sense as long as the operators $\hat{\mathbf{K}}'_j$ act on analytical functions of matrix elements S_{kj}. It must be pointed out that Louck's results cannot be applied to Eulerian angles, in contrast with the relation (A.12) of Tennyson and Sutcliffe.[283‡]

In order to include electrons in considerations one should just separate the motion of the center of mass of the whole molecule[‡‡] and then express the Hamiltonian in terms of molecule-fixed coordinates $\boldsymbol{\eta}_b$ of electrons. The last step means nothing but to reexpress the operators $\hat{K}'_j$ in terms of the

‡As explained by Professor B. T. Sutcliffe one should treat relation (A12) of Ref. 283 as a result of continuation of the Euler angles from the SO(3) group to the full real linear group. As an example one can consider factorization of nonsingular matrices into product of orthogonal and positively definite matrices parametrizing the orthogonal matrices by the Euler angles.

‡‡In particular, as suggested in our work,[8] one can make use of Radau coordinates[284] constructed from radius vectors $\mathbf{x}_b$ $(b = 1, 2, \ldots, n)$ and $\mathbf{z}$ of electrons and the center of nuclear mass respectively; unfortunately we learned of the existence of Radau's work[284] only from Smith's independent discovery of these coordinates.[285]

new variables:

$$\hat{K}_j' = \hat{K}_j + \sum_{b,i,k} \left(\hat{K}_j' S_{ik}\right) x_{bi} \frac{\partial}{\partial \eta_{bk}} = \hat{K}_j - \hat{M}_j \tag{A.11}$$

where $\hat{M}_j$ are molecule-fixed components of the total angular momentum of the electrons, and the only difference between the operators $\hat{K}_j'$ and $\hat{K}_j$ is that for the new operators one must calculate derivatives with respect to θ_k in (A.4) fixing the variables $\boldsymbol{\eta}_b$ instead of the laboratory coordinates $\mathbf{x}_b$. Again we obtain this result directly from the relation (A.9), and we do not need the particular realization of matrices **S** used by Nielsen[286] to prove the assertion.

An extension of the work of Wilson and Howard[75] to nonrigid molecules was done by Hougen et al.[287] for the least-squares set of large- and small-amplitude variables. In a similar way one can express the operator of the nuclear kinetic energy in terms of *arbitrary* large- and small-amplitude variables[15]

$$-\tfrac{1}{2}\Delta_N = \tfrac{1}{2}\tilde{g}^{-1/2} \sum_{\beta} \hat{p}_\beta \tilde{G}_{\beta\gamma}(\rho, q) \tilde{g}^{1/2} \hat{p}_\gamma$$
$$+ \tfrac{1}{2}\tilde{g}^{-1/2} \sum_{s,t=1}^{p+3} (\hat{N}_s - \hat{l}_s - \lambda_s) \tilde{g}^{1/2} \mu_{st} (\hat{N}_t - \hat{l}_t + \lambda_t) \tag{A.12}$$

where the operators $\hat{N}_s$ are defined in Section III, $\hat{p}_\gamma \equiv -i\,\partial/\partial q_\gamma$, the function $\tilde{g}(\rho, q)$ is connected with the Jacobian $g^{1/2}(\rho, q, \theta)$ by Eq. (A.3), and $\tilde{\mathbf{G}}(\rho, q)$ is the inverse of the matrix $\mathbf{T}^{\text{vib}}(\rho, q)$ with elements

$$T_{\beta\gamma}^{\text{vib}}(\rho, q) \equiv \sum_A m_A \frac{\partial \mathbf{r}_A}{\partial q_\beta} \cdot \frac{\partial \mathbf{r}_A}{\partial q_\gamma}, \qquad \beta, \gamma = 1, 2, \ldots, \Gamma \tag{A.13}$$

$$\hat{l}_s \equiv \frac{1}{2} \sum_{\beta=1}^{\Gamma} (\hat{p}_\beta r_{s,\beta} + r_{s,\beta} \hat{p}_\beta), \qquad s \le p+3 \tag{A.14}$$

$$\lambda_{s,\beta}(\rho, q) \equiv -\sum_{\gamma=1}^{\Gamma} \tilde{g}_{s,\gamma+p+3} \tilde{G}_{\beta\gamma}, \qquad s \le p+3 \tag{A.15}$$

$$\tilde{g}_{\mu\nu}(\rho, q) \equiv \sum_{A=1}^{N} m_A \mathbf{g}_A^\mu \cdot \mathbf{g}_A^\nu, \qquad \mu, \nu = 1, 2, \ldots, (3N-3) \tag{A.16}$$

$$\mathbf{g}_A^\mu(\rho, q) \equiv \begin{cases} \mathbf{A}_\mu \mathbf{y}_A, & \mu = 1, 2, 3 \\ \dfrac{\partial \mathbf{y}_A}{\partial Я_{\mu-3}}, & 3 < \mu \le 3N-3; \quad Я \equiv \{\rho, q\} \end{cases} \tag{A.17}$$

$$\lambda_s(\rho, q) \equiv \frac{1}{2} \sum_{\gamma=1}^{\Gamma} (\hat{p}_\gamma \lambda_{s,\gamma}), \qquad s \le p+3 \tag{A.18}$$

These relations involve a slight change of notation and a correction of the misprint in sign of λ_k in Eq. (17) of our work.[15] We also prefer not to employ the Podolsky transformation [Eq. (A.2)]. Applying the algorithm of Gauss[288] to the matrix $\|\tilde{g}_{\mu\nu}\|$, one finds

$$\tilde{g} = \frac{T^{(\mathrm{la})}}{G} \tag{A.19}$$

where $T^{(\mathrm{la})}$ and G are the determinants of the matrices $\mathbf{T}^{(\mathrm{la})}$ and $\mathbf{G}$ with elements

$$T_{st}^{(\mathrm{la})}(\rho, q) \equiv \tilde{g}_{st}(\rho, q), \qquad s, t = 1, 2, \ldots, p+3 \tag{A.20}$$

$$G_{\beta\gamma}(\rho, q) = \sum_A m_A \nabla_A q_\beta \cdot \nabla_A q_A \tag{A.21}$$

We are ready to study the Hamiltonian corresponding to frozen vibrations:

$$\hat{h} = \tilde{g}_0^{-1/2} \sum_{s,t=1}^{p+3} (\hat{N}_s - 2\lambda_s^0) \tilde{g}_0^{1/2} \{\boldsymbol{\mu}_0\}_{st} \hat{N}_t + V(\rho) \tag{A.22}$$

where the index 0 means $q = 0$. Note that $\mathbf{T}_0^{(\mathrm{la})} = \boldsymbol{\mu}_0^{-1}$ and $\mathbf{T}_0^{\mathrm{vib}} = \mathbf{G}_0^{-1}$, that is,

$$\tilde{g}_0 = \frac{T_0^{\mathrm{vib}}}{\mu_0} \tag{A.23}$$

where T_0^{vib} and μ_0 are determinants of the matrices $\mathbf{T}_0^{\mathrm{vib}}$ and $\boldsymbol{\mu}_0$ respectively.

Up to this point we do not specify our variables. Let us now make use of least-squares large-amplitude variables $\theta^{(\mathrm{ls})}$ and $\rho^{(\mathrm{ls})}$, still considering an arbitrary set of vibrational coordinates. Differentiating the global Eckart–Sayvetz conditions [Eqs. (6.2a) and (6.2b)] twice with respect to vibrational coordinates q, one finds

$$\sum_A m_A \mathbf{g}_A^s(\rho, 0) \cdot \frac{\partial^2 \mathbf{y}_A}{\partial q_\beta \, \partial q_\gamma} = 0, \quad s \le p+3 \tag{A.24}$$

Of course, this relation is trivial if q is a set of rectilinear vibrational coordinates. Putting $q = 0$ in Eq. (A.18) and taking into account Eq. (A.24), we have

$$\lambda_s^0(\rho) = \begin{cases} 0, & s = 1, 2, 3 \\ \frac{1}{2} i \displaystyle\sum_{\beta,\gamma=1}^{\Gamma} m_A \mathbf{c}_A^\beta \cdot \frac{\partial \mathbf{c}_A^\gamma}{\partial \rho_\nu} \{\mathbf{G}_0\}_{\beta\gamma}, & s = \nu + 3, \quad \nu > 0 \end{cases} \tag{A.25}$$

where $\mathbf{c}_A^\beta = \partial \mathbf{y}_A/\partial q_\beta|_{q=0}$. To prove that the functions $\lambda_s^0(\rho)$ vanish for $s = 1,2,3$ we also took into account that the matrix with elements $\mathbf{c}_A^\beta \cdot \mathbf{A}_s \mathbf{c}_A^\gamma$ is antisymmetric and hence its convolution with $\mathbf{G}_0$ is equal to zero.

According to (A.13),

$$\{\mathbf{T}_0^{\text{vib}}(\rho)\}_{\beta\gamma} \equiv T_{\beta\gamma}^{\text{vib}}(\rho,0) = \sum_A m_A \mathbf{c}_A^\beta \cdot \mathbf{c}_A^\gamma \tag{A.26}$$

and

$$\mathbf{T}_0^{\text{vib}} \mathbf{G}_0 = \mathbf{1}_\Gamma \tag{A.27}$$

that is, we have

$$\lambda_{\nu+3}^0(\rho) = \tfrac{1}{4}(\hat{N}_{\nu+3} \ln T_0^{\text{vib}}), \qquad \nu = 1,2,\ldots, p \tag{A.28}$$

Substituting (A.23) and (A.28) into (A.22), we are directly led to the Hamiltonian [Eq. (3.10)] discussed in Section III. If, under the action of feasible operations, the vibrational coordinates q undergo linear homogeneous transformations $\mathbf{Q}^g$ independent of ρ, we conclude that the Hamiltonian in Eq. (3.10) must be invariant under elements of the $\mathcal{F}_{\theta,\rho}$ group. Note that in the general case of arbitrary large-amplitude variables the Hamiltonian (A.22) contains an additional term, whose physical nature is unclear to us.

APPENDIX B

For conventional internal variables, in contrast with rectilinear vibrational coordinates, we usually know their derivatives with respect to Cartesian coordinates. Therefore in order to transform the operator of the kinetic energy to external and internal variables one only needs to find derivatives $\nabla_A \theta$.

Let us consider a set of standard configurations $\bar{\mathbf{y}}$ defined by three relations of the form

$$F_k(\bar{\mathbf{y}}) = 0, \qquad k = 1,2,3 \tag{B.1}$$

As stressed by Tennyson and Sutcliffe,[283] the relations (B.1) "can be thought of as a set of constraining equations through which the three Euler angles that parametrize $\mathbf{S}$ are expressed as functions of the $\bar{\mathbf{r}}$" if one substitutes the $\overline{\mathbf{S}^{-1}\mathbf{r}}$ for the $\bar{\mathbf{y}}$ in Eq. (B.1). Representing the operator ∇_A as

$$\nabla_A = -i \sum_{\gamma=1}^{3} \mathbf{f}_{A,j} \hat{K}'_j + \sum_{m=1}^{3N-6} \nabla_A \mathfrak{A}_m \partial_m \tag{B.2}$$

where $\partial_m \equiv \partial/\partial \mathfrak{A}_m$, acting by it on (B.1), and taking into account (A.9), we find the set of linear equations for the vectors $\mathbf{S}^{-1}\mathbf{f}_{A,j}$,

$$\sum_{j=1}^{3} \mathbf{S}^{-1}\mathbf{f}_{A,j}P_{jk} + \frac{\partial}{\partial \mathbf{y}_A}F_k(\bar{\mathbf{y}}) = 0, \qquad k = 1,2,3 \tag{B.3}$$

where the 3×3 matrix $\mathbf{P}$ with elements

$$P_{jk} \equiv F_k\left(\overline{-\mathbf{A}_j\mathbf{y}}\right) \tag{B.4}$$

must be invertible, for otherwise the relations in Eq. (B.1) would be invariant under some infinitesimal rotations, in contradiction with the definition of the standard configurations $\bar{\mathbf{y}}$.

The aforementioned method of calculating the derivatives $\nabla_A\theta_i$ was first suggested by Curtiss, Hirshfelder, and Alder,[289] who constructed a set of standard configurations attaching molecule-fixed axes to three arbitrary atoms of a polyatomic molecule. The special property of this set is that non-vanishing components of the vectors $\mathbf{y}_A$ can be chosen as internal variables. However, in the general case one cannot vary those components in an independent way, in contrast with Eq. (2.13) in Ref. 274 or with the first of the relations of Eq. (A.7) in Ref. 283. If $\Psi(\bar{\mathbf{r}})$ is a function invariant under overall rotations, Eq. (2.13) of Bardo and Wolfsberg[274] should be written in our notation as

$$\nabla_A\Psi = \mathbf{S}\nabla_A\Psi|_{\bar{\mathbf{r}}=\bar{\mathbf{y}}} \tag{B.5}$$

where $\mathbf{y}_A = \mathbf{S}^{-1}\mathbf{r}_A$. Note that Eq. (A.4) of Tennyson and Sutcliffe[283] for $\nabla_A q_k$ has an additional term whose nature is incomprehensible.‡

If a set of standard configurations satisfies linear conditions, we can parametrize it by rectilinear vibrational coordinates

$$q_\gamma = \sum_A \mathbf{s}_A^\gamma \cdot (\mathbf{y}_A - \mathbf{a}_A) \tag{B.6}$$

and find the derivatives $\nabla_A q_\gamma$ substituting $\mathbf{S}^{-1}\mathbf{r}_A$ for the $\mathbf{y}_A$ in (B.6) and differentiating (B.6) with respect to $\mathbf{r}_A$ as suggested by Louck.[272] As an alternative one can express the vectors $\nabla_A q_\gamma$ from general relations

$$\frac{\partial \mathbf{r}_A}{\partial q_\gamma} = \sum_{i=1}^{3} g_{\gamma+3i}\nabla_A\theta_i + \sum_\beta g_{\gamma+3\beta+3}\nabla_A q_\beta \tag{B.7}$$

‡According to comments made by Professor B. T. Sutcliffe, the method discussed in Appendix A of Ref. 283 is restricted to the cases such that there exist analytic expressions of vibrational coordinates in terms of components of the vectors $\mathbf{y}_A$ (using our notation). Those expressions are assumed to be continued analytically beyond the region where these vectors are defined.

We have used this idea in Ref. 15 to represent the molecular Hamiltonian in the form of Eq. (A.12) discussed in Appendix A.

The aforementioned method of calculating derivatives $\nabla_A \theta_i$ was extended to least-squares large-amplitude *internal* variables in Ref. 33. As stressed by Makushkin and Ulenikov,[273] the direct calculation of derivatives of external and internal variables with respect to Cartesian coordinates may be especially useful if an initial Hamiltonian is not represented as a quadratic form over impulses. But in emphasizing the advantage of their method, those authors overlooked that at least for nonlinear molecules the problem had been solved in our work,[15,33] including electronic and moreover *large-amplitude* nuclear internal motions.

It should be mentioned that when we suggested differentiating the global Eckart–Sayvetz conditions with respect to Cartesian coordinates in order to find the derivatives $\nabla_A \theta_i$ and $\nabla_A \rho_j$, we[33] limited ourselves to consideration of rectilinear vibrational coordinates. Neither we nor Louck,[272] who independently suggested the same technique for quasirigid molecules, noticed that the real advantage of those algorithms is connected with the use of conventional internal variables.

Note that to calculate derivatives $\nabla_A \theta_i$ Sorensen[290] formally needs rectilinear vibrational coordinates, namely, he makes use of the fact that derivatives of $\mathbf{r}_A$ with respect to *rectilinear* vibrational coordinates satisfy the relations

$$\sum_A m_A \mathbf{S}\mathbf{g}^{s}_{0,A} \cdot \frac{\partial \mathbf{r}_A}{\partial q_\gamma} = 0 \qquad \text{(B.8)}$$

and hence the vectors $\overline{\nabla \theta_i}$ are linear combinations of the vectors $\overline{m\mathbf{S}\mathbf{g}^{s}_{0}}$. However, after the derivatives $\nabla_A \theta_i$ have been found, they can be used together with any set of internal variables.

Except the calculation of the derivatives $\nabla_A q_\gamma$, our method[15,33] of transforming the molecular Hamiltonian to least-squares variables differs from Louck's[272] by the initial form of the operator of the kinetic energy. Remember that Louck calculates second derivatives with respect to Cartesian coordinates, whereas we[15] make use of the Hermiticity of the operator ∇_A, transcribing Eq. (B.2) into the form

$$\nabla_A = \tilde{g}^{-1/2}\left[-i\sum_{\gamma=1}^{3}\hat{K}'_j \mathbf{f}_{A,j} + \sum_{m=1}^{3N-6}\partial_m \nabla_A \mathfrak{A}_m\right]\tilde{g}^{1/2} \qquad \text{(B.9)}$$

Equations (B.2) and (B.9) immediately lead to the covariant Laplacian of Eq. (A.1). Therefore the idea of Islampour and Kasha[275] to substitute the derivatives $\nabla_A \theta_i$ and $\nabla_A q_\gamma$ directly into the matrix $\|g^{\mu\nu}\|$ has been implicitly realized by us. The only difference between our work[15,33] and Ref. 275 is in the way in which we calculate the derivatives $\nabla_A q$ (in Ref. 275 both $\nabla_A \theta_i$ and $\nabla_A q$ are calculated according to Louck's recipe[272]).

A similar technique can be applied to linear molecules. We can easily calculate derivatives of azimuthal and polar angles φ and θ representing the Eckart conditions in the form

$$\mathbf{n}(\varphi, \theta) = \frac{\sum_A m_A \mathbf{r}_A}{\left|\sum_A m_A \mathbf{r}_A\right|} \tag{B.10}$$

and then differentiating them with respect to Cartesian coordinates $\mathbf{r}_A$[291] (cf. Ref. 292). The extension of this technique to nonrigid linear molecules is discussed in Appendix A of Ref. 17.

Note added in proof. I would like to thank Professor R. G. Woolley for his severe critical remarks on Section II.B although I do not think that I can agree with some of them. Woolley's main contention that molecular structure cannot be derived from quantum mechanics is completely unacceptable to me. I sincerely tried to understand arguments in support of this contention but apparently failed in my efforts. I am also indebted to Professor B. T. Sutcliffe for very helpful comments clarifying his method to express the molecular Hamiltonian in terms of rotational and vibrational variables.

REFERENCES

1. F. Hund, *Z. Phys.* **43**, 805 (1927).
2. C. J. Brester, *Kristallsymmetrie and Reststrahlen*, dissertation, Utrecht, 1923.
3. E. P. Wigner, *Göttinger Nachrichten*, p. 133 (1930).
4. C. Eckart, *Phys. Rev.* **47**, 552 (1935).
5. J. T. Hougen, *J. Chem. Phys.* **37**, 1433 (1962).
6. J. T. Hougen, *J. Chem. Phys.* **39**, 358 (1963).
7. A. A. Kiselev, *J. Phys. B* **3**, 904 (1970).
8. M. N. Adamov and G. A. Natanson, *Vestn. Leningr. Univ.*, No. 4, p. 28 (1973) (in Russian).
9. T. Pedersen, *J. Mol. Spectrosc.* **80**, 229 (1980).
10. L. Lathouwers and P. van Leuven, *Adv. Chem. Phys.* **49**, 116 (1982).
11. M. Born and K. Huang, *Dynamical Theory of Crystal Lattices*, Clarendon Press, Oxford, 1956.
12. M. Born and R. Oppenheimer, *Ann. Phys.* **84**, 457 (1927).
13. B. J. E. Mayer, S. Brunauer, and M. G. Mayer, *J. Amer. Chem. Soc.* **55**, 37 (1933).

14. E. B. Wilson, Jr., *J. Chem. Phys.* **6**, 740 (1938).
15. G. A. Natanson and M. N. Adamov, *Vestn. Leningr. Univ.*, No. 4, p. 22 (1974) (in Russian).
16. M. N. Adamov and G. A. Natanson, *Vestn. Leningr. Univ.*, No. 16, p. 25 (1978) (in Russian).
17. G. A. Natanson, *Molec. Phys.* **46**, 481 (1982).
18. E. B. Wilson, Jr., *J. Chem. Phys.* **3**, 276 (1935).
19. E. B. Wilson, Jr., *J. Chem. Phys.* **3**, 818 (1935).
20. E. B. Wilson, Jr., C. C. Lin, and D. R. Lide, Jr., *J. Chem. Phys.* **23**, 136 (1955).
21. R. J. Myers and E. B. Wilson, *J. Chem. Phys.* **33**, 186 (1960).
22. T. Kasuya, *Sci. Papers Inst. Phys. Chem. Res.* **56**, 1 (1962).
23. T. Kasuya and T. Kojima, *J. Phys. Soc. Japan* **18**, 364 (1963).
24. H. Dreizler, *Z. Naturforsch. A* **16**, 477 (1961).
25. H. Dreizler, *Z. Naturforsch. A* **16**, 1354 (1961).
26. K. D. Moller and H. G. Andresen, *J. Chem. Phys.* **39**, 17 (1963).
27. R. S. Berry, *Rev. Mod. Phys.* **32**, 447 (1960).
28. M. N. Adamov and G. A. Natanson, *Fiz. Molek.* **2**, 3 (1976) (in Russian; see Los Alamos Translation LA-TR-77-32).
29. H. C. Longuet-Higgins, *Molec. Phys.* **6**, 445 (1963).
30. J. K. G. Watson, *Can. J. Phys.* **43**, 1996 (1965).
31. J. K. G. Watson, *Molec. Phys.* **21**, 753 (1971).
32. P. R. Bunker, *Molecular Symmetry and Spectroscopy*, Academic Press, 1979.
33. G. A. Natanson and M. N. Adamov, *Vestn. Leningr. Univ.* No. 10, p. 24 (1974) (in Russian).
34. M. N. Adamov and G. A. Natanson, Reprint of Inst. Theor. Phys., Acad. Sci. Ukr. SSR, ITF-76-82R, Kiev, 1976 (in Russian).
35. G. A. Natanson, *Fiz. Molek.* **6**, 3 (1978) (in Russian).
36. T. Pedersen, *J. Chem. Phys.* **54**, 4028 (1971).
37. G. A. Natanson, Abstracts of Fourteenth Annual Midwest Theoretical Chemistry Conference, Chicago, 1981, A6 (unpublished).
38. P. R. Bunker, *Vibrational Spectroscopy and Structure*, Vol. 3, J. R. Durig (ed.), Marcel Dekker, 1975.
39. R. S. Berry, in *Quantum Dynamics of Molecules*, R. G. Woolley (ed.), Plenum, New York, 1980, p. 143.
40. J. T. Hougen, *Can. J. Phys.* **42**, 1920 (1964).
41. J. T. Hougen, *Can. J. Phys.* **43**, 935 (1965).
42. J. T. Hougen, *Pure Appl. Chem.* **11**, 481 (1965).
43. J. T. Hougen, *Can. J. Phys.* **44**, 1169 (1966).
44. B. J. Dalton, *Molec. Phys.* **11**, 265 (1966).
45. R. S. Berry, *Lect. Notes in Chem.* **12**, 753 (1979).
46. A. Bauder, R. Meyer, and Hs. H. Günthard, *Molec. Phys.* **28**, 1305 (1974).
47. H. Frei, P. Groner, A. Bauder, and Hs. H. Günthard, *Molec. Phys.* **3**, 443 (1976).
48. H. Frei and Hs. H. Günthard, *Lect. Notes in Phys.* **79**, 92 (1978).

49. H. Frei and Hs. H. Günthard, *Topics Curr. Chem.* **81**, 7 (1979).
50. H. Frei, P. Groner, A. Bauder, and Hs. H. Günthard, *Molec. Phys.* **36**, 1469 (1978).
51. H. Frei, A. Bauder, and Hs. H. Günthard, *Molec. Phys.* **43**, 785 (1981).
52. Hs. H. Günthard, *Studies in Phys. & Theor. Chem.* (Elsevier, Amsterdam) **23**, 133 (1983).
53. J. D. Louck and H. W. Galbraith, *Rev. Mod. Phys.* **48**, 69 (1976).
54. K. Pedersen, Unpublished report, The H. C. Orsted Institute, Copenhagen, 1977.
55. H. Frei and Hs. H. Günthard, *Chem. Phys.* **15**, 155 (1976).
56. M. Hollenstein, A. Bauder, and Hs. H. Günthard, *Chem. Phys.* **47**, 269 (1980).
57. Z. Slanina, *Adv. Quant. Chem.* **13**, 89 (1981).
58. S. L. Altmann, *Proc. Roy. Soc. A* **298**, 1841 (1967).
59. S. L. Altmann, *Mol. Phys.* **21**, 587 (1971).
60. S. L. Altmann, *Induced Representations in Crystals and Molecules. Point Space and Non-rigid Molecule Groups*, Academic, London, 1977.
61. S. L. Altmann, *Studies in Phys. & Theor. Chem.* (Elsevier, Amsterdam) **23**, 95 (1983).
62. G. S. Ezra, *Molec. Phys.* **38**, 863 (1979).
63. G. S. Ezra, in *Energy Storage and Redistribution in Molecules*, J. Hinze (ed.), Plenum, New York & London, 1983, p. 293.
64. G. S. Ezra, *Lect. Notes in Chem.* (Springer, Berlin) **28** (1982).
65. R. L. Flurry, Jr., *J. Molec. Spectrosc.* **56**, 88 (1975).
66. R. L. Flurry, Jr., *Symmetry Groups: Theory and Chemical Applications*, Prentice-Hall, Englewood Cliffs, N.J., 1980.
67. R. L. Flurry, Jr., *J. Phys. Chem.* **80**, 778 (1976).
68. R. L. Flurry, Jr. and S. F. Abdulnur, *J. Molec. Spectrosc.* **63**, 33 (1976).
69. R. L. Flurry, Jr., *Studies in Phys. & Theor. Chem.* (Elsevier, Amsterdam) **23**, 113 (1983).
70. Y. G. Smeyers, *Fol. Chim. Theoret. Lat.* **6**, 139 (1978).
71. C. M. Woodman, *Molec. Phys.* **19**, 753 (1970).
72. G. S. Ezra, *Molec. Phys.* **43**, 773 (1981).
73. G. S. Ezra, *Studies in Phys. & Theor. Chem.* (Elsevier, Amsterdam) **23**, 81 (1983).
74. J. Maruani, Y. G. Smeyers, and A. Hernandez-Laguna, *J. Chem. Phys.* **76**, 3123 (1982).
75. E. B. Wilson and J. B. Howard, *J. Chem. Phys.* **4**, 260 (1936).
76. P. R. Bunker, *Bull. Soc. Chem. Belg.* **89**, 565 (1980).
77. G. Dellepiane, *Z. Naturforsch.* **34a**, 1230 (1979).
78. P. Groner, *Molec. Phys.* **43**, 415 (1981).
79. G. A. Natanson, *Studies in Phys. & Theor. Chem.* (Elsevier, Amsterdam) **23**, 201 (1983).
80. B. L. Crawford and E. B. Wilson, Jr., *J. Chem. Phys.* **9**, 323 (1941).
81. P. R. Bunker, *J. Chem. Phys.* **47**, 718 (1967).
82. J. W. Fleming and C. N. Banwell, *J. Molec. Spectrosc.* **31**, 378 (1969).
83. G. S. Ezra, private communication.
84. Y. G. Smeyers and M. N. Bellido, *Int. J. Quant. Chem.* **19**, 553 (1981).
85. G. A. Natanson, *Optics Spectrosc. Wash.* **41**, 18 (1976).
86. G. A. Natanson, *Optics Spectrosc. Wash.* **47**, 139 (1979).
87. C. Trindle, *Israel J. Chem.* **19**, 47 (1980).

88. J. M. F. Gilles and J. Philippot, *Int. J. Quant. Chem.* **6**, 225 (1972).
89. M. I. Petrashen and E. D. Trifonov, *Applications of Group Theory in Quantum Mechanics*, M.I.T. Press, Cambridge, Mass., 1969.
90. B. J. Dalton, *Studies in Phys. & Theor. Chem.* (Elsevier, Amsterdam) **23**, 169 (1983).
91. J. Brocas and D. Fastenakel, *Molec. Phys.* **30**, 193 (1975).
92. B. J. Dalton, *J. Chem. Phys.* **54**, 4745 (1971).
93. M. M. Cheron and J. Borde, *J. de Phys.* **35**, 641 (1974).
94. B. J. Dalton and P. D. Nicholson, *Int. J. Quant. Chem.* **9**, 325 (1975).
95. D. Fastenakel and J. Brocas, *Bull. Soc. Chim. Belg.* **84**, 1093 (1975).
96. B. J. Dalton, J. Brocas, and D. Fastenakel, *Molec. Phys.* **31**, 1887 (1976).
97. C. Trindle, S. N. Datta, and T. D. Bouman, *Int. J. Quant. Chem.* **11**, 627 (1977).
98. J. Brocas and C. Rusu, *Int. J. Quant. Chem.* **22**, 331 (1982).
99. J. Brocas and C. Rusu, *Studies in Phys. & Theor. Chem.* (Elsevier, Amsterdam) **23**, 257 (1983).
100. D. Fastenakel and J. Brocas, *Mol. Phys.* **40**, 361 (1980).
101. J. Brocas, D. Fastenakel, and J. Buschen, *Mol. Phys.* **41**, 1163 (1980).
102. J. Brocas, *J. Stat. Phys.* **24**, 199 (1981).
103. A. A. Kiselev and K. Luders, *Vopr. Kvant. Teor. Atom. Molek.* **2**, 56 (1981) (in Russian).
104. A. G. Zhilich, A. A. Kiselev, and V. P. Smirnov, *Soviet Phys. Solid State* **23**, 459 (1981).
105. W. G. Harter, C. W. Patterson, and F. J. Paixao, *Rev. Mod. Phys.* **50**, 37 (1978).
106. J. Borde, C. J. Borde, C. Salamon, A. Van Lerberghe, M. Ouhayoun, and C. D. Cantrell, *Phys. Rev. Lett.* **45**, 14 (1980).
107. J. Borde and C. J. Borde, *Chem. Phys.* **71**, 417 (1982).
108. L. C. Biedenharn and J. D. Louck, *Angular Momentum in Quantum Physics. Theory and Application*, Addison-Wesley, (1981).
109. G. A. Natanson and R. S. Berry, *Ann. Phys. (N.Y.)*, **155**, 178 (1984).
110. G. A. Natanson, *Rev. Mod. Phys.* to be submitted.
111. L. D. Landau and E. M. Lifshitz, *Quantum Mechanics (Non-Relativistic Theory)*, Pergamon, 1976.
112. T. Oka, *J. Mol. Spectrosc.* **48**, 503 (1973).
113. H. Berger, *J. de Phys.* **38**, 1371 (1977).
114. F. Michelot, B. Bobin, and J. Moret-Bailly, *J. Mol. Spectrosc.* **76**, 314 (1979).
115. W. G. Harter and C. W. Patterson, *Lecture Notes in Chem.* (Springer, Berlin) **22**, 306 (1979).
116. R. G. Woolley, *Adv. Phys.* **25**, 27 (1976).
117. R. G. Woolley, *J. Amer. Chem. Soc.* **100**, 1073 (1978).
118. R. G. Woolley, *Israel J. Chem.* **19**, 30 (1980).
119. M. N. Adamov and G. A. Natanson, *Vestn. Leningr. Univ.*, No. 4, p. 50 (1970) (in Russian).
120. F. Jørgensen, *Int. J. Quant. Chem.* **14**, 55 (1978).
121. M. N. Adamov and G. A. Natanson, *Vestn. Leningr. Univ.*, No. 22, p. 28 (1978) (in Russian).
122. A. A. Kiselev and A. N. Petelin, *Vestn. Leningr. Univ.*, No. 22, p. 16 (1970).

123. M. Bixon, *Chem. Phys.* **70**, 199 (1982).
124. E. B. Wilson, *Int. J. Quant. Chem. Symp.* **13**, 5 (1979).
125. P. Claverie and S. Diner, *Israel J. Chem.* **19**, 54 (1980).
126. P. Claverie, *Studies in Phys. & Theor. Chem.* (Elsevier, Amsterdam) **23**, 13 (1983).
127. R. G. Woolley, *Int. J. Quant. Chem.* **12**, Suppl. 1, 301 (1977).
128. R. G. Woolley, *Chem. Phys. Lett.* **55**, 443 (1978).
129. R. G. Wooley, in *Structure and Bonding*, vol. 52.1, Springer, Berlin & Heidelberg, (1982).
130. J. M. F. Gilles and J. Philippot, *Int. J. Quant. Chem.* **14**, 299 (1978).
131. M. Simonius, *Phys. Rev. Lett.* **40**, 980 (1978).
132. R. A. Harris and L. Stodolsky, *Phys. Lett.* **78B**, 313 (1978).
133. R. A. Harris, in *Quantum Dynamics of Molecules*, R. G. Wooley (ed.), Plenum, New York, 1980, p. 357.
134. R. A. Harris and L. Stodolsky, *Phys. Lett.* **116B**, 464 (1982).
135. R. A. Harris and R. Silbey, *J. Chem. Phys.* **78**, 7330 (1983).
136. R. A. Harris and L. Stodolsky, *J. Chem. Phys.* **74**, 2145 (1981).
137. B. Fain, *Phys. Lett.* **89A**, 455 (1982).
138. P. Pfeifer, *Studies in Phys. & Theor. Chem.* (Elsevier, Amsterdam) **23**, 379 (1983).
139. D. J. R. Lloyd-Evans, *Mol. Phys.* **10**, 377 (1966).
140. G. A. Natanson, F. Amar, and R. S. Berry, *J. Chem. Phys.* **78**, 399 (1983).
141. I. G. Kaplan, *Symmetry of Many-Electron Systems*, Academic, New York, (1975).
142. E. Nathier, D. Welti, A. Bauder, and Hs. H. Günthard, *J. Mol. Spectrosc.* **37**, 63 (1971).
143. A. Attanasio, A. Bauder, and Hs. H. Günthard, *Mol. Phys.* **23**, 827 (1972).
144. J. Philippot and I. Senders, *Int. J. Quant. Chem.* **15**, 713 (1979).
145. K. Mislow, *Introduction to Stereochemistry*, Benjamin, New York, 1966.
146. D. Papousek, J. M. R. Stone, and V. Spirko, *J. Mol. Spectrosc.* **48**, 17 (1973).
147. D. Papousek and V. Spirko, *Top. Curr. Chem.* **68**, 59 (1976).
148. Lord Kelvin, Baltimore Lectures, 1884, in *Baltimore Lectures*, C. J. Clay, London, 1904, pp. 436, 619.
149. H. Eyring, J. Walter, and G. E. Kimball, *Quantum Chemistry*, Wiley, New York, 1949.
150. R. S. Berry, *J. Chem. Phys.* **32**, 933 (1960).
151. K. C. Kulander and C. Bottcher, *Chem. Phys.* **29**, 141 (1978).
152. K. C. Kulander and E. J. Heller, *J. Chem. Phys.* **69**, 2439 (1978).
153. P. J. Hay, R. T. Pack, R. B. Walker, and E. J. Heller, *J. Phys. Chem.* **86**, 862. (1982).
154. L. S. Bernstein, J. J. Kim, and K. S. Pitzer, *J. Chem. Phys.* **62**, 3671 (1975).
155. L. S. Bernstein, S. Abramowitz, and I. W. Levin, *J. Chem. Phys.* **64**, 3228 (1976).
156. A. V. Demidov, A. A. Ivanov, L. S. Ivashkevich, A. A. Ischenko, V. P. Spiridonov, J. Almlöf, and T. G. Strand, *Chem. Phys. Lett.* **64**, 528 (1979).
157. V. P. Spiridonov, A. A. Ischenko, and L. S. Ivashkevich, *J. Mol. Struct.* **72**, 165 (1981).
158. M. E. Kellman, *J. Phys. Chem.* **87**, 2161 (1983).
159. M. E. Kellman, *Chem. Phys. Lett.* **94**, 331 (1983).
160. F. Baltagi, A. Bauder, T. Ueda, and Hs. H. Günthard, *J. Molec. Spectrosc.* **42**, 112 (1972).
161. F. Baltagi, A. Bauder, P. Henrici, T. Ueda, and Hs. H. Günthard, *Mol. Phys.* **24**, 945 (1972).

162. J. Maruani, A. Hernandez Laguna, and V. G. Smeyers, *J. Chem. Phys.* **63**, 4515 (1975).
163. J. Maruani, V. G. Smeyers, and A. Hernandez Laguan, *J. Chem. Phys.* **76**, 3123 (1982).
164. A. T. Labbe and J. Maruani, *Int. J. Quant. Chem.* **22**, 115 (1982).
165. G. Herzberg and H. C. Longuet-Higgins, *Discuss. Faraday Soc.* **35**, 77 (1963).
166. H. C. Longuet-Higgins, *Proc. Roy. Soc. London* **A344**, 147 (1975).
167. C. A. Mead and D. G. Truhlar, *J. Chem. Phys.* **70**, 2284 (1979).
168. C. A. Mead, *J. Chem. Phys.* **78**, 807 (1983).
169. J. C. D. Brand, D. R. Williams, and T. J. Cook, *J. Mol. Spectrosc.* **20**, 359 (1966).
170. T. Itoh, *J. Phys. Soc. Japan* **11**, 264 (1956).
171. M. Kreglewski, *J. Mol. Spectrosc.* **72**, 1 (1978).
172. E. S. Kryachko and Yu. A. Kruglyak, *J. Struct. Chem. Wash.* **21**, 299 (1980).
173. E. S. Kryachko, Yu. A. Kruglyak, and B. J. Dalton, *Int. J. Quant. Chem.* **17**, 1229 (1980).
174. M. Tsuboi, A. Y. Hirakawa, T. Ino, T. Sasaki, and K. Tamagake, *J. Chem. Phys.* **41**, 2721 (1964).
175. R. Cervellati, G. Corbelli, A. D. Borgo, and D. G. Lister, *J. Mol. Struct.* **73**, 31 (1981).
176. R. A. Kydd and A. R. C. Dunham, *J. Mol. Struct.* **98**, 39 (1983).
177. P. Russegger and J. Brickmann, *J. Chem. Phys.* **62**, 1086 (1975).
178. J. T. Hougen, *J. Mol. Spectrosc.* **82**, 92 (1980).
179. J. T. Hougen, *J. Mol. Spectrosc.* **89**, 296 (1981).
180. S. Tsunekawa, T. Kojima, and J. T. Hougen, *J. Mol. Spectrosc.* **95**, 133 (1982).
181. M. R. Aliev, *Sov. Phys. Usp.* **19**, 627 (1976).
182. I. B. Bersuker and V. Z. Polinger, *Adv. Quant. Chem.* **15**, 85 (1982).
183. J. Moret-Bailly, *Cahier Phys.* **178**, 253 (1965).
184. R. E. Miller and J. C. Decius, *J. Chem. Phys.* **59**, 4871 (1973).
185. A. A. Kiselev and K. Luders, *Phys. Stat. Sol. (b)* **93**, 285 (1979).
186. A. A. Kiselev and K. Luders, *Vestn. Leningr. Univ.*, No. 16, p. 31 (1979) (in Russian).
187. H. Nichols and R. M. Hexter, *J. Chem. Phys.* **75**, 3126 (1981).
188. H. Nichols and R. M. Hexter, *Surf. Sci.* **118**, 597 (1982).
189. G. A. Natanson, *Phys. Rev. Let.*, to be submitted.
190. M. I. Petrashen and E. M. Ledovskaja, *Vestn. Leningr. Univ.*, No. 4, p. 34 (1980) (in Russian).
191. H. D. Bist, J. C. D. Brand, A. R. Moy, V. T. Jones, and R. J. Pirkle, *J. Mol. Spectrosc.* **66**, 411 (1977).
192. L. Fredin and B. Nelander, *Chem. Phys.* **60**, 181 (1981).
193. K. D. Moller, *J. Mol. Struct.* **22**, 183 (1974).
194. M. Gut, A. Bauder, and Hs. H. Günthard, *Chem. Phys.* **8**, 252 (1975).
195. A. Bauder, E. Mathier, R. Meyer, M. Ribeaud, and Hs. H. Günthard, *Mol. Phys.* **15**, 597 (1968).
196. P. Nösberger, A. Bauder, and Hs. H. Günthard, *Chem. Phys.* **4**, 196 (1974).
197. T. S. G. K. Murty, L. S. R. K. Prasad, and M. K. Rao, *Int. J. Pure Appl. Phys.* **18**, 713 (1980).
198. G. D. Renkes, *Chem. Phys.* **57**, 261 (1981).
199. J. Suskind, *J. Chem. Phys.* **53**, 2492 (1970).

200. D. C. McKean and R. A. Watt, *J. Mol. Spectrosc.* **61**, 184 (1976).
201. G. O. Sørensen, T. Pedersen, M. Dreizler, A. Guarnieri, and A. P. Cox, *J. Mol. Struct.* **97**, 77 (1983).
202. G. O. Sorensen and T. Pedersen, *Studies in Phys. & Theor. Chem.* (Elsevier, Amsterdam) **23**, 219 (1983).
203. G. A. Natanson, *Chem. Phys. Lett.*, to be submitted.
204. J. M. R. Stone and I. M. Mills, *Mol. Phys.* **18**, 631. (1970).
205. I. Mills, *Mol. Phys.* **20**, 127 (1971).
206. F. A. Miller, R. J. Capwell, R. C. Lord, and D. G. Rea, *Spectrochim. Acta* **28A**, 603 (1972).
207. J. E. Leonard, G. S. Hammond, and H. E. Simmons, *J Am. Chem. Soc.* **97**, 5052 (1975).
208. H. M. Pickett and H. L. Strauss, *J. Chem. Phys.* **55**, 324 (1971).
209. H. M. Pickett and H. L. Strauss, *J. Amer. Chem. Soc.* **92**, 7281 (1970).
210. D. F. Bocian, H. M. Pickett, T. C. Rounds, and H. L. Strauss, *J. Am. Chem. Soc.* **97**, 687 (1975).
211. T. C. Rounds and H. L. Strauss, *J. Chem. Phys.* **69**, 268 (1978).
212. H. L. Strauss, *J. Chem. Educ.* **48**, 221 (1971).
213. J. D. Louck, *Lect. Notes in Chem.* **12**, 57 (1979).
214. A. Metropoulos and Y. N. Chiu, *J. Chem. Phys.* **68**, 5607 (1978).
215. P. R. Bunker, *Can. J. Phys.* **57**, 2099 (1979).
216. A. Metropoulos, *Chem. Phys. Lett.* **83**, 357 (1981).
217. J. Tennyson and A. van der Avoird, *J. Chem. Phys.* **77**, 5664 (1982).
218. T. R. Dyke, B. J. Howard, and W. Klemperer, *J. Chem. Phys.* **56**, 2442 (1972).
219. A. E. Barton and B. J. Howard, *Faraday Discuss. Chem. Soc.* **73**, 45 (1982).
220. A. Metropoulos, *Studies in Phys. & Theor. Chem.* (Elsevier, Amsterdam) **23**, 419 (1983).
221. D. Papousek, K. Sarka, V. Spirko, and B. Jordanov, *Collect. Czech. Chem. Commun.* **36**, 890 (1971).
222. A. J. Merer and J. K. G. Watson, *J. Mol. Spectrosc.* **47**, 499 (1973).
223. G. Dellepiane, M. Gussoni, and J. T. Hougen, *J. Mol. Spectrosc.* **47**, 515 (1973).
224. J. T. Hougen, *J. Chem. Phys.* **56**, 6245 (1972).
225. A. A. Kiselev and A. V. Lyaptsev, *Opt. Spectrosc. Wash.* **35**, 251 (1973).
226. A. A. Kiselev and A. V. Lyaptsev, *Opt. Spectrosc. Wash.* **38**, 159 (1975).
227. T. R. Dyke, *J. Chem. Phys.* **66**, 492 (1977).
228. J. A. Odutola, D. L. Alvis, C. W. Curtis, and T. R. Dyke, *Mol. Phys.* **42**, 267 (1981).
229. J. A. Odutola, T. R. Dyke, B. J. Howard and J. S. Muenter, *J. Chem. Phys.* **70**, 4884 (1979).
230. E. B. Wilson, J. C. Decius, and P. C. Cross, *Molecular Vibrations*, McGraw-Hill, New York, (1955).
231. A. V. Lyaptsev and A. A. Kiselev, *Opt. Spectrosc. Wash.* **39**, 262 (1975).
232. Yu. S. Makushkin, O. N. Ulenikov, and A. E. Cheglokov, *Opt. Spectrosc. Wash.* **45**, 266 (1978).
233. O. N. Ulenikov and A. E. Cheglokov, *Opt. Spectrosc. Wash.* **46**, 483 (1979).
234. G. L. Hofacker, *Z. Naturforsch. (a)* **18**, 607 (1963).

235. R. A. Marcus, *J. Chem. Phys.* **41**, 610 (1964).
236. C. R. Quade, *J. Chem. Phys.* **65**, 700 (1976).
237. D. H. Cress and C. R. Quade, *J. Chem. Phys.* **67**, 5695 (1977).
238. C. R. Quade, *J. Chem. Phys.* **73**, 2107 (1980).
239. H.-Y. Sheng, E. F. Barker, and D. M. Dennison, *Phys. Rev.* **60**, 786 (1941).
240. R. R. Newton and L. H. Thomas, *J. Chem. Phys.* **16**, 310 (1948).
241. D. R. Lide, Jr. and D. Christensen, *J. Chem. Phys.* **35**, 1374 (1961).
242. M. K. Kemp and W. H. Flygare, *J. Am. Chem. Soc.* **91**, 3163 (1969).
243. J. E. Wollrab, *J. Chem. Phys.* **53**, 1543 (1970).
244. A. P. Cox and S. Waring, *J. Chem. Soc. Faraday Trans. II* **68**, 1060 (1972).
245. R. Meyer and E. B. Wilson, Jr., *J. Chem. Phys.* **53**, 3969 (1970).
246. A. Bauder and Hs. H. Günthard, *J. Mol. Spectrosc.* **60**, 290 (1976).
247. R. Meyer, *J. Mol. Spectrosc.* **76**, 266 (1979).
248. A. Veillard, *Chem. Phys. Lett.* **4**, 51 (1969).
249. A. Veillard, *Theor. Chim. Acta* **18**, 21 (1970).
250. T.-K. Ha, R. Meyer, P. N. Ghosh, A. Bauder, and Hs. H. Günthard, *Chem. Phys. Lett.* **81**, 610 (1981).
251. R. A. Hegstrom, D. W. Rein, and P. G. H. Sandars, *J. Chem. Phys.* **73**, 2329 (1980).
252. N. J. D. Lucas, *J. Phys. B. Atom. Molec. Phys.* **6**, 155 (1973).
253. L. A. Gribov, *Opt. Spectrosc. Wash.* **31**, 842 (1971).
254. L. A. Gribov and G. V. Khovrin, *Opt. Spectrosc. Wash.* **36**, 274 (1974).
255. C. R. Quade, *J. Chem. Phys.* **64**, 2783 (1976).
256. A. R. Hoy, I. M. Mills, and G. Srey, *Molec. Phys.* **24**, 1265 (1972).
257. A. I. Pavlyuchko, G. F. Lozenko, G. V. Khovrin, and L. A. Gribov, *Opt. Spectrosc. Wash.* **52**, 37 (1982).
258. R. Meyer and Hs. H. Günthard, *J. Chem. Phys.* **49**, 1510 (1968).
259. P. Groner, Thesis, Swiss Fed. Inst. Technology, Zurich, Diss. No. 5394, (1974).
260. F. Lichene, G. Dellepiane, and M. Gussoni, *Chem. Phys.* **40**, 163 (1979).
261. J. Mink and B. Gellai, *J. Organomet. Chem.* **66**, 1 (1974).
262. S. Kato, H. Kato, and K. Fukui, *J. Amer. Chem. Soc.* **99**, 684 (1977).
263. K. Yamashita, T. Yamabe, and K. Fukui, *Chem. Phys. Lett.* **84**, 123 (1981).
264. K. Yamashita, T. Yamabe, and K. Fukui, *Theor. Chim. Acta* **60**, 523 (1982).
265. S. K. Ganguly and P. N. Ghosh, *Chem. Phys. Lett.* **90**, 140 (1982).
266. H. D. Bist, J. C. D. Brand, and D. R. Williams, *J. Mol. Spectrosc.* **21**, 76 (1966).
267. J. B. Howard, *J. Chem. Phys.* **5**, 442 (1937).
268. R. K. Harris and M. Thorley, *J. Mol. Spectrosc.* **42**, 407 (1972).
269. G. A. Natanson, *Opt. Spectrosc. Wash.* **38**, 375 (1975).
270. J. T. Hougen, private communication.
271. B. T. Darling, *J. Mol. Spectrosc.* **11**, 67 (1963).
272. J. D. Louck, *J. Mol. Spectrosc.* **61**, 107 (1976).
273. Yu. S. Makushkin and O. N. Ulenikov, *J. Mol. Spectrosc.* **68**, 1 (1977).
274. R. D. Bardo and R. D. Wolfsberg, *J. Chem. Phys.* **67**, 593 (1977).

275. R. Islampour and M. Kasha, *Chem. Phys.* **74**, 67 (1983).
276. B. T. Sutcliffe, in *Quantum Dynamics of Molecules*, R. G. Woolley (ed.), Plenum, New York, (1980), p. 1.
277. J. K. G. Watson, *Mol. Phys.* **19**, 465 (1970).
278. B. T. Darling and D. M. Dennison, *Phys. Rev.* **57**, 128 (1940).
279. B. Podolsky, *Phys. Rev.* **32**, 812 (1928).
280. E. Bodewig, *Matrix Calculus*, North-Holland, Amsterdam, 1956.
281. F. Webster, M.-J. Huang, and M. Wolfsberg, *J. Chem. Phys.* **75**, 2306 (1981).
282. N. M. Marques and J. Borde, *Mol. Phys.* **22**, 809 (1971).
283. J. Tennyson and B. T. Sutcliffe, *J. Chem. Phys.* **77**, 4061 (1982).
284. R. Radau, *Ann. Sci. Ecole Normale Superior* **5**, 311 (1868).
285. F. T. Smith, *Phys. Rev. Lett.* **45**, 1157 (1980).
286. H. H. Nielsen, *Rev. Mod. Phys.* **23**, 90 (1951).
287. J. T. Hougen, P. R. Bunker, and J. W. C. Johns, *J. Mol. Spectrosc.* **34**, 136 (1970).
288. F. R. Gantmacher, *The Theory of Matrices*, Chelsea, New York, (1977).
289. C. F. Curtiss, J. O. Hirschfelder, and F. T. Adler, *J. Chem. Phys.* **18**, 1638 (1950).
290. G. O. Sorensen, *Top. Current Chem.* **82**, 99 (1979).
291. M. N. Adamov and G. A. Natanson, *Vestn. Leningr. Univ.*, No. 22, p. 30 (1970) (in Russian).
292. R. Islampour and M. Kasha, *Chem. Phys.* **75**, 157 (1983).

COLLISIONAL DISSOCIATION AND CHEMICAL RELAXATION OF RUBIDIUM AND CESIUM HALIDE MOLECULES

JOSEPH N. WEBER* AND R. STEPHEN BERRY

Department of Chemistry and the James Franck Institute
The University of Chicago
Chicago, Illinois 60637

CONTENTS

*Present address: Christina Laboratory, E. I. duPont de Nemours and Co., Inc., Wilmington, Delaware 19898.

I. INTRODUCTION

The kinetics of chemical relaxation of alkali halide systems has been an object of study by several groups, by shock-wave studies[1-11] and in flames.[12-15] These systems are attractive for several reasons. When a shock wave is used to produce dissociation, it generates a fast temperature, pressure, and density jump, after which concentrations can be monitored for individual species. The alkali halide systems are simple enough that all reasonable binary and ternary processes can be included explicitly in a kinetic model. Moreover the number of independent species is small enough, enough is known about the properties of the systems at equilibrium, enough species may be monitored, and enough internal consistency is demanded among any set of such systems that there is little room for ambiguity in the interpretation of the results. The chemical relaxation of dissociated alkali halide vapors is probably as close as any kinetic process may be to a problem in which the mechanism can be inferred from the kinetics.

The shock-tube studies were reviewed by Mandl[7] through 1978; his discussion included some of the data from the study presented here and in the preceding paper.[11] Kinetic studies of alkali halides in flames[12-15] have contributed a number of rate coefficients that are also obtained from the shock-tube experiments.

In addition to the review and integration of previous experiments of different sorts, the principal new contributions of this work are: (1) the extension of the kinetic-measurements to temperatures well above any reached in previous kinetic studies, so that the temperature range in which these processes have been studied now covers about a factor of 4, up to the temperature region in which oscillatory emission and light amplification have been reported for NaI;[16,17] and (2) the critical comparison of the rate coefficients obtained by the various groups and methods for *all* the reactions that play a significant role in the chemical relaxation of the rubidium and cesium halides.

In order to combine previous work with our own, it is most convenient to go first into the new studies and their results, and then examine all the data together. From this sum we can make our best inferences about the microscopic processes and phenomenological rates.

II. EXPERIMENTAL METHOD

The shock tube used in this work has been described in previous publications and in detail in the doctoral dissertation from which this presentation is drawn.[18] The experimental procedure differs in some ways from that of the preceding paper, in large part because of the higher temperatures encountered in this work.

The gas being studied consisted of argon with a mole fraction of 2.5×10^{-4} to 2.0×10^{-3} of salt behind the incident shock and 3.5×10^{-4} to 3.0×10^{-3} behind the reflected shock, where the relaxation process was studied. The driver gas was hydrogen or helium.

A. Velocity Measurement

Most of the measurements of the shock-wave velocity were made by one of two methods. One was the detection of emitted light from the shock front passing three successive windows with very narrow apertures. These windows consisted of polished quartz rods (diam. $\cong 1$ mm, length $\cong 25$ mm) sealed to metal plugs held by bolted vacuum seals to the shock-tube wall. This method was capable in principle of giving time resolution to 1 μsec. However, the luminous fronts encountered in this work were intense enough that the sensors sometimes responded to scattered light before the shock front passed the narrow apertures of the measuring gates.

The second method for determining the shock velocity is based on measuring the time interval Δt_l between the arrival of the incident and reflected shock fronts at the monitoring point. The time interval was taken from the film record of the rotating drum camera, which gave a precision of 0.1–0.2 μsec. The apparatus was calibrated to account for the nonideality of the velocity of the reflected shock; the correction to be subtracted from the ideal velocity was linear in the speed u_I of the incident shock. Details of the procedure are given in Ref. 18. The shock velocities determined this way agreed with those based on light emission for the runs in which both methods were employed. For the more intense shocks, scattered light was enough of a problem that only the second method was used.

B. Concentrations

Salts were introduced into the argon carrier gas by vaporization from a filament mounted on a plug in the bottom wall of the shock tube. The current through the filament was chosen to produce significant convective mixing of the aerosol formed from the salt vapor with the argon. The salt left a fairly uniform deposit on the $\cong 25$ cm of the tube walls from the end wall upstream. Estimates of the total salt concentration were made from the amounts of material so deposited.

The principal method of measuring concentrations was absorption spectroscopy on film, with a rotating-drum camera spectrograph. Spectra were taken primarily of the halide ions' photodetachment continua and of neutral alkalis, for which two or more principal series lines were measured. In the work done at lower temperatures,[11] the concentrations of neutral molecules were also measured via their diffuse absorption. In this work, the concentrations of molecules were too low to be determined in this way, particularly because the molecular absorptions occur in regions where the negative-ion continua are moderately intense and the negative-ion concentrations are, for the most part, very much higher than the molecular concentrations.

The negative ions always behaved as optically thin samples. Moreover the intensity of the source $I_0(\lambda)$ and the optical absorption cross sections $\sigma(\lambda)$ of the halides are essentially constant over the range $\Delta\lambda$ used to monitor their concentrations. Hence the number density of halide ions could be found directly from Beer's law.

In most of the work done at lower temperatures, and reported in Ref. 11, the alkali atomic lines used to monitor the neutral-metal-atom concentrations were also optically thin. However, in a few of the shocks run at higher temperatures in that work, the lines used were not strictly thin, and in the higher-temperature studies reported here, we encountered many instances in which the samples were not optically thin in the lines we used. Consequently we determined the alkali concentrations from the curves of growth of the relevant lines. These curves were constructed from the equivalent widths of the lines by assuming that the weakest usefully measurable lines, the fourth in the principal series for both Rb and Cs, were optically thin, so that the concentrations of alkalis could be determined and the curves of growth could be constructed for other, more intense lines recorded on the same film. In no case was there a very large deviation from the thin-line approximation. The curves of growth for the Cs $6\,^2S_{1/2} \rightarrow 7\,^2P$ line and the Rb $5\,^2S_{1/2} \rightarrow 6\,^2P^0$ and $7\,^2P^0$ lines are shown in Fig. 1*a* and *b*, respectively. These curves of growth are completely consistent with those reported by James and Sugden[19] from flame studies.

The alkali concentrations were determined from a line-shape profile approximating the Voigt profile as

$$P(\lambda) = \frac{0.40\,\Delta\lambda}{(\Delta\lambda)^2 + (\Delta\lambda_{1/2})^2} \tag{2.1}$$

where $\Delta\lambda$ is the separation in wavelength from the line center and $\Delta\lambda_{1/2}$ is the separation of the half-height wavelength from the line center. Typically, $\Delta\lambda_{1/2}$ was 2 Å for the wider lines we used. The concentrations determined this way from the curve-of-growth method were typically 18% higher than those one would infer from the naive assumption of Beer's law for the thickest lines of Rb and 32% higher for the thickest lines of Cs. The concentrations at equilibrium were determined from the averages based on as many

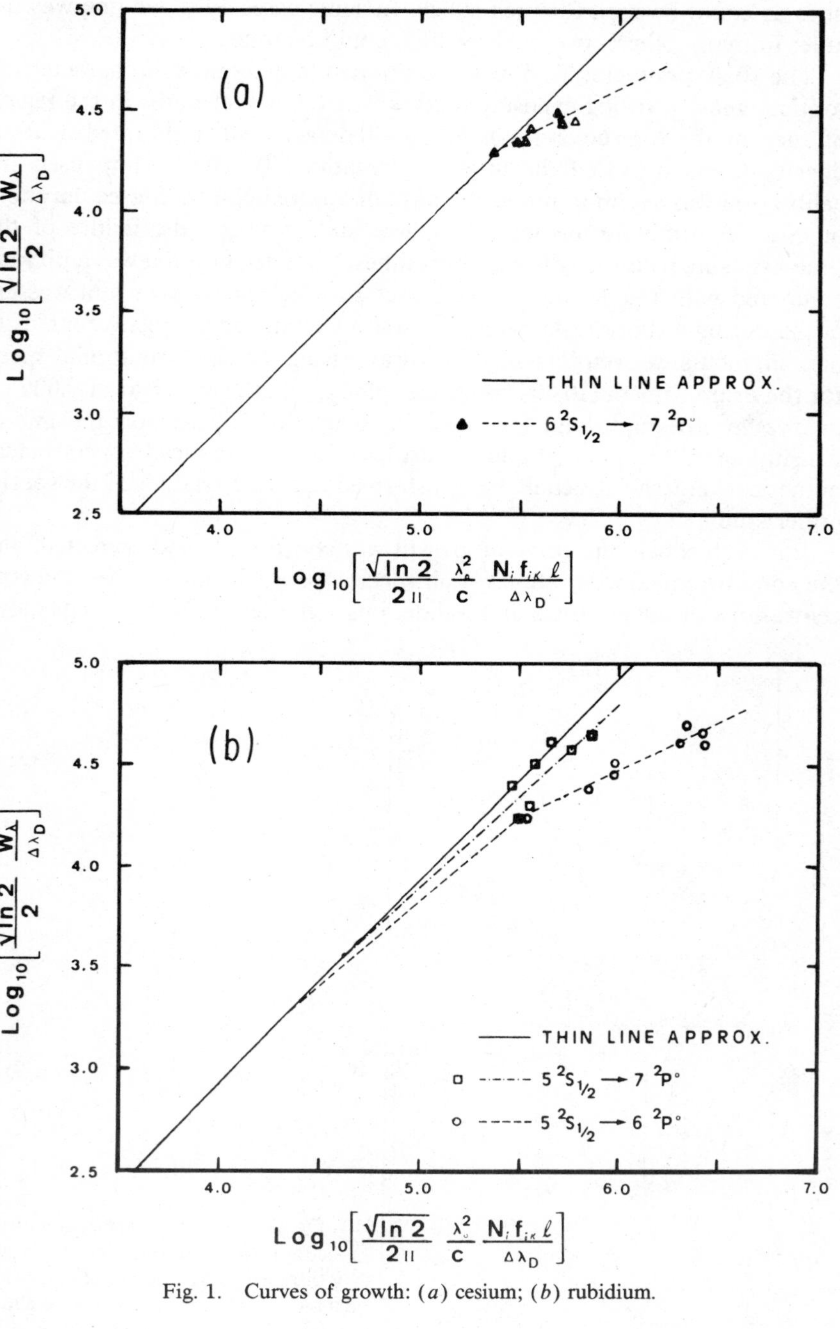

Fig. 1. Curves of growth: (*a*) cesium; (*b*) rubidium.

lines as could be used for each shock. In some cases only one line was usable; in many others, two or three lines could be used.

The 103-0 photographic film was calibrated to determine its characteristic (optical density vs. log exposure) curves[20] at two wavelengths in the region of each of the thresholds for halide photodetachment. Calibrated neutral-density filters provided the intensity variation. The flash lamp used for calibration was set up to produce a flash of approximately 25-μsec duration, in order to simulate the spectral source and minimize the failure of the time-exposure reciprocity law at short times.[20,21] This duration was a bit long compared with the duration of $\cong 1$ μsec to which an area of film was exposed during a shock experiment, but was adequate for the relative intensity measurements we were making. By always using the same rotational speed for the drum camera (10,404 rpm), the same entrance slit height of 0.005 in. on the spectrograph, and the same slit height of 200 μm on the microdensitometer, the photographic densities could be reduced to consistent numerical values which could be transformed to concentrations of the species under study.

For each shock, the curve of growth was constructed and corrected, and the equilibrium concentrations of all species were calculated. Then the concentrations of alkali atoms and halide ions were determined at steps along

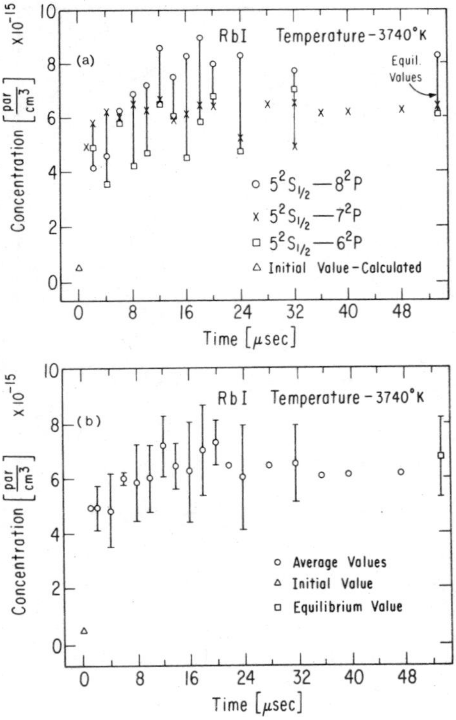

Fig. 2. Profiles of concentration vs. time for rubidium, showing (*a*) the results for three individual lines, and (*b*) the same data as in (*a*), but averaged. Also shown are error bars for ± 1 standard deviation where applicable.

the film separated by intervals corresponding to 2 μsec in the hot gas, for the entire period from passage of the reflected shock until equilibrium was attained. The profiles of concentrations vs. time were then evaluated; for the alkalis, all the usable lines were evaluated independently and the results were averaged. Figure 2 shows an example of one shock for which three lines of Rb were used. The oscillator strengths of the relevant rubidium and cesium lines and the photodetachment cross sections of the halide ions at the selected wavelengths are given in Table I.

The upper limit of temperature for these experiments was about 6000 K, because at higher temperatures the spectral lines of excited argon cause significant interference with the alkali lines used for monitoring concentrations. Interference with the system's chemistry from Ar^* and Ar^+ is not a serious problem at temperatures below 7000 K. In practice, we did make corrections for argon interference with Rb lines in two of the highest-temperature cases, by estimating the concentration of Ar^* from an isolated, thin Ar line near the interfering line.

The most important impurities were probably residual N_2 and O_2. The upper bound for their concentrations, based on the highest leak rates we ever

TABLE I
Oscillator Strengths and Photodetachment Cross Sections

(a) *Oscillator Strengths of Rubidium*

λ weighted average	Oscillator strengths Ref. 22	Ref. 23	Ref. 24	Average
3349.5	7.28×10^{-4}	7.21×10^{-4}	5.52×10^{-4}	6.67×10^{-4}
3588.0	2.46×10^{-3}	2.15×10^{-3}	1.87×10^{-3}	2.16×10^{-3}
4205.5	1.34×10^{-2}	1.37×10^{-2}	1.30×10^{-2}	1.34×10^{-2}

(b) *Oscillator Strengths of Cesium*

λ weighted average	Oscillator strengths Ref. 24	Ref. 25	Ref. 26	Ref. 27	Average
3477.0	2.23×10^{-4}	2.15×10^{-4}	2.4×10^{-4}	2.24×10^{-4}	2.26×10^{-4}
3611.5	5.12×10^{-4}	5.16×10^{-4}	6.0×10^{-4}	5.15×10^{-4}	5.36×10^{-4}
3877.5	1.64×10^{-3}	1.78×10^{-3}	1.7×10^{-3}	1.64×10^{-3}	1.89×10^{-3}

(c) *Halide Negative-Ion Photodetachment Cross Sections*[a]

Species	$\sigma(\lambda)$ [cm^2]	λ [Å]
Cl^-	9.00×10^{-18}	3310.5 ± 20
Br^-	9.29×10^{-18}	3400 ± 100
I^-	1.90×10^{-17}	3600 ± 300

[a]Refs. 7 and 28.

observed under working conditions, was about 1×10^{-4} Torr of air in the unshocked gas, and corresponded to about 1 part in 10^6 of the argon or 1 part in 10^3 of the salt. To avoid contamination of the reflected shock by impurities from the end wall, the observation point was chosen several millimeters from the end wall, further in fact than in the work of Ref. 11. When one salt was being studied, the shock tube was cleaned after three to five shocks. The tube was cleaned carefully and shocked several times with argon when experiments with a new salt were begun. We were never able to detect lines of more than one alkali in the absorption spectra. (Salt samples were of purity 99.9% or higher.)

C. Calculations of Equilibrium Concentrations

Equilibrium concentrations behind each reflected shock are required for constructing the curves of growth and for checking the theoretical models of the time evolution. The equilibrium concentrations following the incident shock, multiplied by the density ratio across the reflected shock, are the initial conditions immediately behind the reflected shock.

The species assumed present are M^0, X^0, M^+, X^-, MX, X^+, e, and Ar. In fact $[X^+]$, the concentrations of positive ions of halogens, are almost always negligible under our conditions. The Ar is chemically inactive except as a "solvent," so there are really seven species present whose concentrations vary because of reactions. There are two mass-balance equations, one for M^0 and one for X^0, and a charge-balance equation, so we need four independent equilibrium constants to determine the entire set of concentrations. We found it most convenient to use the following relations (the subscripts correspond to the reactions as indexed below):

$$K_1 = \frac{[M^+][X^-]}{[MX]} \tag{2.2}$$

$$K_2 = \frac{[X^0][e]}{[X^-]} \tag{2.3}$$

$$K_5 = \frac{[M^+][e]}{[M^0]} \tag{2.4}$$

$$K_9 = \frac{[X^+][e]}{[X^0]} \tag{2.5}$$

The other three equations are

$$[MX]+[M^0]+[M^+]=a_0 \tag{2.6}$$

$$[MX]+[X^-]+[X^0]+[X^+]=a_0 \tag{2.7}$$

$$[M^+]+[X^+]=[X^-]+[e] \tag{2.8}$$

where a_0 is the initial total concentration of MX.

The concentration a_0 is known behind the incident shock from temperature and spectroscopic measurements as described below. The concentrations $[M^0]$ and $[X^-]$ are both measured behind the reflected shock, so the system is overdetermined at equilibrium there.

The equilibrium constants K_1, K_2, K_5, and K_9 were derived from the corresponding partition functions:

$$K_1 = \frac{Q(M^+)Q(X^-)}{Q(MX)} S^*(MX) \tag{2.9}$$

$$K_2 = \frac{Q(X^0)Q(e)}{Q(X^-)} S^*(X^-) \tag{2.10}$$

$$K_5 = \frac{Q(M^+)Q(e)}{Q(M^0)} S^*(M^0) \tag{2.11}$$

$$K_9 = \frac{Q(X^+)Q(e)}{Q(X^0)} S^*(X^0) \tag{2.12}$$

Here the Q's are the internal partition functions; $Q(M^+)$ was identically 1.000 for all of our conditions, $Q(e)$ is identically 2, and $Q(X^-)$ is exactly 1. We evaluated the atomic Q's from published tables;[29] where our temperatures were below those of the tables for $Q(X^0)$ and $Q(X^+)$, we computed the partition functions from the first two levels for X^0 and first five levels for X^+, enough levels to give the tabulated values at the lowest temperatures for which published values are given. The molecular partition functions were evaluated using the standard anharmonic-oscillator, nonrigid form.[30-32]

The functions S^* are these, in units of cm^{-3}:

$$S^*(MX) = 1.879 \times 10^{20} (\mu_{MX} T)^{3/2} \exp\left[\frac{-D_0^{ion}(MX)}{kT}\right] \tag{2.13}$$

$$S^*(X^-) = 2.415 \times 10^{15} T^{3/2} \exp\left[\frac{-EA(X^0)}{kT}\right] \tag{2.14}$$

$$S^*(M^0) = 2.415 \times 10^{15} T^{3/2} \exp\left[\frac{-[IP(M^0) - \Delta IP]}{kT}\right] \tag{2.15}$$

$$S^*(X^0) = 2.415 \times 10^{15} T^{3/2} \exp\left[\frac{-[IP(X^0) - \Delta IP]}{kT}\right] \tag{2.16}$$

The dissociation energy $D_0^{ion}(MX)$ is the dissociation energy to ions:

$$D_0^{ion}(MX) = D_0(MX) + IP(M^0) - EA(X^0); \tag{2.17}$$

the reduced mass is

$$\mu_{MX} = \frac{m(M^+)m(X^-)}{m(M^+)+m(X^-)} \tag{2.18}$$

and ΔIP is the shift of ionization potential due to the plasma. Under our conditions ΔIP was about 0.03 eV behind the incident shocks and about 0.05 eV behind the reflected shocks. The relevant energies of reaction are given in Table II. In Table III are values of the harmonic, rigid-rotor, and correction factors for the internal partition function, the entire internal partition function, and the factor S^* for CsCl.

The equations were manipulated to give a fourth-order equation for $[M^0]$ in terms of $[e]$;

$$\begin{aligned}&[e]^2\{[e]^2+[M^0][e]+K_2[e]+K_2K_9\}\\&\quad -K_2[M^0]\{K_2[e]+K_9[e]+2K_5K_9\}=0\end{aligned} \tag{2.19}$$

This equation was solved iteratively for measured values of $[M^0]$. The mass-balance conditions were used to check both for self-consistency and for roundoff errors, and to fix a_0.

When the concentration of X^- was used as the observable, the electron concentration was determined from

$$[e]^2\{[e]^2+[e](K_5+[X^-])\}-\{[X^-]K_2(2K_5K_9+[e](K_5+K_9))\}=0 \tag{2.20}$$

TABLE II
Energies of Reaction of Monatomic Species

Species	Symbol	ε [eV]
Rb	IP(Rb)	4.177[a]
Cs	IP(Cs)	3.894[a]
Cl	IP(Cl)	12.967[a]
Br	IP(Br)	11.814[a]
I	IP(I)	10.451[a]
Cl^-	EA(Cl)	3.615[b]
Br^-	EA(Br)	3.364[b]
I^-	EA(I)	3.061[b]
Ar	IP(Ar)	15.759[a]

[a] Ref. 33.
[b] Ref. 34.

TABLE III
Partition Function for CsCl

T [K]	Q_{vib}	Q_{rot}	$Q_{correction}$	$Q_{internal}$	S^* [cm^{-3}]
0	1	1	1	1	0
500	2.18422	4831.87	1.01642	1.07271×10^{4}	6.407×10^{-23}
1000	3.79251	9663.74	1.04051	3.81344×10^{4}	4.002×10^{2}
1500	5.41755	14495.6	1.06631	8.37383×10^{4}	9.560×10^{10}
2000	7.04683	19327.5	1.09342	1.48921×10^{5}	1.678×10^{15}
2500	8.67780	24159.4	1.12174	2.35173×10^{5}	6.374×10^{17}
3000	10.3096	28991.2	1.15125	3.44094×10^{5}	3.515×10^{19}
3500	11.9419	33823.1	1.18192	4.77392×10^{5}	6.391×10^{20}
4000	13.5745	38655.0	1.21375	6.36883×10^{5}	5.780×10^{21}
4500	15.2073	43486.8	1.24674	8.24495×10^{5}	3.272×10^{22}
5000	16.8403	48318.7	1.28089	1.04226×10^{6}	1.331×10^{23}
5500	18.4734	53150.6	1.31619	1.29233×10^{6}	4.256×10^{23}
6000	20.1065	57982.5	1.35264	1.57694×10^{6}	1.133×10^{24}
6500	21.7396	62814.3	1.39025	1.89847×10^{6}	2.623×10^{24}
7000	23.3729	67646.2	1.42900	2.25938×10^{6}	5.427×10^{24}
7500	25.0061	72478.1	1.46891	2.66225×10^{6}	1.026×10^{25}
8000	26.6395	77309.9	1.50997	3.10978×10^{6}	1.804×10^{25}

The equilibrium conditions behind the incident shock were determined from the value of total salt concentration, in turn obtained from the equilibrium conditions behind the reflected shock as just described. The value of a_0(inc) is just a_0(refl)$\cdot(\rho^{inc}/\rho^{refl})$, the value behind the reflected shock multiplied by the ratio of incident to reflected shock densities. The electron concentration was obtained by iterative solution of the eighth-order equation

$$[e]^4BK_5(K_5+[e])+[e]ABK_1(K_5+[e])-a_0A^2K_1=0 \quad (2.21)$$

where

$$A=2K_2K_5K_9+[e]K_2(K_5+K_9)-[e]^3 \quad (2.22)$$

and

$$B=K_2([e]+K_9)+[e]^2 \quad (2.23)$$

The initial guessed value of [e] must be rather close to the correct value in order for a Newton–Raphson solution of (2.21) to converge, especially in the lower ranges of temperatures we studied.

Initial concentrations behind the reflected shock were calculated from the density-ratio-renormalized concentrations at equilibrium behind the incident shock. These values, rather than the observed values, were used because scattering of the probe light by the shock front made the signal too weak to be used for 2–4 μsec following the passage of the reflected shock.

Independent rough estimates of a_0 were made by determining the surface area over which solid salt was found after the tube was opened. These estimates, albeit somewhat uncertain, were entirely consistent with the values estimated from equilibrium conditions in the reflected-shock zone. The range of total weights of samples in the gas inferred from the collected residue was 1.5–3.5 mg, obtained from 2–15 mg of salt initially put on the heater filament.

III. KINETIC MODEL

The alkali halide systems reported here are particularly favorable for kinetic study because all the significant species are known, and all the unimolecular and bimolecular processes can be included in a straightforward kinetic model. All the termolecular processes can be tabulated as well, but we shall not include all of these processes in our model. We shall see, in the concluding discussion, that one process which is successfully modeled as bimolecular in these and other shock-tube experiments is probably more complicated in reality; this is the collisional detachment of electrons from negative halide ions.

In this section, we first list the species that must be considered, then select the elementary reactions to be included, and transform the quantitative description into a set of coupled differential equations. Then we describe the solution of these equations by use of an analog computer.

In the temperature range of our experiments, the most important species involving the argon carrier gas and the salt MX are the diatomic MX molecules, the neutral atoms M^0, X^0, and Ar, the positive ions M^+ and X^+, the negative ions X^-, and electrons, Negative alkali ions M^- are not thermally stable in the temperature range of our experiments, 2000–6500 K, nor are dimers or higher polymers of the alkali halides. The positive halogen ions X^+ are only present at our very highest temperatures; the ions of argon would appear at still higher temperatures.

Details of the justification for dropping the dimers $(MX)_2$, the halide molecules X_2 and their ions X_2^- and X_2^+, the alkali halide ions MX^+ and MX^-, and the species Ar^+, Ar_2^+, ArX^+, and ArM^+ are given elsewhere.[18] The neutral diatomic species among these have dissociation energies of 2.5 eV or less, so that at 2000 K the equilibrium constant for dissociation of AB to A + B is of order 10^{-3} in pressure units of atmospheres; with samples of order 1 mg/l, the undissociated fraction of such diatomic species under the most favorable of our conditions is less than 2% of the total possible stoichiometric concentration. For higher temperatures and more easily dissociated species (among them all, only Cl_2 has a $D_e > 2$ eV), the undissociated fractions are of course smaller. The diatomic halogen ions X_2^+ do have larger dissociation energies, but their ionization potentials are high, comparable with the atomic halogens. The ionization potential of I_2 is the lowest, about 9.3 eV,[33] but its dissociation energy is only about 2.68 eV;[33] these, together with the large mass of iodine, make I_2^+ a negligible species in our system.

The relevant ionization potentials, electron affinities, and dissociation energies used for our calculations are given in Table IV.

Next we turn to the elementary reactions of MX, M^+, X^-, M^0, X^0, and e. There are 51 reactions of these species, plus the reverse reactions, which are unimolecular, bimolecular, or termolecular. We considered no reactions of higher molecularity in either direction. We also neglect all photoreactions, a valid assumption under our conditions but not under the hotter conditions of Gait and Berry.[16,17] From the 51 reactions, we selected eight, with their reverse reactions, giving us 16 in all. These were the criteria for selection:

1. Reactions with argon as a third body were included because of its high concentration.

TABLE IV
Energies of Reaction of Diatomic Species of Relevant Alkali Halides and Related Substances[a]

Symbol	Species and Values [eV]					
		Cs		Rb		
$IP(M_2)$		3.7		3.75		
$D(M_2)$		0.394		0.49		
$D(M_2^+)$		0.61		0.72		
		I		Br		Cl
$EA(X_2)$		2.56[b]		2.55[b]		2.39[b]
$IP(X_2)$		(9.31)[c]		10.52		11.50
$D(X_2)$		1.542		1.971		2.479
$D(X_2^-)$		1.04		1.15		1.26
$D(X_2^+)$		(2.683)[c]		3.26		3.95
	CsI	CsBr	CsCl	RbI	RbBr	RbCl
IP(MX)	7.25	7.72	8.32	7.12	7.75	8.26
D(MX)	3.56	4.17	4.58	3.30	3.90	4.34
$D(MX^+)$	0.21	0.34	—	0.36	0.33	0.26

[a]See Ref. 35.
[b]See Ref. 34
[c]Values are uncertain.

2. Reactions with electrons as third bodies were kept because of the high velocities of thermal electrons.
3. Reactions involving oppositely charged species were kept because of the long range of the Coulomb interaction (or, alternatively, the large Debye radii).
4. Unimolecular processes were neglected because none of the species have enough internal degrees of freedom available to stabilize collision complexes; thus a process $A+B \rightarrow C^*$ was not considered, but if $C^* \rightarrow D+E$, then $A+B \rightarrow D+E$ might be considered.
5. Processes involving two or three neutral or like-charged alkali- or halide-containing species in forward and reverse directions were excluded as being too infrequent. (In fact, one reaction, $e+X^- \rightarrow 2e+X^0$, was kept.)

The final set of reactions comprises the following eight forward and eight reverse processes:

$$\mathrm{Ar}+\mathrm{MX} \underset{k_{1r}}{\overset{k_{1f}}{\rightleftarrows}} \mathrm{Ar}+\mathrm{M}^+ +\mathrm{X}^- \tag{R1}$$

$$\mathrm{Ar}+\mathrm{X}^- \underset{k_{2r}}{\overset{k_{2f}}{\rightleftarrows}} \mathrm{Ar}+\mathrm{X}^0 + e \tag{R2}$$

$$\mathrm{X}^- + e \underset{k_{3r}}{\overset{k_{3f}}{\rightleftarrows}} \mathrm{X}^0 + 2e \tag{R3}$$

$$\mathrm{MX}+ e \underset{k_{4r}}{\overset{k_{4f}}{\rightleftarrows}} \mathrm{M}^0 +\mathrm{X}^- \tag{R4}$$

$$\mathrm{M}^0 + e \underset{k_{5r}}{\overset{k_{5f}}{\rightleftarrows}} \mathrm{M}^+ + 2e \tag{R5}$$

$$\mathrm{MX}+ e \underset{k_{6r}}{\overset{k_{6f}}{\rightleftarrows}} \mathrm{M}^0 +\mathrm{X}^0 + e \tag{R6}$$

$$\mathrm{Ar}+\mathrm{MX} \underset{k_{7r}}{\overset{k_{7f}}{\rightleftarrows}} \mathrm{Ar}+\mathrm{M}^0 +\mathrm{X}^0 \tag{R7}$$

$$\mathrm{Ar}+\mathrm{M}^0 +\mathrm{X}^0 \underset{k_{8r}}{\overset{k_{8f}}{\rightleftarrows}} \mathrm{Ar}+\mathrm{M}^+ +\mathrm{X}^- \tag{R8}$$

In practice, because of the very large concentration of Ar, the forward reactions R1, R2, and R7 are always in the quasi-first-order regime and the forward reaction R8 is in the quasi-second-order regime. Hence their inclusion implicity includes any contributions of the corresponding unimolecular counterparts of R1, R2, and R7 and bimolecular counterpart of R8.

Mandl, in his early work,[3-5,36-38] included $M^+ + X^- \rightleftarrows M^+ + X^0 + e^-$. This process was disregarded in his later studies.[6,39] In his system, it represented less than a 10% effect on the concentration of F^-. We disregard it as a perturbation, within our experimental uncertainty, on the much more important detachment processes R2 and R3.

Reaction R4 plays a catchall role much like R1, R2, R7, and R8. Reaction R4 subsumes effects of $Ar + MX + e \rightleftarrows Ar + M^0 + X^-$ and of $2e + MX \rightleftarrows e + M^0 + X^-$. One related reaction that was not included is $MX + e \rightleftarrows M^+ + X^- + e$. While this is probably the most likely of the omitted reactions to play any role, we excluded it on grounds that the dissociative attachment process is far more likely than polar dissociation when the electrons are only thermal.[40] We also excluded $X^0 + e \rightleftarrows X^+ + 2e$.[18]

The next stage of constructing the kinetic model was the reduction of the differential equations to computable form. The sixteen reactions R1–R8 and their reverses involve six concentrations. As mentioned, there are two mass-conservation relations and one equation of electric-charge balance. Hence there are three independent concentrations. We regularly monitored two of these, M^0 and X^-. In some cases we were able to follow MX as well, but because the temperature dependence of the molecular absorption coefficients is not known, we did not depend on the MX concentrations for our analysis.

In addition to the three balance equations, there are several other constraints, which overdetermine the system with the concentrations of two species measured as functions of time. These are the additional constraints:

1. The concentrations at the arrival of the reflected shock are fixed by the equilibrium conditions behind the incident shock.
2. The concentrations at times well after the passage of the reflected shock must also be equilibrium concentrations, based on the final reflected-shock conditions.
3. The rate constant for each reaction appearing in more than one system must be the same in all the systems where it appears.
4. The rate constants must vary systematically from alkali to alkali and from halogen to halogen.
5. All sixteen rate constants must fit reasonable Arrhenius plots.
6. The concentration profiles from the kinetic model must reproduce the experimental concentration profiles of M^0 and X^-.

The first criterion is met by using the appropriate equilibrium concentrations as the initial conditions for the solution of the differential equations. The second and last criteria are the most stringent in most cases. The third, tied to the fifth, means, for example, that the data for $Ar + Br^- \rightarrow Ar + Br^0 + e$ from RbBr and CsBr must lie on the same Arrhenius curve. The criterion of smooth variation from salt to salt is predicated on the supposition that alkalis, especially heavy alkalis, are rather like one another, and that chlorine, bromine, and iodine also are similar and differ systematically and smoothly in all the properties and rates relevant to our systems.

The kinetics of reactions R1–R8 can be expressed in the following six differential equations:

$$\frac{d[\mathrm{M}^0]}{dt} = k_{4f}[\mathrm{MX}][e] - k_{4r}[\mathrm{M}^0][\mathrm{X}^-] - k_{5f}[\mathrm{M}^0][e] + k_{5r}[\mathrm{M}^+][e]^2 + k_{6f}[\mathrm{MX}][e] - k_{6r}[\mathrm{M}^0][\mathrm{X}^0][e] + k_{7f}[\mathrm{Ar}][\mathrm{MX}] - k_{7r}[\mathrm{Ar}][\mathrm{M}^0][\mathrm{X}^0] - k_{8f}[\mathrm{Ar}][\mathrm{M}^0][\mathrm{X}^0] + k_{8r}[\mathrm{Ar}][\mathrm{M}^+][\mathrm{X}^-] \quad (3.1)$$

$$\frac{d[\mathrm{X}^-]}{dt} = k_{1f}[\mathrm{Ar}][\mathrm{MX}] - k_{1r}[\mathrm{Ar}][\mathrm{M}^+][\mathrm{X}^-] - k_{2f}[\mathrm{Ar}][\mathrm{X}^-] + k_{2r}[\mathrm{Ar}][\mathrm{X}^0][e] - k_{3f}[\mathrm{X}^-][e] + k_{3r}[\mathrm{X}^0][e]^2 + k_{4f}[\mathrm{MX}][e^-] - k_{4r}[\mathrm{M}^0][X^-] + k_{8f}[\mathrm{Ar}][\mathrm{M}^0][\mathrm{X}^0] - k_{8r}[\mathrm{Ar}][\mathrm{M}^+][\mathrm{X}^-] \quad (3.2)$$

$$\frac{d[\mathrm{MX}]}{dt} = -k_{1f}[\mathrm{Ar}][\mathrm{MX}] + k_{1r}[\mathrm{Ar}][\mathrm{M}^+][\mathrm{X}^-] - k_{4f}[\mathrm{MX}][e] + k_{4r}[\mathrm{M}^0][\mathrm{X}^-] - k_{6f}[\mathrm{MX}][e] + k_{6r}[\mathrm{M}^0][\mathrm{X}^0][e] - k_{7f}[\mathrm{Ar}][\mathrm{MX}] + k_{7r}[\mathrm{Ar}][\mathrm{M}^0][\mathrm{X}^0] \quad (3.3)$$

$$\frac{d[\mathrm{M}^+]}{dt} = k_{1f}[\mathrm{Ar}][\mathrm{MX}] - k_{1r}[\mathrm{Ar}][\mathrm{M}^+][\mathrm{X}^-] + k_{5f}[\mathrm{M}^0][e] - k_{5r}[\mathrm{M}^+][e]^2 + k_{8f}[\mathrm{Ar}][\mathrm{M}^0][\mathrm{X}^0] - k_{8r}[\mathrm{Ar}][\mathrm{M}^+][\mathrm{X}^-] \quad (3.4)$$

$$\frac{d[\mathrm{X}^0]}{dt} = k_{2f}[\mathrm{Ar}][\mathrm{X}^-] - k_{2r}[\mathrm{Ar}][\mathrm{X}^0][e] + k_{3f}[\mathrm{X}^-][e] - k_{3r}[\mathrm{X}^0][e]^2 + k_{6f}[\mathrm{MX}][e] - k_{6r}[\mathrm{M}^0][\mathrm{X}^0][e] + k_{7f}[\mathrm{Ar}][\mathrm{MX}] - k_{7r}[\mathrm{Ar}][\mathrm{M}^0][\mathrm{X}^0] - k_{8f}[\mathrm{Ar}][\mathrm{M}^0][\mathrm{X}^0] + k_{8r}[\mathrm{Ar}][\mathrm{M}^+][\mathrm{X}^-] \quad (3.5)$$

$$\frac{d[e]}{dt} = k_{2f}[\mathrm{Ar}][\mathrm{X}^-] - k_{2r}[\mathrm{Ar}][\mathrm{X}^0][e] + k_{3f}[\mathrm{X}^-][e] - k_{3r}[\mathrm{X}^0][e]^2 - k_{4f}[\mathrm{MX}][e] + k_{4r}[\mathrm{M}^0][X^-] + k_{5f}[\mathrm{M}^0][e] - k_{5r}[\mathrm{M}^+][e]^2 \quad (3.6)$$

These equations are reduced further by introducing the relation between the unknown rate coefficients and the calculable equilibrium constants,

$$K_j = \frac{k_{jf}}{k_{jr}}, \qquad j = 1, \ldots, 8 \quad (3.7)$$

and the time-derivative forms of the mass and charge conservation equations,

$$\frac{d[\mathrm{MX}]}{dt}+\frac{d[\mathrm{M}^+]}{dt}+\frac{d[\mathrm{M}^0]}{dt}=0 \tag{3.8}$$

$$\frac{d[\mathrm{MX}]}{dt}+\frac{d[\mathrm{X}^0]}{dt}+\frac{d[\mathrm{X}^-]}{dt}=0 \tag{3.9}$$

$$\frac{d[\mathrm{M}^+]}{dt}-\frac{d[\mathrm{X}^-]}{dt}-\frac{d[e]}{dt}=0 \tag{3.10}$$

The system of six differential equations in six variable concentrations which require solution by computer is

$$\begin{aligned}\frac{d[\mathrm{M}^0]}{dt}&=k_{4f}([\mathrm{MX}][e]-K_4^{-1}[\mathrm{M}^0][\mathrm{X}^-])+k_{5r}([\mathrm{M}^+][e]-K_5[\mathrm{M}^0])[e]\\&\quad+k_{6f}([\mathrm{MX}]-K_6^{-1}[\mathrm{M}^0][\mathrm{X}^0])[e]\\&\quad+k_{7f}[\mathrm{Ar}]([\mathrm{MX}]-K_7^{-1}[\mathrm{M}^0][\mathrm{X}^0])\\&\quad+k_{8r}[\mathrm{Ar}]([\mathrm{M}^+][\mathrm{X}^-]-K_8[\mathrm{M}^0][\mathrm{X}^0])\end{aligned} \tag{3.11}$$

$$\begin{aligned}\frac{d[\mathrm{X}^-]}{dt}&=k_{1f}[\mathrm{Ar}]([\mathrm{MX}]-K_1^{-1}[\mathrm{M}^+][\mathrm{X}^-])\\&\quad-k_{2f}[\mathrm{Ar}]([\mathrm{X}^-]-K_2^{-1}[\mathrm{X}^0][e])\\&\quad+k_{4f}([\mathrm{MX}][e]-K_4^{-1}[\mathrm{M}^0][\mathrm{X}^-])\\&\quad-k_{3f}([\mathrm{X}^-]-K_3^{-1}[\mathrm{X}^0][e])[e]\\&\quad-k_{8r}[\mathrm{Ar}]([\mathrm{M}^+][\mathrm{X}^-]-K_8[\mathrm{M}^0][\mathrm{X}^0])\end{aligned} \tag{3.12}$$

$$\begin{aligned}\frac{d[\mathrm{MX}]}{dt}&=-k_{1f}[\mathrm{Ar}]([\mathrm{MX}]-K_1^{-1}[\mathrm{M}^+][\mathrm{X}^-])\\&\quad-k_{4f}([\mathrm{MX}][e]-K_4^{-1}[\mathrm{M}^0][X^-])\\&\quad-k_{6f}([\mathrm{MX}]-K_6^{-1}[\mathrm{M}^0][\mathrm{X}^0])[e]\\&\quad-k_{7f}[\mathrm{Ar}]([\mathrm{MX}]-K_7^{-1}[\mathrm{M}^0][\mathrm{X}^0])\end{aligned} \tag{3.13}$$

$$\frac{-d[\mathrm{M}^+]}{dt}=\frac{d[\mathrm{M}^0]}{dt}+\frac{d[\mathrm{MX}]}{dt} \tag{3.14}$$

$$\frac{-d[\mathrm{X}^0]}{dt}=\frac{d[\mathrm{X}^-]}{dt}+\frac{d[\mathrm{MX}]}{dt} \tag{3.15}$$

$$\frac{-d[e]}{dt}=\frac{d[\mathrm{MX}]}{dt}+\frac{d[\mathrm{M}^0]}{dt}+\frac{d[\mathrm{X}^-]}{dt} \tag{3.16}$$

Of the 51 reactions for the entire scheme, the only ones not included either effectively or explicitly are

$$MX \rightleftarrows M^+ + X^0 + e$$
$$MX + M^+ + e \rightleftarrows 2M^0 + X^0$$
$$MX + X^0 + e \rightleftarrows M^+ + 2X^-$$

The kinetic effects of the others are all essentially buried in small fluctuations of the rate coefficients $k_{1f} - k_{8f}$ and $k_{1r} - k_{8r}$.[18] Furthermore, two-body contributions cannot be distinguished from three-body contributions when the third body is Ar.

The six differential equations (3.11)–(3.16) and the six initial concentrations are the basis for the computation of profiles of concentration as functions of time, from which we infer the sixteen unknown rate coefficients. The system of equations is stiff,[41] and, while it could now be solved with modern digital methods,[42] lends itself especially to solution by analog procedures because it appears to have many shallow local minima. The machine used for solving our equations was an Applied Dynamics/4 analog-hybrid computer. The one aspect of the analog calculation that required care was the scaling of equations and independent variables. The equations were scaled directly to machine units for dependent variables; the equations themselves were rescaled to accommodate the independent variables.[43] Further details of the setup for analog computation are given in Ref. 18.

Experimental data were entered digitally to the hybrid and transferred to the display screen of the analog component. Initial estimates of the rate coefficients were made by extrapolating the values of Ref. 11. After a mock solution for static input voltages was checked against calculated values, the scaled trial rate coefficients were entered as potentiometer settings and the analog computer was put into a repetitive solution mode. The solution was displayed with the experimental data on the screen, and the potentiometers were varied until the best visual fit was found. The potentiometer settings were recorded and then varied to test the sensitivity of the solution to the values of the rate constants.

A total of 34 shocks were fitted and refitted to experimental data, almost all of them several times in order to get good simultaneous agreement among all the reactions, according to the criteria given earlier. The average number of computer runs done to fit an experiment was 3, and the range was 1 to 5 trials.

IV. RESULTS: OBSERVATIONS AND ARRHENIUS GRAPHS

Typical fitted solutions of the concentration profiles are shown in Fig. 3 for an example of RbCl, in Fig. 4 for RbBr, and in Fig. 5 for an example of

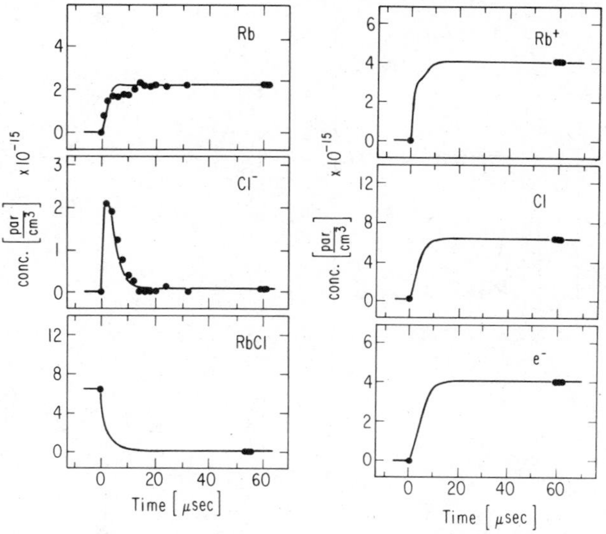

Fig. 3. Experimental and/or theoretical profiles of concentration versus time for RbCl at 4230 K: Rb^0, Cl^-, RbCl, Rb^+, Cl^0, *e*. The data point at $t = 0$ μsec and the three at $t \cong 60$ μsec were calculated from theoretical values of the equilibrium constants. The data points at all other times were determined experimentally. The solid curve is the analog fit to the experimental data points.

CsI. The data points for the neutral metal atoms and the halide ions are taken directly from observations. The concentrations of the other four species were computed from the assumption of chemical equilibrium in the gases of the incident shock just prior to passage of the reflected shock, and again of equilibrium well behind the reflected shock front.

As in our previous observations,[10,11] the metal-atom concentrations normally rise monotonically or show a small overshoot, The halide concentrations show a large overshoot immediately following dissociation and then relax back to their new equilibrium concentrations. The concentrations of diatomic species fall smoothly and rather rapidly to near zero. The results obtained in this work differed from those at lower temperatures[10,11] in two ways, mainly. The conditions of this work brought the system to approximate equilibrium in 20 μsec or less, considerably shorter than we observed at lower temperatures. Second, the gases behind the incident shocks in this work were hot enough to cause significant dissociation in that zone, prior to passage of the reflected shock and its temperature and density jump. In the studies at lower temperatures,[11] the kinetics in the reflected shock were dominated by molecular dissociation. The consequence is that the high-temperature experiments are more severe tests of the overall kinetic model.

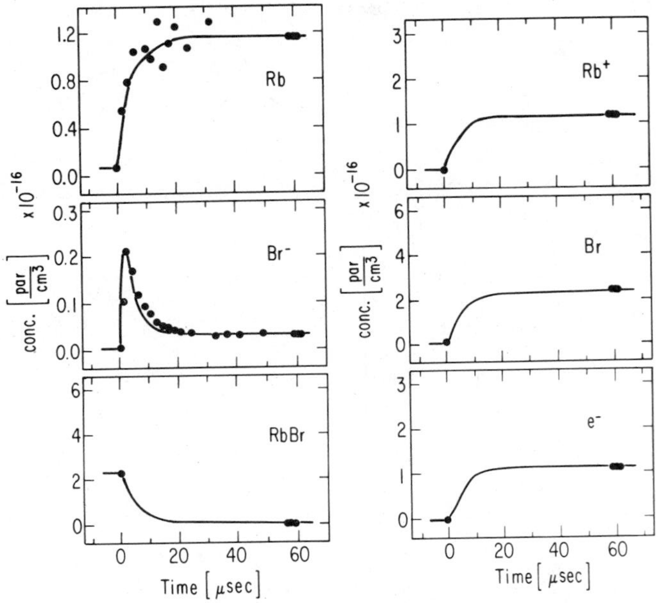

Fig. 4. Experimental and/or theoretical concentration profiles for RbBr at 4370 K: Rb^0, Br^-, RbBr, Rb^+, Br^0, e; see caption of Fig. 3 for explanations.

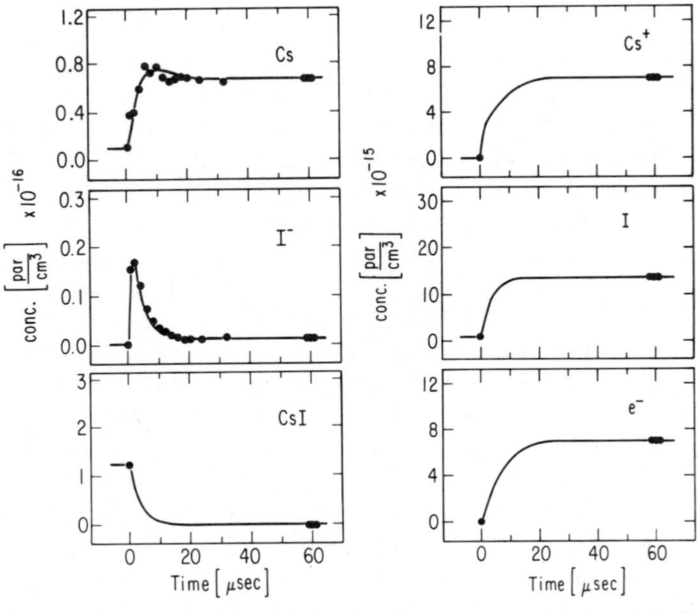

Fig. 5. Experimental and/or theoretical concentration profiles for CsI at 3975 K; Cs^0, I^-, CsI, Cs^+, I^0, e; see caption of Fig. 3 for explanations.

The small overshoots sometimes seen in the computed concentrations of alkali atoms were sometimes within the scatter and uncertainty of the measurements, but in other cases both the measured concentrations and the analog fits implied that these small overshoots in $[M^0]$ are real. Sometimes there were small systematic deviations, especially between halide concentrations observed and computed. The effect of these deviations on the resulting rate coefficients is discussed below; the variations will be seen to fall within the best estimates of the uncertainties of the two sets of data.

The sensitivities were estimated by finding how much the potentiometers had to be varied to produce changes of 2–5% in any portion of the analog curves. The minimum changes observable on the screen were about $\frac{1}{10}$ of this amount, so we used as our estimate of the uncertainty in the potentiometer setting $\frac{1}{10}$ of the sensitivity as just defined. Typical fractional fitting uncertainties in the rate constants $\Delta k_j/k_j$ for all eight reactions are these, for CsCl at 4010 K:

$$\Delta k_{1f}/k_{1f} = 0.023$$
$$\Delta k_{2f}/k_{2f} = 0.015$$
$$\Delta k_{3f}/k_{3f} = 170$$
$$\Delta k_{4f}/k_{4f} = 160$$
$$\Delta k_{5f}/k_{5f} = 0.10$$
$$\Delta k_{6f}/k_{6f} = 1000$$
$$\Delta k_{7f}/k_{7f} = 0.10$$
$$\Delta k_{8f}/k_{8f} = 0.043$$

As we shall see shortly, reactions 3, 4, and 6 contribute so little to the overall kinetics that we could only put upper bounds on their values.

A. Rate Coefficients

The forward rate coefficients, equilibrium constants, and reverse rate constants for all six salts, for reaction R1, are given in Table V. In this and subsequent tables, only about 60% of the measured rate coefficients are given, for the sake of brevity. They are all available in Ref. 18 and are all included in the figures. Table VI gives the rate coefficients k_{2f} for collisional detachment, $Ar + X^- \rightleftarrows Ar + X^0 + e$. We shall return to reactions R3 and R4 below. The ionization of alkali atoms by collisions with electrons (reaction R5, $M^0 + e \rightleftarrows M^+ + 2e$) is described by the rate coefficients k_{5f}, equilibrium constants K_5, and rate coefficients k_{5r} given in Table VII. Again, we shall return to reaction R6. The rate coefficients and equilibrium constants for

TABLE V

Collisional Dissociation into Ions and Ionic Recombination Rate Constants:

$$Ar + MX \underset{k_r}{\overset{k_f}{\rightleftarrows}} Ar + M^+ + X^-$$

T [K]	k_f [cm^3/sec] Fit[a]	k_f [cm^3/sec] Calculated[a]	K_{eq} [cm^{-3}]	k_r [cm^6/sec] Fit[a]	k_r [cm^6/sec] Calculated[a]
		$Ar + RbCl \underset{k_r}{\overset{k_f}{\rightleftarrows}} Ar + Rb^+ + Cl^-$			
2860	$2.98 \pm 6.78 \times 10^{-15}$	$3.04 \pm 1.80 \times 10^{-15}$	2.758×10^{13}	$1.08 \pm 2.46 \times 10^{-28}$	$1.09 \pm 0.63 \times 10^{-28}$
3360	$5.45 \pm 1.25 \times 10^{-15}$	$7.36 \pm 2.44 \times 10^{-15}$	5.005×10^{14}	$1.09 \pm 0.25 \times 10^{-29}$	$1.48 \pm 0.48 \times 10^{-29}$
3765	$1.17 \pm 0.99 \times 10^{-14}$	$1.20 \pm 0.35 \times 10^{-14}$	2.937×10^{15}	$3.98 \pm 3.37 \times 10^{-30}$	$4.13 \pm 1.17 \times 10^{-30}$
4230	$1.62 \pm 5.00 \times 10^{-14}$	$1.81 \pm 0.66 \times 10^{-14}$	1.460×10^{16}	$1.11 \pm 3.42 \times 10^{-30}$	$1.24 \pm 0.44 \times 10^{-30}$
5810	$2.99 \pm 4.03 \times 10^{-14}$	$3.56 \pm 2.67 \times 10^{-14}$	4.691×10^{17}	$6.37 \pm 8.59 \times 10^{-32}$	$7.35 \pm 5.31 \times 10^{-32}$
		$Ar + RbBr \underset{k_r}{\overset{k_f}{\rightleftarrows}} Ar + Rb^+ + Br^-$			
3795	$2.02 \pm 0.92 \times 10^{-15}$	$1.99 \pm 0.63 \times 10^{-15}$	4.377×10^{15}	$4.62 \pm 2.10 \times 10^{-31}$	$4.53 \pm 1.46 \times 10^{-31}$
3960	$2.90 \pm 1.33 \times 10^{-15}$	$2.49 \pm 0.63 \times 10^{-15}$	7.881×10^{15}	$3.68 \pm 1.69 \times 10^{-31}$	$3.18 \pm 0.81 \times 10^{-31}$
4370	$3.51 \pm 2.72 \times 10^{-15}$	$3.97 \pm 0.80 \times 10^{-15}$	2.754×10^{16}	$1.27 \pm 0.99 \times 10^{-31}$	$1.44 \pm 0.29 \times 10^{-31}$
4560	$3.59 \pm 1.50 \times 10^{-15}$	$4.73 \pm 1.06 \times 10^{-15}$	4.556×10^{16}	$7.88 \pm 3.29 \times 10^{-32}$	$1.04 \pm 0.24 \times 10^{-31}$
5135	$7.43 \pm 2.74 \times 10^{-15}$	$7.23 \pm 2.65 \times 10^{-15}$	1.641×10^{17}	$4.53 \pm 1.67 \times 10^{-32}$	$4.38 \pm 1.63 \times 10^{-32}$
		$Ar + RbI \underset{k_r}{\overset{k_f}{\rightleftarrows}} Ar + Rb^+ + I^-$			
3320	$1.66 \pm 1.29 \times 10^{-15}$	$1.22 \pm 0.95 \times 10^{-15}$	9.708×10^{14}	$1.71 \pm 1.33 \times 10^{-30}$	$1.24 \pm 0.98 \times 10^{-30}$
3740	$2.00 \pm 1.96 \times 10^{-15}$	$2.34 \pm 1.10 \times 10^{-15}$	5.433×10^{15}	$3.68 \pm 3.61 \times 10^{-31}$	$4.31 \pm 2.06 \times 10^{-31}$
4240	$2.52 \pm 2.62 \times 10^{-15}$	$4.10 \pm 1.67 \times 10^{-15}$	2.670×10^{16}	$9.44 \pm 9.81 \times 10^{-32}$	$1.54 \pm 0.64 \times 10^{-31}$
4840	$7.75 \pm 5.54 \times 10^{-15}$	$6.51 \pm 3.76 \times 10^{-15}$	1.151×10^{17}	$6.73 \pm 4.81 \times 10^{-32}$	$5.68 \pm 3.34 \times 10^{-32}$
5945	$1.13 \pm 1.20 \times 10^{-14}$	$1.07 \pm 1.11 \times 10^{-14}$	7.570×10^{17}	$1.49 \pm 1.59 \times 10^{-32}$	$1.39 \pm 1.46 \times 10^{-32}$
		$Ar + CsCl \underset{k_r}{\overset{k_f}{\rightleftarrows}} Ar + Cs^+ + Cl^-$			
3145	$6.12 \pm 5.00 \times 10^{-15}$	$4.56 \pm 2.34 \times 10^{-15}$	2.333×10^{14}	$2.62 \pm 2.14 \times 10^{-29}$	$1.95 \pm 1.01 \times 10^{-29}$
3300	$5.37 \pm 7.91 \times 10^{-15}$	$5.18 \pm 2.08 \times 10^{-15}$	5.228×10^{14}	$1.03 \pm 1.51 \times 10^{-29}$	$9.91 \pm 4.00 \times 10^{-30}$
4010	$7.61 \pm 1.60 \times 10^{-15}$	$7.60 \pm 2.06 \times 10^{-15}$	9.379×10^{15}	$8.11 \pm 1.71 \times 10^{-31}$	$8.15 \pm 2.21 \times 10^{-31}$
4155	$7.50 \pm 7.03 \times 10^{-15}$	$7.98 \pm 2.42 \times 10^{-15}$	1.485×10^{16}	$5.05 \pm 4.73 \times 10^{-31}$	$5.37 \pm 1.64 \times 10^{-31}$
4695	$1.12 \pm 0.47 \times 10^{-14}$	$9.02 \pm 4.32 \times 10^{-15}$	6.433×10^{16}	$1.74 \pm 0.73 \times 10^{-31}$	$1.39 \pm 0.67 \times 10^{-31}$
		$Ar + CsBr \underset{k_r}{\overset{k_f}{\rightleftarrows}} Ar + Cs^+ + Br^-$			
2990	$2.66 \pm 4.01 \times 10^{-15}$	$2.88 \pm 1.36 \times 10^{-15}$	1.759×10^{14}	$1.51 \pm 2.28 \times 10^{-29}$	$1.63 \pm 0.96 \times 10^{-29}$
3835	$6.48 \pm 1.06 \times 10^{-15}$	$6.37 \pm 1.20 \times 10^{-15}$	7.806×10^{15}	$8.30 \pm 1.36 \times 10^{-30}$	$8.20 \pm 1.89 \times 10^{-31}$
4090	$9.74 \pm 3.52 \times 10^{-15}$	$7.34 \pm 1.24 \times 10^{-15}$	1.793×10^{16}	$5.43 \pm 1.96 \times 10^{-31}$	$4.12 \pm 0.85 \times 10^{-31}$
4310	$7.45 \pm 2.78 \times 10^{-15}$	$8.10 \pm 1.39 \times 10^{-15}$	3.374×10^{16}	$2.21 \pm 0.82 \times 10^{-31}$	$2.41 \pm 0.50 \times 10^{-31}$
4840	$9.19 \pm 1.63 \times 10^{-15}$	$9.58 \pm 2.14 \times 10^{-15}$	1.216×10^{17}	$7.56 \pm 1.34 \times 10^{-32}$	$7.85 \pm 2.15 \times 10^{-32}$

TABLE V (*Continued*)

T [K]	k_f [cm^3/sec] Fit[a]	k_f [cm^3/sec] Calculated[a]	K_{eq} [cm^{-3}]	k_r [cm^6/sec] Fit[a]	k_r [cm^6/sec] Calculated[a]
			$Ar + CsI \underset{k_r}{\overset{k_f}{\rightleftarrows}} Ar + Cs^+ + I^-$		
3105	$3.73 \pm 4.96 \times 10^{-15}$	$3.01 \pm 3.26 \times 10^{-15}$	1.143×10^{15}	$3.26 \pm 4.34 \times 10^{-30}$	$2.62 \pm 2.90 \times 10^{-30}$
3590	$3.54 \pm 3.49 \times 10^{-15}$	$5.43 \pm 2.78 \times 10^{-15}$	8.796×10^{15}	$4.02 \pm 3.97 \times 10^{-31}$	$6.19 \pm 3.23 \times 10^{-31}$
3975	$6.99 \pm 2.64 \times 10^{-15}$	$7.51 \pm 2.99 \times 10^{-15}$	3.089×10^{16}	$2.26 \pm 0.86 \times 10^{-31}$	$2.44 \pm 0.99 \times 10^{-31}$
4475	$1.08 \pm 0.55 \times 10^{-14}$	$1.01 \pm 0.55 \times 10^{-14}$	1.126×10^{17}	$9.59 \pm 4.88 \times 10^{-32}$	$8.93 \pm 4.96 \times 10^{-32}$
4500	$1.11 \pm 0.29 \times 10^{-14}$	$1.02 \pm 0.57 \times 10^{-14}$	1.193×10^{17}	$9.30 \pm 2.43 \times 10^{-32}$	$8.53 \pm 4.84 \times 10^{-32}$

[a] The $\pm$ uncertainties strictly apply to the positive side only. On the negative side the uncertainty is the same percentage of the rate coefficient as on the positive side.

Reaction R7 ($Ar + MX \rightleftarrows Ar + M^0 + X^0$, the collisional dissociation to neutrals) are given in Table VIII. The equilibrium constants and rate coefficients for three-body ion-pair formation (reaction R8, $Ar + M^0 + X^0 \rightleftarrows Ar + M^+ + X^-$) are given in Table IX. In Table X are the rate coefficients for ion-pair formation if it is interpreted as a two-body process. The equilibrium constants of Table X are the same as those of Table IX, of course. Later we shall discuss whether to interpret $M^0 + X^0 \rightleftarrows M^+ + X^-$ as bimolecular or termolecular with Ar as the third body. Throughout Tables V–X, the values under "Fit" are taken from the potentiometer readings of the analog computer; the values under "Calculated" were obtained by statistical analyses of the data described below.

The next set of tables give upper bounds for four forward and four reverse reactions for which we could obtain no precise values. None of these —associative detachment, either two-body or three-body; collisional dissociation by electrons; or collisional detachment by electrons—was significant enough to play a measurable role in our observed kinetics. Table XI gives the upper bounds for dissociative attachment, $e + MX \to M^0 + X^-$, interpreted as a bimolecular process. Table XII gives upper bounds for the same process interpreted as a three-body reaction, $Ar + e + MX \to Ar + M^0 + X^-$. Table XIII shows upper bounds for the rates of the related collisional dissociation by electrons, $e + MX \to e + M^0 + X^0$. Finally among these tables of rates, Table XIV shows upper bounds for collisional detachment by electrons, $e + X^- \to 2e + X^0$.

Note that the forward reactions are all endothermic and their rate coefficients show the expected increase with temperature. The rate coefficients of the reverse reactions were computed from $k_r = k_f/K_{eq}$, and are therefore not independently measured. The estimates of uncertainties of the reverse reactions were taken as $\Delta k_r = \Delta k_f/K_{eq}$. The limits given in Tables XI–XIV were

TABLE VI
Collisional Detachment and Three-Body Attachment Rate Constants:

$$\mathrm{Ar} + \mathrm{X}^- \underset{k_r}{\overset{k_f}{\rightleftarrows}} \mathrm{Ar} + \mathrm{X}^0 + e$$

T [K]	k_f [cm^3/sec] Fit[a]	k_f [cm^3/sec] Calculated[a]	K_{eq} [cm^{-3}]	k_r [cm^6/sec] Fit[a]	k_r [cm^6/sec] Calculated[a]
		$\mathrm{Ar} + \mathrm{Cl}^- \underset{k_r}{\overset{k_f}{\rightleftarrows}} \mathrm{Ar} + \mathrm{Cl}^0 + e$			
2860	$3.54 \pm 5.00 \times 10^{-16}$	$4.21 \pm 3.22 \times 10^{-16}$	1.664×10^{15}	$2.13 \pm 3.00 \times 10^{-31}$	$2.62 \pm 2.02 \times 10^{-31}$
3145	$9.47 \pm 5.00 \times 10^{-16}$	$1.19 \pm 0.59 \times 10^{-15}$	7.321×10^{15}	$1.29 \pm 0.68 \times 10^{-31}$	$1.63 \pm 0.82 \times 10^{-31}$
3260	$2.16 \pm 5.00 \times 10^{-15}$	$1.71 \pm 0.73 \times 10^{-15}$	1.241×10^{16}	$1.74 \pm 4.03 \times 10^{-31}$	$1.38 \pm 0.59 \times 10^{-31}$
3300	$2.62 \pm 5.00 \times 10^{-15}$	$1.93 \pm 0.78 \times 10^{-15}$	1.478×10^{16}	$1.77 \pm 1.91 \times 10^{-31}$	$1.30 \pm 0.53 \times 10^{-31}$
3360	$3.16 \pm 4.15 \times 10^{-15}$	$2.31 \pm 0.86 \times 10^{-15}$	1.908×10^{16}	$1.66 \pm 2.18 \times 10^{-31}$	$1.20 \pm 0.46 \times 10^{-31}$
3765	$1.06 \pm 2.48 \times 10^{-14}$	$6.59 \pm 1.90 \times 10^{-15}$	8.728×10^{16}	$1.21 \pm 2.34 \times 10^{-31}$	$7.40 \pm 2.15 \times 10^{-32}$
4010	$1.17 \pm 0.18 \times 10^{-14}$	$1.12 \pm 0.35 \times 10^{-14}$	1.903×10^{17}	$6.15 \pm 0.95 \times 10^{-32}$	$5.80 \pm 1.84 \times 10^{-32}$
4230	$1.00 \pm 1.64 \times 10^{-14}$	$1.71 \pm 0.63 \times 10^{-14}$	3.565×10^{17}	$2.81 \pm 4.60 \times 10^{-32}$	$4.77 \pm 1.76 \times 10^{-32}$
4695	$3.61 \pm 3.46 \times 10^{-14}$	$3.69 \pm 1.89 \times 10^{-14}$	1.122×10^{18}	$3.22 \pm 3.08 \times 10^{-32}$	$3.35 \pm 1.73 \times 10^{-32}$
5810	$1.56 \pm 3.14 \times 10^{-13}$	$1.41 \pm 1.23 \times 10^{-13}$	8.736×10^{18}	$1.79 \pm 3.59 \times 10^{-32}$	$1.81 \pm 1.60 \times 10^{-32}$
		$\mathrm{Ar} + \mathrm{Br}^- \underset{k_r}{\overset{k_f}{\rightleftarrows}} \mathrm{Ar} + \mathrm{Br}^0 + e$			
2990	$1.18 \pm 4.51 \times 10^{-15}$	$1.21 \pm 1.61 \times 10^{-15}$	7.310×10^{15}	$1.61 \pm 6.17 \times 10^{-31}$	$1.74 \pm 2.29 \times 10^{-31}$
3795	$1.09 \pm 1.85 \times 10^{-14}$	$1.11 \pm 0.49 \times 10^{-14}$	1.720×10^{17}	$6.34 \pm 5.00 \times 10^{-32}$	$6.34 \pm 2.75 \times 10^{-32}$
3960	$1.25 \pm 0.88 \times 10^{-14}$	$1.57 \pm 0.59 \times 10^{-14}$	2.833×10^{17}	$4.41 \pm 0.31 \times 10^{-32}$	$5.42 \pm 2.01 \times 10^{-32}$
4370	$1.45 \pm 1.82 \times 10^{-14}$	$3.30 \pm 1.16 \times 10^{-14}$	8.411×10^{17}	$1.72 \pm 2.16 \times 10^{-32}$	$3.86 \pm 1.35 \times 10^{-32}$
4560	$7.10 \pm 4.00 \times 10^{-14}$	$4.44 \pm 1.72 \times 10^{-14}$	1.310×10^{18}	$5.46 \pm 3.08 \times 10^{-32}$	$3.37 \pm 1.30 \times 10^{-32}$
4840	$5.92 \pm 2.56 \times 10^{-14}$	$6.61 \pm 3.11 \times 10^{-14}$	2.375×10^{18}	$2.49 \pm 1.08 \times 10^{-32}$	$2.81 \pm 1.31 \times 10^{-32}$
5135	$1.59 \pm 0.78 \times 10^{-13}$	$9.58 \pm 5.52 \times 10^{-14}$	4.170×10^{18}	$3.81 \pm 1.87 \times 10^{-32}$	$2.37 \pm 1.35 \times 10^{-32}$
		$\mathrm{Ar} + \mathrm{I}^- \underset{k_r}{\overset{k_f}{\rightleftarrows}} \mathrm{Ar} + \mathrm{I}^0 + e$			
3105	$4.85 \pm 5.00 \times 10^{-15}$	$4.52 \pm 5.23 \times 10^{-15}$	3.660×10^{16}	$1.33 \pm 1.37 \times 10^{-31}$	$1.30 \pm 1.46 \times 10^{-31}$
3320	$7.18 \pm 5.00 \times 10^{-15}$	$8.27 \pm 6.88 \times 10^{-15}$	8.525×10^{16}	$8.42 \pm 5.87 \times 10^{-32}$	$9.88 \pm 8.02 \times 10^{-32}$
3590	$9.53 \pm 5.00 \times 10^{-15}$	$1.59 \pm 0.91 \times 10^{-14}$	2.155×10^{17}	$4.42 \pm 2.32 \times 10^{-32}$	$7.36 \pm 4.13 \times 10^{-32}$
3740	$3.55 \pm 5.05 \times 10^{-14}$	$2.20 \pm 1.08 \times 10^{-14}$	3.417×10^{17}	$1.04 \pm 1.48 \times 10^{-31}$	$6.37 \pm 3.04 \times 10^{-32}$
3975	$1.62 \pm 1.06 \times 10^{-14}$	$3.49 \pm 1.50 \times 10^{-14}$	6.594×10^{17}	$2.46 \pm 1.61 \times 10^{-32}$	$5.18 \pm 2.18 \times 10^{-32}$
4240	$7.80 \pm 6.22 \times 10^{-14}$	$5.50 \pm 2.50 \times 10^{-14}$	1.277×10^{18}	$6.11 \pm 4.87 \times 10^{-32}$	$4.23 \pm 1.88 \times 10^{-32}$
4500	$6.19 \pm 1.14 \times 10^{-14}$	$8.16 \pm 4.38 \times 10^{-14}$	2.276×10^{18}	$2.72 \pm 0.50 \times 10^{-32}$	$3.54 \pm 1.86 \times 10^{-32}$
4840	$1.62 \pm 1.19 \times 10^{-13}$	$1.28 \pm 0.88 \times 10^{-13}$	4.448×10^{18}	$3.64 \pm 2.68 \times 10^{-32}$	$2.89 \pm 1.94 \times 10^{-32}$
5945	$7.15 \pm 1.20 \times 10^{-13}$	$3.91 \pm 4.95 \times 10^{-13}$	2.429×10^{19}	$2.94 \pm 0.49 \times 10^{-32}$	$1.76 \pm 2.17 \times 10^{-32}$

[a] The $\pm$ uncertainties strictly apply to the positive side only. On the negative side the uncertainty is the same percentage as on the positive side.

TABLE VII

Ionization and Electronic Collisional–Radiative Recombination Rate Constants:

$$e + M^0 \underset{k_r}{\overset{k_f}{\rightleftarrows}} 2e + M^+$$

	k_f [cm^3/sec]			k_r [cm^6/sec]	
T [K]	Fit[a]	Calculated[a]	K_{eq} [cm^{-3}]	Fit[a]	Calculated[a]
			$Rb^0 + e \underset{k_r}{\overset{k_f}{\rightleftarrows}} Rb^+ + 2e$		
2860	$7.87 \pm 5.00 \times 10^{-14}$	$8.54 \pm 20.73 \times 10^{-14}$	1.961×10^{13}	$4.01 \pm 2.55 \times 10^{-27}$	$4.26 \pm 10.36 \times 10^{-27}$
3360	$1.34 \pm 5.00 \times 10^{-12}$	$1.02 \pm 1.12 \times 10^{-12}$	2.986×10^{14}	$4.49 \pm 5.00 \times 10^{-27}$	$3.37 \pm 3.70 \times 10^{-27}$
3550	$3.99 \pm 5.00 \times 10^{-12}$	$2.19 \pm 1.85 \times 10^{-12}$	6.912×10^{14}	$5.77 \pm 7.23 \times 10^{-27}$	$3.13 \pm 2.66 \times 10^{-27}$
3765	$5.01 \pm 5.00 \times 10^{-12}$	$4.71 \pm 3.16 \times 10^{-12}$	1.617×10^{15}	$3.10 \pm 3.09 \times 10^{-27}$	$2.91 \pm 1.95 \times 10^{-27}$
3960	$3.33 \pm 5.00 \times 10^{-12}$	$8.79 \pm 5.18 \times 10^{-12}$	3.230×10^{15}	$1.03 \pm 1.55 \times 10^{-27}$	$2.74 \pm 1.62 \times 10^{-27}$
4230	$8.63 \pm 9.41 \times 10^{-11}$	$1.89 \pm 1.10 \times 10^{-11}$	7.568×10^{15}	$1.14 \pm 1.24 \times 10^{-26}$	$2.55 \pm 1.48 \times 10^{-27}$
4370	$2.37 \pm 4.54 \times 10^{-11}$	$2.72 \pm 1.66 \times 10^{-11}$	1.129×10^{16}	$2.10 \pm 4.02 \times 10^{-27}$	$2.47 \pm 1.51 \times 10^{-27}$
4560	$1.56 \pm 1.75 \times 10^{-11}$	$4.29 \pm 2.91 \times 10^{-11}$	1.863×10^{16}	$8.37 \pm 9.39 \times 10^{-28}$	$2.36 \pm 1.61 \times 10^{-27}$
4840	$4.28 \pm 5.00 \times 10^{-11}$	$7.86 \pm 6.42 \times 10^{-11}$	3.611×10^{16}	$1.19 \pm 1.38 \times 10^{-27}$	$2.30 \pm 1.82 \times 10^{-27}$
5135	$7.56 \pm 5.00 \times 10^{-11}$	$1.38 \pm 1.37 \times 10^{-10}$	6.636×10^{17}	$1.14 \pm 0.75 \times 10^{-27}$	$2.11 \pm 2.09 \times 10^{-27}$
5810	$7.28 \pm 5.00 \times 10^{-10}$	$4.07 \pm 5.79 \times 10^{-10}$	2.023×10^{17}	$3.60 \pm 2.47 \times 10^{-27}$	$1.91 \pm 2.72 \times 10^{-27}$
5945	$2.43 \pm 2.05 \times 10^{-10}$	$4.91 \pm 7.43 \times 10^{-10}$	2.426×10^{17}	$1.00 \pm 0.85 \times 10^{-27}$	$1.87 \pm 2.84 \times 10^{-27}$
			$Cs^0 + e \underset{k_r}{\overset{k_f}{\rightleftarrows}} Cs^+ + 2e$		
2990	$2.57 \pm 5.00 \times 10^{-13}$	$3.59 \pm 2.81 \times 10^{-13}$	1.287×10^{14}	$2.00 \pm 3.89 \times 10^{-27}$	$2.77 \pm 2.15 \times 10^{-27}$
3145	$8.64 \pm 5.00 \times 10^{-13}$	$7.35 \pm 4.54 \times 10^{-13}$	2.879×10^{14}	$3.00 \pm 1.74 \times 10^{-27}$	$2.53 \pm 1.55 \times 10^{-27}$
3260	$4.99 \pm 5.00 \times 10^{-13}$	$1.20 \pm 0.62 \times 10^{-12}$	4.985×10^{14}	$1.00 \pm 1.00 \times 10^{-27}$	$2.39 \pm 1.23 \times 10^{-27}$
3300	$2.39 \pm 5.00 \times 10^{-12}$	$1.41 \pm 0.69 \times 10^{-12}$	5.981×10^{14}	$4.00 \pm 4.38 \times 10^{-27}$	$2.34 \pm 1.14 \times 10^{-27}$
3590	$3.97 \pm 5.00 \times 10^{-12}$	$4.07 \pm 1.41 \times 10^{-12}$	1.986×10^{15}	$2.00 \pm 2.52 \times 10^{-27}$	$2.06 \pm 0.71 \times 10^{-27}$
3835	$1.33 \pm 1.24 \times 10^{-11}$	$8.81 \pm 2.68 \times 10^{-12}$	4.748×10^{15}	$2.80 \pm 2.61 \times 10^{-27}$	$1.87 \pm 0.57 \times 10^{-27}$
3975	$1.03 \pm 1.94 \times 10^{-11}$	$1.31 \pm 0.41 \times 10^{-11}$	7.443×10^{15}	$1.38 \pm 2.61 \times 10^{-27}$	$1.78 \pm 0.55 \times 10^{-27}$
4090	$2.10 \pm 2.20 \times 10^{-11}$	$1.79 \pm 0.52 \times 10^{-11}$	1.050×10^{16}	$2.00 \pm 2.10 \times 10^{-27}$	$1.72 \pm 0.56 \times 10^{-27}$
4155	$2.53 \pm 4.15 \times 10^{-11}$	$2.11 \pm 0.72 \times 10^{-11}$	1.265×10^{16}	$2.00 \pm 3.28 \times 10^{-27}$	$1.68 \pm 0.57 \times 10^{-27}$
4310	$2.75 \pm 2.78 \times 10^{-11}$	$3.07 \pm 1.17 \times 10^{-11}$	1.923×10^{16}	$1.43 \pm 1.45 \times 10^{-27}$	$1.61 \pm 0.61 \times 10^{-27}$
4500	$3.37 \pm 4.29 \times 10^{-11}$	$4.70 \pm 2.07 \times 10^{-11}$	3.078×10^{16}	$1.09 \pm 1.39 \times 10^{-27}$	$1.52 \pm 0.67 \times 10^{-27}$
4840	$4.71 \pm 3.03 \times 10^{-11}$	$9.25 \pm 5.18 \times 10^{-11}$	6.441×10^{16}	$7.31 \pm 4.70 \times 10^{-28}$	$1.40 \pm 0.78 \times 10^{-27}$

[a] The ± uncertainties strictly apply to the positive side only. On the negative side the uncertainty is the same percentage as on the positive side.

taken from the sensitivities; the first limit corresponds to the lower limit of the temperature range, and the second to the upper limit. The bounds on the reverse rate coefficients were computed from bounds on forward rate coefficients and equilibrium constants. We should emphasize that the bounds in Tables XI–XIV may be larger, possibly much larger, than the real rate coefficients.

TABLE VIII

Collisional Dissociation into Atoms and Atomic Recombination Rate Constants:

$$Ar + MX \underset{k_r}{\overset{k_f}{\rightleftarrows}} Ar + M^0 + X^0$$

T [K]	k_f [cm^3/sec] Fit[a]	k_f [cm^3/sec] Calculated[a]	K_{eq} [cm^{-3}]	k_r [cm^6/sec] Fit[a]	k_r [cm^6/sec] Calculated[a]
		$Ar + RbCl \underset{k_r}{\overset{k_f}{\rightleftarrows}} Ar + Rb^0 + Cl^0$			
2860	$5.60 \pm 5.18 \times 10^{-16}$	$6.53 \pm 5.15 \times 10^{-16}$	2.340×10^{15}	$2.39 \pm 2.21 \times 10^{-31}$	$3.00 \pm 2.29 \times 10^{-31}$
3360	$1.58 \pm 1.87 \times 10^{-15}$	$1.14 \pm 0.50 \times 10^{-15}$	3.198×10^{16}	$4.94 \pm 5.85 \times 10^{-32}$	$3.58 \pm 1.54 \times 10^{-32}$
3765	$1.96 \pm 7.94 \times 10^{-15}$	$1.52 \pm 0.51 \times 10^{-15}$	1.585×10^{17}	$1.24 \pm 5.01 \times 10^{-32}$	$9.28 \pm 3.00 \times 10^{-32}$
3935	$1.54 \pm 5.00 \times 10^{-15}$	$1.66 \pm 0.54 \times 10^{-15}$	2.828×10^{17}	$5.45 \pm 5.00 \times 10^{-33}$	$5.66 \pm 1.77 \times 10^{-33}$
4230	$1.54 \pm 6.95 \times 10^{-15}$	$1.89 \pm 0.64 \times 10^{-15}$	6.878×10^{17}	$2.24 \pm 5.00 \times 10^{-33}$	$2.60 \pm 0.85 \times 10^{-33}$
5810	$2.69 \pm 5.00 \times 10^{-15}$	$2.48 \pm 1.61 \times 10^{-15}$	2.026×10^{19}	$1.33 \pm 2.47 \times 10^{-34}$	$1.32 \pm 0.83 \times 10^{-34}$
		$Ar + RbBr \underset{k_r}{\overset{k_f}{\rightleftarrows}} Ar + Rb^0 + Br^0$			
3795	$2.00 \pm 4.15 \times 10^{-15}$	$2.02 \pm 0.85 \times 10^{-15}$	4.169×10^{17}	$4.80 \pm 5.00 \times 10^{-33}$	$4.87 \pm 2.07 \times 10^{-33}$
3960	$2.45 \pm 4.42 \times 10^{-15}$	$2.25 \pm 0.75 \times 10^{-15}$	6.912×10^{17}	$3.54 \pm 6.39 \times 10^{-33}$	$3.25 \pm 1.10 \times 10^{-33}$
4370	$1.87 \pm 4.09 \times 10^{-15}$	$2.77 \pm 0.70 \times 10^{-15}$	2.052×10^{18}	$9.11 \pm 5.00 \times 10^{-34}$	$1.33 \pm 0.34 \times 10^{-33}$
4560	$3.59 \pm 4.75 \times 10^{-15}$	$2.99 \pm 0.81 \times 10^{-15}$	3.204×10^{18}	$1.12 \pm 1.48 \times 10^{-33}$	$9.16 \pm 2.51 \times 10^{-34}$
5000	$3.76 \pm 5.00 \times 10^{-15}$	$3.40 \pm 1.34 \times 10^{-15}$	8.009×10^{18}	$4.69 \pm 6.24 \times 10^{-34}$	$4.26 \pm 1.69 \times 10^{-34}$
5135	$2.94 \pm 5.00 \times 10^{-15}$	$3.50 \pm 1.54 \times 10^{-15}$	1.031×10^{19}	$2.85 \pm 4.85 \times 10^{-34}$	$3.44 \pm 1.52 \times 10^{-34}$
		$Ar + RbI \underset{k_r}{\overset{k_f}{\rightleftarrows}} Ar + Rb^0 + I^0$			
3320	$9.97 \pm 5.00 \times 10^{-15}$	$1.01 \pm 1.06 \times 10^{-14}$	3.347×10^{17}	$2.98 \pm 1.49 \times 10^{-32}$	$3.28 \pm 3.11 \times 10^{-32}$
3550	$1.50 \pm 5.00 \times 10^{-14}$	$1.10 \pm 0.87 \times 10^{-14}$	7.226×10^{17}	$2.08 \pm 6.92 \times 10^{-32}$	$1.63 \pm 1.13 \times 10^{-32}$
3740	$8.87 \pm 7.85 \times 10^{-15}$	$1.16 \pm 0.78 \times 10^{-14}$	1.261×10^{18}	$7.03 \pm 6.23 \times 10^{-33}$	$9.62 \pm 5.58 \times 10^{-32}$
4240	$1.70 \pm 0.98 \times 10^{-14}$	$1.25 \pm 0.80 \times 10^{-14}$	4.374×10^{18}	$3.89 \pm 2.24 \times 10^{-33}$	$2.93 \pm 1.71 \times 10^{-33}$
4840	$7.30 \pm 5.00 \times 10^{-15}$	$1.27 \pm 1.14 \times 10^{-14}$	1.418×10^{19}	$5.15 \pm 3.53 \times 10^{-34}$	$9.28 \pm 8.27 \times 10^{-34}$
5945	$1.79 \pm 3.08 \times 10^{-14}$	$1.16 \pm 1.83 \times 10^{-14}$	7.579×10^{19}	$2.36 \pm 4.06 \times 10^{-34}$	$1.86 \pm 3.05 \times 10^{-34}$
		$Ar + CsCl \underset{k_r}{\overset{k_f}{\rightleftarrows}} Ar + Cs^0 + Cl^0$			
3145	$1.04 \pm 2.50 \times 10^{-15}$	$8.97 \pm 6.15 \times 10^{-16}$	5.933×10^{15}	$1.75 \pm 4.21 \times 10^{-31}$	$1.52 \pm 1.00 \times 10^{-31}$
3260	$1.10 \pm 1.69 \times 10^{-15}$	$1.04 \pm 0.60 \times 10^{-15}$	1.064×10^{16}	$1.03 \pm 1.59 \times 10^{-31}$	$9.75 \pm 5.39 \times 10^{-32}$
3300	$8.67 \pm 7.74 \times 10^{-16}$	$1.09 \pm 0.59 \times 10^{-15}$	1.292×10^{16}	$6.71 \pm 5.99 \times 10^{-32}$	$8.41 \pm 4.40 \times 10^{-32}$
4010	$1.30 \pm 5.00 \times 10^{-15}$	$2.05 \pm 1.13 \times 10^{-15}$	2.155×10^{17}	$6.03 \pm 2.32 \times 10^{-33}$	$9.37 \pm 4.94 \times 10^{-33}$
4155	$2.86 \pm 5.00 \times 10^{-15}$	$2.24 \pm 1.41 \times 10^{-15}$	3.402×10^{17}	$8.41 \pm 5.00 \times 10^{-33}$	$6.48 \pm 3.90 \times 10^{-33}$
4695	$3.63 \pm 2.20 \times 10^{-15}$	$2.88 \pm 2.81 \times 10^{-15}$	1.513×10^{18}	$2.40 \pm 1.45 \times 10^{-33}$	$1.96 \pm 1.83 \times 10^{-33}$

TABLE VIII (*Continued*)

T [K]	k_f [cm^3/sec] Fit[a]	k_f [cm^3/sec] Calculated[a]	K_{eq} [cm^{-3}]	k_r [cm^6/sec] Fit[a]	k_r [cm^6/sec] Calculated[a]
		$Ar + CsBr \underset{k_r}{\overset{k_f}{\rightleftarrows}} Ar + Cs^0 + Br^0$			
2990	$8.86 \pm 7.52 \times 10^{-16}$	$9.00 \pm 7.70 \times 10^{-16}$	9.991×10^{15}	$8.87 \pm 7.53 \times 10^{-32}$	$9.30 \pm 7.95 \times 10^{-32}$
3835	$3.46 \pm 3.00 \times 10^{-15}$	$2.75 \pm 0.87 \times 10^{-15}$	3.203×10^{17}	$1.08 \pm 0.94 \times 10^{-32}$	$8.31 \pm 2.64 \times 10^{-33}$
4090	$2.76 \pm 4.84 \times 10^{-15}$	$3.40 \pm 1.04 \times 10^{-15}$	6.981×10^{17}	$3.95 \pm 6.93 \times 10^{-33}$	$4.74 \pm 1.45 \times 10^{-33}$
4310	$3.86 \pm 8.34 \times 10^{-15}$	$3.96 \pm 1.32 \times 10^{-15}$	1.274×10^{18}	$3.03 \pm 6.55 \times 10^{-33}$	$3.05 \pm 1.01 \times 10^{-33}$
4840	$5.51 \pm 4.19 \times 10^{-15}$	$5.23 \pm 2.43 \times 10^{-15}$	4.484×10^{18}	$1.23 \pm 0.93 \times 10^{-33}$	$1.21 \pm 0.57 \times 10^{-33}$
		$Ar + CsI \underset{k_r}{\overset{k_f}{\rightleftarrows}} Ar + Cs^0 + I^0$			
3105	$1.49 \pm 5.00 \times 10^{-15}$	$1.44 \pm 0.28 \times 10^{-15}$	1.775×10^{17}	$8.39 \pm 5.00 \times 10^{-33}$	$8.27 \pm 1.21 \times 10^{-33}$
3590	$2.21 \pm 5.00 \times 10^{-15}$	$2.18 \pm 0.22 \times 10^{-15}$	9.545×10^{17}	$2.32 \pm 5.24 \times 10^{-33}$	$2.26 \pm 0.17 \times 10^{-33}$
3975	$2.57 \pm 1.76 \times 10^{-15}$	$2.70 \pm 0.22 \times 10^{-15}$	2.737×10^{18}	$9.39 \pm 6.43 \times 10^{-34}$	$9.76 \pm 0.58 \times 10^{-34}$
4475	$3.39 \pm 5.00 \times 10^{-15}$	$3.24 \pm 0.38 \times 10^{-15}$	8.379×10^{18}	$4.05 \pm 5.97 \times 10^{-34}$	$3.92 \pm 0.34 \times 10^{-34}$
4500	$3.38 \pm 4.86 \times 10^{-15}$	$3.27 \pm 0.39 \times 10^{-15}$	8.822×10^{18}	$3.83 \pm 5.51 \times 10^{-34}$	$3.76 \pm 0.33 \times 10^{-34}$

[a]The $\pm$ uncertainties strictly apply to the positive side only. On the negative side the uncertainty is the same percentage as on the positive side.

TABLE IX

Three-Body Ion-Pair Formation and Neutralization Rate Constants:

$$Ar + M^0 + X^0 \underset{k_r}{\overset{k_f}{\rightleftarrows}} Ar + M^+ + X^-$$

T [K]	k_f [cm^6/sec] Fit[b]	k_f [cm^6/sec] Calculated[b]	K_{eq}[a]	k_r [cm^6/sec] Fit[b]	k_r [cm^6/sec] Calculated[b]
		$Ar + Rb^0 + Cl^0 \underset{k_r}{\overset{k_f}{\rightleftarrows}} Ar + Rb^+ + Cl^-$			
2860	$6.62 \pm 5.00 \times 10^{-32}$	$6.30 \pm 3.14 \times 10^{-32}$	1.179×10^{-2}	$5.61 \pm 5.00 \times 10^{-30}$	$5.00 \pm 3.08 \times 10^{-30}$
3360	$5.73 \pm 5.36 \times 10^{-32}$	$7.13 \pm 2.04 \times 10^{-32}$	1.565×10^{-2}	$3.66 \pm 3.42 \times 10^{-30}$	$4.56 \pm 1.59 \times 10^{-30}$
3765	$1.01 \pm 1.74 \times 10^{-31}$	$7.69 \pm 1.79 \times 10^{-32}$	1.853×10^{-2}	$5.45 \pm 9.39 \times 10^{-30}$	$4.31 \pm 1.22 \times 10^{-30}$
3935	$7.34 \pm 5.00 \times 10^{-32}$	$7.90 \pm 1.86 \times 10^{-32}$	1.960×10^{-2}	$3.74 \pm 8.52 \times 10^{-30}$	$4.23 \pm 1.21 \times 10^{-30}$
4230	$7.34 \pm 9.32 \times 10^{-32}$	$8.24 \pm 2.16 \times 10^{-32}$	2.123×10^{-2}	$3.46 \pm 4.39 \times 10^{-30}$	$4.10 \pm 1.31 \times 10^{-30}$
5810	$1.05 \pm 5.23 \times 10^{-31}$	$9.60 \pm 4.93 \times 10^{-32}$	2.316×10^{-2}	$4.53 \pm 5.00 \times 10^{-30}$	$3.66 \pm 2.33 \times 10^{-30}$

TABLE IX (*Continued*)

T [K]	k_f [cm^6/sec] Fit[b]	k_f [cm^6/sec] Calculated[b]	K_{eq}[a]	k_r [cm^6/sec] Fit[b]	k_r [cm^6/sec] Calculated[b]
			$Ar + Rb^0 + Br^0 \underset{k_r}{\overset{k_f}{\rightleftarrows}} Ar + Rb^+ + Br^-$		
3795	$5.32 \pm 6.56 \times 10^{-32}$	$5.11 \pm 1.28 \times 10^{-32}$	1.050×10^{-2}	$5.07 \pm 6.25 \times 10^{-30}$	$4.82 \pm 1.37 \times 10^{-30}$
3960	$5.64 \pm 7.12 \times 10^{-32}$	$5.34 \pm 1.07 \times 10^{-32}$	1.140×10^{-2}	$4.95 \pm 6.25 \times 10^{-30}$	$4.71 \pm 1.07 \times 10^{-30}$
4370	$4.66 \pm 6.62 \times 10^{-32}$	$5.87 \pm 0.92 \times 10^{-32}$	1.342×10^{-2}	$3.47 \pm 4.93 \times 10^{-30}$	$4.47 \pm 0.80 \times 10^{-30}$
4560	$5.63 \pm 8.60 \times 10^{-32}$	$6.10 \pm 1.05 \times 10^{-32}$	1.422×10^{-2}	$3.96 \pm 6.05 \times 10^{-30}$	$4.38 \pm 0.85 \times 10^{-30}$
5000	$6.71 \pm 8.66 \times 10^{-32}$	$6.60 \pm 1.63 \times 10^{-32}$	1.563×10^{-2}	$4.29 \pm 5.54 \times 10^{-30}$	$4.20 \pm 1.18 \times 10^{-30}$
5135	$7.72 \pm 5.00 \times 10^{-32}$	$6.74 \pm 1.85 \times 10^{-32}$	1.591×10^{-2}	$4.85 \pm 7.23 \times 10^{-30}$	$4.15 \pm 1.30 \times 10^{-30}$
			$Ar + Rb^0 + I^0 \underset{k_r}{\overset{k_f}{\rightleftarrows}} Ar + Rb^+ + I^-$		
3320	$1.87 \pm 5.00 \times 10^{-32}$	$2.06 \pm 0.65 \times 10^{-32}$	2.901×10^{-3}	$6.45 \pm 5.00 \times 10^{-30}$	$6.78 \pm 2.24 \times 10^{-30}$
3550	$2.97 \pm 5.00 \times 10^{-32}$	$2.86 \pm 0.69 \times 10^{-32}$	3.650×10^{-3}	$8.14 \pm 5.00 \times 10^{-30}$	$7.75 \pm 1.97 \times 10^{-30}$
3740	$3.55 \pm 5.00 \times 10^{-32}$	$3.63 \pm 0.74 \times 10^{-32}$	4.308×10^{-3}	$8.24 \pm 5.00 \times 10^{-30}$	$8.54 \pm 1.83 \times 10^{-30}$
4240	$7.58 \pm 5.00 \times 10^{-32}$	$6.15 \pm 1.19 \times 10^{-32}$	6.105×10^{-3}	$1.24 \pm 5.00 \times 10^{-29}$	$1.06 \pm 0.22 \times 10^{-29}$
4840	$8.55 \pm 5.00 \times 10^{-32}$	$1.00 \pm 0.27 \times 10^{-31}$	8.117×10^{-3}	$1.05 \pm 5.70 \times 10^{-29}$	$1.30 \pm 0.37 \times 10^{-29}$
5945	$1.96 \pm 5.00 \times 10^{-31}$	$1.91 \pm 0.86 \times 10^{-31}$	9.990×10^{-3}	$1.96 \pm 5.00 \times 10^{-29}$	$1.69 \pm 0.80 \times 10^{-29}$
			$Ar + Cs^0 + Cl^0 \underset{k_r}{\overset{k_f}{\rightleftarrows}} Ar + Cs^+ + Cl^-$		
3145	$1.43 \pm 5.09 \times 10^{-31}$	$1.35 \pm 0.50 \times 10^{-31}$	3.932×10^{-2}	$3.64 \pm 5.00 \times 10^{-30}$	$3.39 \pm 1.43 \times 10^{-30}$
3260	$1.59 \pm 5.00 \times 10^{-31}$	$1.36 \pm 0.42 \times 10^{-31}$	4.018×10^{-2}	$3.96 \pm 5.00 \times 10^{-30}$	$3.37 \pm 1.19 \times 10^{-30}$
3300	$1.22 \pm 3.79 \times 10^{-31}$	$1.36 \pm 0.40 \times 10^{-31}$	4.047×10^{-2}	$3.01 \pm 5.00 \times 10^{-30}$	$3.36 \pm 1.12 \times 10^{-30}$
4010	$1.40 \pm 0.53 \times 10^{-31}$	$1.39 \pm 0.28 \times 10^{-31}$	4.354×10^{-2}	$3.22 \pm 1.22 \times 10^{-30}$	$3.25 \pm 0.75 \times 10^{-30}$
4155	$1.27 \pm 1.22 \times 10^{-31}$	$1.40 \pm 0.32 \times 10^{-31}$	4.365×10^{-2}	$2.91 \pm 2.79 \times 10^{-30}$	$3.23 \pm 0.84 \times 10^{-30}$
4695	$1.87 \pm 2.46 \times 10^{-31}$	$1.41 \pm 0.50 \times 10^{-31}$	4.252×10^{-2}	$4.40 \pm 5.79 \times 10^{-30}$	$3.18 \pm 1.27 \times 10^{-30}$
			$Ar + Cs^0 + Br^0 \underset{k_r}{\overset{k_f}{\rightleftarrows}} Ar + Cs^+ + Br^-$		
2990	$8.84 \pm 5.00 \times 10^{-32}$	$9.25 \pm 5.41 \times 10^{-32}$	1.761×10^{-2}	$5.02 \pm 2.84 \times 10^{-30}$	$5.03 \pm 3.05 \times 10^{-30}$
3835	$1.62 \pm 0.77 \times 10^{-31}$	$1.37 \pm 0.31 \times 10^{-31}$	2.437×10^{-2}	$6.65 \pm 3.16 \times 10^{-30}$	$5.83 \pm 1.38 \times 10^{-30}$
4090	$1.28 \pm 0.63 \times 10^{-31}$	$1.49 \pm 0.32 \times 10^{-31}$	2.569×10^{-2}	$4.98 \pm 2.45 \times 10^{-30}$	$6.02 \pm 1.33 \times 10^{-30}$
4310	$1.47 \pm 1.67 \times 10^{-31}$	$1.59 \pm 0.36 \times 10^{-31}$	2.649×10^{-2}	$5.55 \pm 6.30 \times 10^{-30}$	$6.17 \pm 1.44 \times 10^{-30}$
4840	$1.84 \pm 4.19 \times 10^{-31}$	$1.82 \pm 0.55 \times 10^{-31}$	2.712×10^{-2}	$6.78 \pm 5.00 \times 10^{-30}$	$6.49 \pm 2.04 \times 10^{-30}$
			$Ar + Cs^0 + I^0 \underset{k_r}{\overset{k_f}{\rightleftarrows}} Ar + Cs^+ + I^-$		
3105	$3.36 \pm 8.46 \times 10^{-32}$	$3.92 \pm 2.42 \times 10^{-32}$	6.439×10^{-3}	$5.22 \pm 5.00 \times 10^{-30}$	$5.95 \pm 3.27 \times 10^{-30}$
3590	$6.81 \pm 5.00 \times 10^{-32}$	$5.32 \pm 1.60 \times 10^{-32}$	9.217×10^{-3}	$7.39 \pm 5.42 \times 10^{-30}$	$5.86 \pm 1.58 \times 10^{-30}$
3975	$6.43 \pm 5.66 \times 10^{-32}$	$6.43 \pm 1.54 \times 10^{-32}$	1.129×10^{-2}	$5.70 \pm 5.01 \times 10^{-30}$	$5.81 \pm 1.25 \times 10^{-30}$
4475	$7.52 \pm 5.00 \times 10^{-32}$	$7.83 \pm 2.68 \times 10^{-32}$	1.344×10^{-2}	$5.60 \pm 5.00 \times 10^{-30}$	$5.75 \pm 1.76 \times 10^{-30}$
4500	$7.51 \pm 5.00 \times 10^{-32}$	$7.90 \pm 2.76 \times 10^{-32}$	1.352×10^{-2}	$5.55 \pm 5.00 \times 10^{-30}$	$5.75 \pm 1.80 \times 10^{-30}$

[a] Dimensionless.

[b] The $\pm$ uncertainties strictly apply to the positive side only. On the negative side the uncertainty is the same percentage as on the positive side.

TABLE X

Two-Body Ion-Pair Formation and Ion–Ion Mutual Neutralization Rate Constants:

$$M^0 + X^0 \underset{k_r}{\overset{k_f}{\rightleftarrows}} M^+ + X^-$$

	k_f [cm^3/sec]			k_r [cm^3/sec]	
[K]	Fit[b]	Calculated[b]	K_{eq}[a]	Fit[b]	Calculated[b]
			$Rb^0 + Cl^0 \underset{k_r}{\overset{k_f}{\rightleftarrows}} Rb^+ + Cl^-$		
360	$1.89 \pm 7.75 \times 10^{-12}$	$1.75 \pm 0.98 \times 10^{-12}$	1.179×10^{-2}	$1.60 \pm 6.57 \times 10^{-10}$	$1.39 \pm 1.10 \times 10^{-10}$
360	$2.00 \pm 1.87 \times 10^{-12}$	$1.99 \pm 0.64 \times 10^{-12}$	1.565×10^{-2}	$1.28 \pm 1.19 \times 10^{-10}$	$1.28 \pm 0.56 \times 10^{-10}$
765	$2.59 \pm 4.46 \times 10^{-12}$	$2.16 \pm 0.56 \times 10^{-12}$	1.853×10^{-2}	$1.40 \pm 2.41 \times 10^{-10}$	$1.21 \pm 0.43 \times 10^{-10}$
935	$1.90 \pm 4.32 \times 10^{-12}$	$2.23 \pm 0.59 \times 10^{-12}$	1.960×10^{-2}	$9.69 \pm 5.00 \times 10^{-11}$	$1.19 \pm 0.43 \times 10^{-10}$
230	$1.93 \pm 2.45 \times 10^{-12}$	$2.33 \pm 0.68 \times 10^{-12}$	2.123×10^{-2}	$9.09 \pm 5.00 \times 10^{-11}$	$1.16 \pm 0.46 \times 10^{-10}$
310	$3.50 \pm 5.00 \times 10^{-12}$	$2.75 \pm 1.59 \times 10^{-12}$	2.316×10^{-2}	$1.51 \pm 5.00 \times 10^{-10}$	$1.05 \pm 0.86 \times 10^{-10}$
			$Rb^0 + Br^0 \underset{k_r}{\overset{k_f}{\rightleftarrows}} Rb^+ + Br^-$		
795	$1.50 \pm 1.85 \times 10^{-12}$	$1.51 \pm 0.16 \times 10^{-12}$	1.050×10^{-2}	$1.43 \pm 1.76 \times 10^{-10}$	$1.43 \pm 0.18 \times 10^{-10}$
960	$1.75 \pm 2.21 \times 10^{-12}$	$1.59 \pm 0.14 \times 10^{-12}$	1.140×10^{-2}	$1.54 \pm 1.94 \times 10^{-10}$	$1.40 \pm 0.15 \times 10^{-10}$
370	$1.60 \pm 2.27 \times 10^{-12}$	$1.75 \pm 0.12 \times 10^{-12}$	1.342×10^{-2}	$1.19 \pm 1.69 \times 10^{-10}$	$1.34 \pm 0.11 \times 10^{-10}$
560	$1.80 \pm 2.75 \times 10^{-12}$	$1.83 \pm 0.14 \times 10^{-12}$	1.422×10^{-2}	$1.27 \pm 1.93 \times 10^{-10}$	$1.31 \pm 0.12 \times 10^{-10}$
00	$2.00 \pm 2.58 \times 10^{-12}$	$1.99 \pm 0.21 \times 10^{-12}$	1.563×10^{-2}	$1.28 \pm 1.65 \times 10^{-10}$	$1.27 \pm 0.16 \times 10^{-10}$
135	$2.10 \pm 3.13 \times 10^{-12}$	$2.03 \pm 0.24 \times 10^{-12}$	1.591×10^{-2}	$1.32 \pm 1.97 \times 10^{-10}$	$1.25 \pm 0.18 \times 10^{-10}$
			$Rb^0 + I^0 \underset{k_r}{\overset{k_f}{\rightleftarrows}} Rb^+ + I^-$		
320	$6.50 \pm 5.00 \times 10^{-13}$	$6.39 \pm 0.84 \times 10^{-13}$	2.901×10^{-3}	$2.24 \pm 5.00 \times 10^{-10}$	$2.10 \pm 0.35 \times 10^{-10}$
550	$9.00 \pm 5.00 \times 10^{-13}$	$8.80 \pm 0.90 \times 10^{-13}$	3.650×10^{-3}	$2.47 \pm 5.00 \times 10^{-10}$	$2.39 \pm 0.31 \times 10^{-10}$
740	$1.00 \pm 5.00 \times 10^{-12}$	$1.11 \pm 0.10 \times 10^{-12}$	4.308×10^{-3}	$2.32 \pm 5.00 \times 10^{-10}$	$2.62 \pm 0.29 \times 10^{-10}$
240	$2.00 \pm 5.00 \times 10^{-12}$	$1.86 \pm 0.15 \times 10^{-12}$	6.105×10^{-3}	$3.28 \pm 5.00 \times 10^{-10}$	$3.22 \pm 0.34 \times 10^{-10}$
340	$3.00 \pm 5.00 \times 10^{-12}$	$3.01 \pm 0.34 \times 10^{-12}$	8.117×10^{-3}	$3.70 \pm 5.00 \times 10^{-10}$	$3.89 \pm 0.57 \times 10^{-10}$
945	$5.50 \pm 5.00 \times 10^{-12}$	$5.64 \pm 1.02 \times 10^{-12}$	9.990×10^{-3}	$5.51 \pm 5.00 \times 10^{-10}$	$5.00 \pm 1.17 \times 10^{-10}$
			$Cs^0 + Cl^0 \underset{k_r}{\overset{k_f}{\rightleftarrows}} Cs^+ + Cl^-$		
145	$3.50 \pm 5.00 \times 10^{-12}$	$3.47 \pm 0.45 \times 10^{-12}$	3.932×10^{-2}	$8.90 \pm 5.00 \times 10^{-11}$	$8.72 \pm 1.52 \times 10^{-11}$
260	$3.54 \pm 5.00 \times 10^{-12}$	$3.57 \pm 0.39 \times 10^{-12}$	4.018×10^{-2}	$8.81 \pm 5.00 \times 10^{-11}$	$8.85 \pm 1.32 \times 10^{-11}$
300	$3.75 \pm 5.00 \times 10^{-12}$	$3.60 \pm 0.38 \times 10^{-12}$	4.047×10^{-2}	$9.27 \pm 5.00 \times 10^{-11}$	$8.89 \pm 1.25 \times 10^{-11}$
910	$4.00 \pm 1.51 \times 10^{-12}$	$4.08 \pm 0.30 \times 10^{-12}$	4.354×10^{-2}	$9.19 \pm 3.47 \times 10^{-11}$	$9.53 \pm 0.95 \times 10^{-11}$
155	$4.01 \pm 3.83 \times 10^{-12}$	$4.16 \pm 0.34 \times 10^{-12}$	4.365×10^{-2}	$9.20 \pm 8.77 \times 10^{-11}$	$9.63 \pm 1.07 \times 10^{-11}$
595	$5.02 \pm 6.61 \times 10^{-12}$	$4.44 \pm 0.55 \times 10^{-12}$	4.252×10^{-2}	$1.18 \pm 1.55 \times 10^{-10}$	$9.98 \pm 1.66 \times 10^{-11}$

TABLE X (*Continued*)

T [K]	k_f [cm^3/sec] Fit[b]	k_f [cm^3/sec] Calculated[b]	K_{eq}[a]	k_r [cm^3/sec] Fit[b]	k_r [cm^3/sec] Calculated[b]
		$Cs^0 + Br^0 \underset{k_r}{\overset{k_f}{\rightleftarrows}} Cs^+ + Br^-$			
2990	$2.99 \pm 7.02 \times 10^{-12}$	$2.86 \pm 0.60 \times 10^{-12}$	1.761×10^{-2}	$1.70 \pm 3.99 \times 10^{-10}$	$1.56 \pm 0.62 \times 10^{-10}$
3835	$3.75 \pm 1.77 \times 10^{-12}$	$3.86 \pm 0.35 \times 10^{-12}$	2.437×10^{-2}	$1.54 \pm 0.73 \times 10^{-10}$	$1.65 \pm 0.27 \times 10^{-10}$
4090	$4.00 \pm 1.98 \times 10^{-12}$	$4.12 \pm 0.35 \times 10^{-12}$	2.569×10^{-2}	$1.56 \pm 0.77 \times 10^{-10}$	$1.67 \pm 0.25 \times 10^{-10}$
4310	$4.00 \pm 4.55 \times 10^{-12}$	$4.33 \pm 0.38 \times 10^{-12}$	2.649×10^{-2}	$1.51 \pm 1.72 \times 10^{-10}$	$1.68 \pm 0.27 \times 10^{-10}$
4840	$5.00 \pm 5.00 \times 10^{-12}$	$4.80 \pm 0.56 \times 10^{-12}$	2.712×10^{-2}	$1.84 \pm 4.20 \times 10^{-10}$	$1.71 \pm 0.37 \times 10^{-10}$
		$Cs^0 + I^0 \underset{k_r}{\overset{k_f}{\rightleftarrows}} Cs^+ + I^-$			
3105	$9.00 \pm 5.00 \times 10^{-13}$	$1.03 \pm 0.58 \times 10^{-12}$	6.439×10^{-3}	$1.40 \pm 3.52 \times 10^{-10}$	$1.56 \pm 0.70 \times 10^{-10}$
3590	$1.54 \pm 7.97 \times 10^{-12}$	$1.43 \pm 0.40 \times 10^{-12}$	9.217×10^{-3}	$1.67 \pm 8.65 \times 10^{-10}$	$1.58 \pm 0.35 \times 10^{-10}$
3975	$2.00 \pm 1.76 \times 10^{-12}$	$1.76 \pm 0.39 \times 10^{-12}$	1.129×10^{-2}	$1.77 \pm 1.56 \times 10^{-10}$	$1.59 \pm 0.29 \times 10^{-10}$
4475	$2.00 \pm 4.92 \times 10^{-12}$	$2.18 \pm 0.69 \times 10^{-12}$	1.344×10^{-2}	$1.49 \pm 3.66 \times 10^{-10}$	$1.60 \pm 0.41 \times 10^{-10}$
4500	$2.00 \pm 4.00 \times 10^{-12}$	$2.20 \pm 0.71 \times 10^{-12}$	1.352×10^{-2}	$1.48 \pm 2.96 \times 10^{-10}$	$1.60 \pm 0.42 \times 10^{-10}$

[a] Dimensionless.

[b] The $\pm$ uncertainties strictly apply to the positive side only. On the negative side the uncertainty is the same percentage as on the positive side.

TABLE XI

Upper Bounds on the Two-Body Dissociative Attachment and Associative Detachment Rate Constants:

$$e + MX \underset{k_r}{\overset{k_f}{\rightleftarrows}} M^0 + X^-$$

Species	k_f [cm^3/sec]	k_r [cm^3/sec]
RbCl	$< 0.3\text{–}1 \times 10^{-11}$	$< 2\text{–}4 \times 10^{-12}$
RbBr	$< 3\text{–}5 \times 10^{-12}$	$< 1\text{–}2 \times 10^{-12}$
RbI	$< 2\text{–}1 \times 10^{-11}$	$< 5\text{–}3 \times 10^{-12}$
CsCl	$< 2\text{–}0.2 \times 10^{-10}$	$< 2\text{–}0.1 \times 10^{-10}$
CsBr	$< 2\text{–}1 \times 10^{-12}$	$< 1\text{–}0.5 \times 10^{-12}$
CsI	$< 2\text{–}0.3 \times 10^{-11}$	$< 4\text{–}0.8 \times 10^{-12}$

TABLE XII

Upper Bounds on the Three-Body Dissociative Attachment and Associative Detachment Rate Constants:

$$Ar + e + MX \underset{k_r}{\overset{k_f}{\rightleftarrows}} Ar + M^0 + X^-$$

Species	k_f [cm^6/sec]	k_r [cm^6/sec]
RbCl	$< 2\text{–}5 \times 10^{-31}$	$< 1\text{–}2 \times 10^{-31}$
RbBr	$< 2\text{–}3 \times 10^{-31}$	$< 0.8\text{–}1 \times 10^{-31}$
RbI	$< 1\text{–}5 \times 10^{-31}$	$< 0.3\text{–}2 \times 10^{-31}$
CsCl	$< 1\text{–}0.1 \times 10^{-29}$	$< 1\text{–}0.8 \times 10^{-30}$
CsBr	$< 1\text{–}5 \times 10^{-32}$	$< 0.7\text{–}3 \times 10^{-32}$
CsI	$< 1\text{–}0.2 \times 10^{-30}$	$< 2\text{–}0.5 \times 10^{-31}$

TABLE XIII

Upper Bounds on the Electron Collisional Dissociation into Atoms and Atomic Recombination Rate Constants:

$$e + \mathrm{MX} \underset{k_r}{\overset{k_f}{\rightleftarrows}} e + \mathrm{M}^0 + \mathrm{X}^0$$

Species	k_f [cm³/sec]	k_r [cm⁶/sec]
RbCl	$< 0.3\text{–}2\times10^{-11}$	$<1\times10^{-27}\text{–}1\times10^{-30}$
RbBr	$< 4\text{–}0.5\times10^{-12}$	$<1\times10^{-29}\text{–}5\times10^{-32}$
RbI	$< 2\text{–}0.2\times10^{-10}$	$<6\times10^{-28}\text{–}3\times10^{-31}$
CsCl	$<1\text{–}2\times10^{-11}$	$<2\times10^{-27}\text{–}1\times10^{-29}$
CsBr	$<2\text{–}3\times10^{-12}$	$<2\times10^{-28}\text{–}7\times10^{-31}$
CsI	$<1\text{–}2\times10^{-11}$	$<6\times10^{-29}\text{–}2\times10^{-30}$

TABLE XIV

Upper Bounds on the Electron Collisional Detachment and Three-Body Attachment Rate Constants:

$$e + \mathrm{X}^- \underset{k_r}{\overset{k_f}{\rightleftarrows}} 2e + \mathrm{X}^0$$

Species	k_f [cm³/sec]	k_r [cm⁶/sec]
Cl^-	$<1\text{–}5\times10^{-11}$	$<5\times10^{-27}\text{–}5\times10^{-29}$
Br^-	$<1\text{–}15\times10^{-11}$	$<1\times10^{-27}\text{–}4\times10^{-29}$
I^-	$<1\text{–}20\times10^{-11}$	$<3\times10^{-28}\text{–}1\times10^{-29}$

B. Arrhenius Graphs and Parameters

The rate coefficients determined from the computer fits to the data are shown, with their total uncertainties, in the Arrhenius plots that follow. These rate coefficients were used to fit Arrhenius curves of the form

$$\ln(k_l T^n) = \ln A_l + \frac{E_{al}}{RT} \tag{4.1}$$

by a weighted least squares method. Here, A_l and E_{al} refer to the Arrhenius constant and energy of activation for the lth reaction. The exponent n was taken as zero or as the ratio of small whole numbers, depending on the approximate preexponential temperature dependence expected for each particular process. The weighting factors were the chi-squares χ^2 calculated by comparing the data points with the fitted analog curves. We now describe briefly how the statistical analysis was done; a more detailed discussion is given in the Appendix.

The weighting factors were these:

$$\omega^{(1)} = \left[\chi^2(\mathrm{M}^0)\right]^{-1/2} \tag{4.2}$$

for processes with M^0 but not X^- in the equations,

$$\omega^{(2)} = \left[\chi^2(\mathrm{X}^-)\right]^{-1/2} \tag{4.3}$$

for processes with X^- but not M^0 in the equations, and

$$\omega^{(3)} = \left[\chi^2(\mathrm{M}^0)\chi^2(\mathrm{X}^-)\right]^{-1/4} \tag{4.4}$$

for processes containing both M^0 and X^- in their equations. For convenience, we define the variables for the lth process:

$$X_i = T_i^{-1}$$
$$Y_i = \ln(k_{li}T_i^{\,n})$$
$$m = E_{al}/R$$
$$b = \ln A_l$$
$$\omega_i = \omega_i^{(j)}$$

with $j=1$, 2, or 3 as in Eqs. (4.2), (4.3), and (4.4). The suffix i refers to the ith data point, that is, the ith temperature and rate coefficient. Then the activation energy and preexponential factor are given by the equations[44,45]

$$m = \frac{\left(\dfrac{\Sigma\omega_i X_i Y_i}{\Sigma\omega_i}\right) - \left(\dfrac{\Sigma\omega_i X_i}{\Sigma\omega_i}\right)\left(\dfrac{\Sigma\omega_i Y_i}{\Sigma\omega_i}\right)}{\left(\dfrac{\Sigma\omega_i X_i^2}{\Sigma\omega_i}\right) - \left(\dfrac{\Sigma\omega_i X_i}{\Sigma\omega_i}\right)^2} \tag{4.5}$$

and

$$b = \frac{\Sigma\omega_i Y_i}{\Sigma\omega_i} - m\frac{\Sigma\omega_i X_i}{\Sigma\omega_i} \tag{4.6}$$

We assume the uncertainties of m and b are distributed normally about the least-squares fit. This allows us to use the t statistic $t(N-2,0.95)$ to calculate the estimated uncertainty at the 95% confidence level; N is the number of points used in making the fit. These uncertainties E are

$$E(m) = \pm t(N-2,0.95)s \tag{4.7}$$

and

$$E(b) = E(m)\left[\frac{\Sigma_i \omega_i X_i^2}{\Sigma \omega_i}\right]^{1/2} \tag{4.8}$$

where

$$s = \left[\frac{\dfrac{\Sigma\omega_i Y_i^2}{\Sigma\omega_i} - \left(\dfrac{\Sigma\omega_i Y_i}{\Sigma\omega_i}\right)^2 - m^2 D}{D(N-2)}\right]^{1/2} \tag{4.9}$$

and

$$D = \frac{\Sigma\omega_i X_i^2}{\Sigma\omega_i} - \left(\frac{\Sigma\omega_i X_i}{\Sigma\omega_i}\right)^2 \tag{4.10}$$

The rate constant of each particular experimental temperature and the estimate of its uncertainty at the 95% confidence level were calculated from

$$\hat{Y}_j = b + m\hat{X}_j \tag{4.11}$$

and

$$E(\hat{Y}_j) = E(m)\left[D + \left(\hat{X}_j - \frac{\Sigma\omega_i X_i}{\Sigma\omega_i}\right)^2\right]^{1/2} \tag{4.12}$$

where $\hat{Y}_j$ refers to the value of the dependent variable calculated from the fitted equation at the jth value of the independent variable, $\hat{X}_j$.

Tables V–X show that the uncertainties in the analog computations based on the sensitivities of the computer fits are generally larger than the uncertainties of the rate constants calculated as just described. Moreover the ranges of the two uncertainties overlap in most cases.

The Arrhenius parameters $\log_{10}A$ and the activation energy E_a and their 95% confidence intervals were obtained from the least-squares fitting. These parameters and their uncertainties are collected in Tables XV–XXVI for forward and reverse reactions. For those reactions whose temperature exponent n was different from zero, the value of n is included. The enthalpies of reaction are also given for the exothermic process, to make it easy to compare the phenomenological activation energies with the heats of reaction. The activation energies for collisional dissociation to atoms and to ions are, not

TABLE XV

Comparison of Arrhenius Parameters for Ionization:

$$M^0 + e \xrightarrow{k} M^+ + 2e$$

Species	Source	$\log_{10} A$	A [cm^3/sec]	E_a [eV]	ΔH [eV]
Rb	W[a]	-5.824 ± 1.150	1.50×10^{-6}	4.11 ± 0.93	4.18
	M[b]	-7.699	2.00×10^{-8}	2.11	
	AH[c]	-8.4 ± 0.3	8.6×10^{-9}	4.05 ± 0.15	
Cs	W[a]	-6.137 ± 0.774	7.30×10^{-7}	3.74 ± 0.59	3.89
	M[b]	-6.462	3.45×10^{-7}	2.61	
	KZ[d]	-3.801	1.58×10^{-4}	3.97	
	AH[c]	-7.8 ± 0.3	1.04×10^{-8}	3.98 ± 0.13	

[a] This work.

[b] Ref. 11.

[c] Ref. 14. The reaction referred to in this work is $Ar + M^0 \rightarrow Ar + M^+ + e$. The preexponential factor A has units of $[cm^{-3} \text{ sec } K^{1/2}]^{-1}$, since there is an explicit temperature dependence of $T^{1/2}$ in the expression for the rate constant.

[d] Calculated from data contained in Ref. 46.

TABLE XVI

Comparison of Arrhenius Parameters for Electronic Collisional–Radiative Recombination:

$$2e + M^+ \xrightarrow{k} e + M^0$$

Species	Source	$\log_{10} A$	A [cm^6/sec]	E_a [eV]	ΔH [eV]
Rb	W[a]	-27.058 ± 1.151	8.74×10^{-28}	-0.39 ± 0.93	< 0
	BKM[b]	-27.138	7.28×10^{-28}	-1.83	
	AH[c]	-23.45 ± 0.3	3.6×10^{-24}	0	
Cs	W[a]	-27.329 ± 0.769	4.69×10^{-28}	-0.46 ± 0.58	< 0
	BKM[b]	-27.178	6.64×10^{-28}	-1.85	
	AH[c]	-23.37 ± 0.3	4.3×10^{-24}	0	
	KZ[d]	—	—	—	

[a] This work.

[b] Calculated from data contained in Ref. 47.

[c] Ref. 14. The reaction referred to in this work is $Ar + M^+ + e \rightarrow Ar + M^0$. The preexponential factor A has units of cm^6/K sec, since there is an explicit temperature dependence of T^{-1} in the expression for the rate constant.

[d] Ref. 46; data give a nonlinear Arrhenius graph.

TABLE XVII
Comparison of Arrhenius Parameters for Collisional Detachment:
$$Ar+X^- \xrightarrow{k} Ar+X^0+e$$

Species	Source	$\log_{10}A$	A [cm³/sec]	E_a [eV]	ΔH [eV]
Cl	W[a]	-10.405 ± 0.716	3.93×10^{-11}	2.82 ± 0.53	3.62
	Ma[b]	-9.921	1.2×10^{-10}	3.62	
	LTW[c]	-9.481	3.3×10^{-10}	3.60	
	Mi[d]	-9.016	9.64×10^{-10}	3.81	
Br	W[a]	-10.371 ± 0.851	4.25×10^{-11}	2.70 ± 0.70	3.36
	Ma[b]	-10.032	9.3×10^{-11}	3.36	
	LTW[c]	-9.456	3.5×10^{-10}	3.38	
	Mi[d]	-10.445	3.59×10^{-11}	2.58	
I	W[a]	-10.291 ± 1.006	5.12×10^{-11}	2.50 ± 0.79	3.06
	Ma[b]	-10.155	7.0×10^{-11}	3.06	
	LTW[c]	-8.921	1.2×10^{-9}	3.06	
	Mi[d]	-10.455	3.51×10^{-11}	2.94	

[a] This work.
[b] Calculated from data contained in Ref. 6.
[c] Calculated from data contained in Ref. 2.
[d] Calculated from data contained in Ref. 11.

surprisingly, smaller than the corresponding dissociation energies. More striking is the comparison in Table XVII of the values of E_a and ΔH: here, for collisional detachment, the activation energies are also somewhat smaller than the electron affinities of the halogens. We shall return to this point later, when we interpret the results. These tables will be discussed in detail in the next section.

The rate coefficients can most easily be grasped as a collective entity from their Arrhenius plots. These are shown in Figs. 6–28 for all the forward re-

TABLE XVIII
Arrhenius Parameters for Three-Body Attachment:
$$Ar+X^0+e \xrightarrow{k} Ar+X^-$$

Species	$\log_{10}A$	A [cm⁶/sec]	E_a [eV]	ΔH [eV]
Cl	-32.86 ± 0.72	1.36×10^{-33}	-1.30 ± 0.53	<0
Br	-32.83 ± 0.84	1.47×10^{-33}	-1.23 ± 0.70	<0
I	-32.70 ± 0.99	1.97×10^{-33}	-1.12 ± 0.78	<0

TABLE XIX
Comparison of Arrhenius Parameters for Two-Body Ion-Pair Formation:
$M^0 + X^0 \xrightarrow{k} M^+ + X^-$

Species	Source	$\log_{10} A$	A [cm³/sec]	E_a [eV]	ΔH [eV]
RbCl	W[a]	-11.371 ± 0.507	4.26×10^{-12}	0.220 ± 0.376	0.562
	BH[b]	< -13.000	$< 1.0 \times 10^{-13}$	0.560	
RbBr	W[a]	-11.329 ± 0.248	4.69×10^{-12}	0.370 ± 0.211	0.813
	BH[b]	-12.959	1.1×10^{-13}	0.808	
RbI	W[a]	-10.053 ± 0.202	8.84×10^{-11}	1.41 ± 0.16	1.12
	BH[b]	-11.721	1.9×10^{-12}	1.11	
CsCl	W[a]	-11.136 ± 0.217	7.31×10^{-12}	0.202 ± 0.161	0.279
	BH[b]	< -13.000	$< 1.0 \times 10^{-13}$	0.269	
CsBr	W[a]	-10.956 ± 0.210	1.11×10^{-11}	0.348 ± 0.168	0.530
	BH[b]	< -13.602	$< 2.5 \times 10^{-14}$	0.518	
CsI	W[a]	-10.917 ± 0.668	1.21×10^{-11}	0.660 ± 0.516	0.833
	BH[b]	-12.921	1.2×10^{-13}	0.819	

[a] This work.
[b] Ref. 13.

actions whose rates were fitted with Arrhenius equations. The "data points" are the rate constants calculated from the analog fit to the concentration profiles; the "error bars" represent the uncertainties derived as described earlier from the sensitivities. As stated earlier, the figures include the rate coefficients for all analyzed runs, although some values have been omitted from the tables for brevity. The solid curves are the curves of the equation

TABLE XX
Arrhenius Parameters for Three-Body Ion-Pair Formation:
$Ar + M^0 + X^0 \xrightarrow{k} Ar + M^+ + X^-$

Species	$\log_{10} A$	A [cm⁶/sec]	E_a [eV]	ΔH [eV]
RbCl	-30.840 ± 0.461	1.45×10^{-31}	0.205 ± 0.342	0.562
RbBr	-30.830 ± 0.535	1.48×10^{-31}	0.347 ± 0.456	0.813
RbI	-29.497 ± 0.449	3.18×10^{-30}	1.44 ± 0.35	1.12
CsCl	-30.810 ± 0.561	1.55×10^{-31}	0.037 ± 0.418	0.279
CsBr	-30.265 ± 0.506	5.43×10^{-31}	0.456 ± 0.404	0.530
CsI	-30.424 ± 0.717	3.76×10^{-31}	0.605 ± 0.554	0.833

TABLE XXI
Comparison of Arrhenius Parameters for Ion–Ion Mutual Neutralization:
$$M^+ + X^- \xrightarrow{k} M^0 + X^0$$

Species	Source	$\log_{10} A$	A [cm³/sec]	E_a [eV]	ΔH [eV]
RbCl	W[a]	-10.097 ± 0.664	7.99×10^{-11}	-0.136 ± 0.492	< 0
	BH[b]	< -12.000	$< 1.0 \times 10^{-12}$	0	
	M[c]	-9.699	2.0×10^{-10}	0	
RbBr	W[a]	-10.064 ± 0.291	8.63×10^{-11}	-0.165 ± 0.248	< 0
	BH[b]	-11.824	1.5×10^{-12}	0	
	M[c]	-9.824	1.5×10^{-10}	0	
RbI	W[a]	-8.825 ± 0.255	1.50×10^{-9}	$+0.562 \pm 0.200$	< 0
	BH[b]	-10.745	1.8×10^{-11}	0	
CsCl	W[a]	-9.882 ± 0.287	1.31×10^{-10}	$+0.111 \pm 0.214$	< 0
	BH[b]	< -12.000	$< 1.0 \times 10^{-12}$	0	
	M[c]	-9.886	3.0×10^{-10}	0	
CsBr	W[a]	-9.701 ± 0.369	1.99×10^{-10}	$+0.062 \pm 0.294$	< 0
	BH[b]	< -12.678	$< 2.1 \times 10^{-13}$	0	
	M[c]	-9.699	2.0×10^{-10}	0	
CsI	W[a]	-9.767 ± 0.554	1.71×10^{-10}	$+0.025 \pm 0.427$	< 0
	BH[b]	-11.824	1.5×10^{-12}	0	
	M[c]	-9.602	2.5×10^{-10}	0	

[a] This work.
[b] Calculated from data contained in Ref. 13.
[c] Calculated from data contained in Ref. 11.

TABLE XXII
Arrhenius Parameters for Three-Body Neutralization:
$$Ar + M^+ + X^- \xrightarrow{k} Ar + M^0 + X^0$$

Species	$\log_{10} A$	A [cm⁶/sec]	E_a [eV]	ΔH [eV]
RbCl	-29.567 ± 0.548	2.71×10^{-30}	-0.151 ± 0.406	< 0
RbBr	-29.566 ± 0.600	2.71×10^{-30}	-0.188 ± 0.511	< 0
RbI	-28.273 ± 0.470	5.33×10^{-29}	$+0.590 \pm 0.370$	< 0
CsCl	-29.556 ± 0.628	2.78×10^{-30}	-0.054 ± 0.468	< 0
CsBr	-29.008 ± 0.520	9.80×10^{-30}	$+0.172 \pm 0.415$	< 0
CsI	-29.274 ± 0.652	5.32×10^{-30}	-0.030 ± 0.503	< 0

resulting from the weighted least-squares fit; the parameters for these curves are those of Tables XV–XXVI. A comparison of these figures, especially Figs. 8, 9, and 10, with those from comparable shock-tube studies[2,6] shows a striking consistency of the scatter.

Ionization of alkali atoms by electron impact is shown for Rb in Fig. 6 and for Cs in Fig. 7. The activation energy agrees well with the ionization potentials; the rate coefficients for Rb are smaller than those for Cs, but the two approach one another as T becomes infinite. The reverse reactions have small negative heats of reaction, reflecting the fact that no temperature dependence was included in the preexponential.

One might expect the rate coefficients for collisional detachment to behave much like those for ionization of the alkali atoms. The straight-line plots of Figs. 8–10, however, have slopes corresponding to activation energies less than the electron affinities. For Br^- and I^-, the 95% confidence limits of the

TABLE XXIII

Comparison of Arrhenius Parameters for Collisional Dissociation into Ions:

$$Ar + MX \xrightarrow{k} Ar + M^+ + X^-$$

Species	Source	$\log_{10} A$	A [cm^3 K$^{7/2}$/sec]	E_a [eV]	ΔH [eV]
RbCl	W[a]	1.806 ± 0.600	63.9	2.40 ± 0.43	4.96
	M[b]	−0.061	0.869	1.56	
	BH[c]	6.799	6.3×10^6	4.95	
RbBr	W[a]	1.736 ± 0.684	54.5	2.94 ± 0.58	4.81
	M[b]	0.925	8.41	2.04	
	BH[c]	6.833	6.8×10^6	4.80	
RbI	W[a]	1.554 ± 0.885	35.8	2.73 ± 0.71	4.59
	BH[c]	6.544	3.5×10^6	4.58	
CsCl	W[a]	0.643 ± 0.740	4.40	1.71 ± 0.55	4.83
	M[b]	1.700	40.7	2.07	
	BH[c]	6.568	3.7×10^6	4.82	
CsBr	W[a]	0.904 ± 0.497	8.01	1.95 ± 0.40	4.63
	M[b]	0.993	9.85	2.11	
	BH[c]	6.431	2.7×10^6	4.59	
CsI	W[a]	1.229 ± 1.046	16.9	2.17 ± 0.81	4.23
	BH[c]	6.00	1.0×10^6	4.22	

[a] This work.
[b] Calculated from data contained in Ref. 11.
[c] Calculated from data contained in Ref. 13.

TABLE XXIV
Comparison of Arrhenius Parameters for Collisional Dissociation into Atoms:

$$\mathrm{Ar} + \mathrm{MX} \xrightarrow{k} \mathrm{Ar} + \mathrm{M}^0 + \mathrm{X}^0$$

Species	Source	$\log_{10} A$	A [cm^3 $K^{7/2}$/sec]	E_a [eV]	ΔH [eV]
RbCl	W[a]	0.175 ± 0.583	1.50	1.85 ± 0.45	4.34
	M[b]	-1.107	7.82×10^{-2}	0.732	
RbBr	W[a]	0.510 ± 0.817	3.24	2.02 ± 0.70	3.90
	M[b]	0.017	1.04	1.39	
RbI	W[a]	0.470 ± 1.117	2.95	1.41 ± 0.88	3.30
CsCl	W[a]	0.574 ± 1.104	3.75	2.11 ± 0.78	4.58
	M[b]	0.796	6.25	2.11	
CsBr	W[a]	1.034 ± 0.714	10.8	2.32 ± 0.56	4.17
	M[b]	0.872	7.45	2.17	
CsI	W[a]	0.349 ± 0.276	2.23	1.83 ± 0.21	3.56

[a] This work.
[b] Calculated from data contained in Ref. 11.

TABLE XXV
Comparison of Arrhenius Parameters for Ionic Recombination:

$$\mathrm{Ar} + \mathrm{M}^+ + \mathrm{X}^- \xrightarrow{k} \mathrm{Ar} + \mathrm{MX}$$

Species	Source	$\log_{10} A$	A [cm^6 K^3/sec]	E_a [eV]	ΔH [eV]
RbCl	W[a]	-22.020 ± 0.583	9.54×10^{-23}	-2.15 ± 0.42	< 0
	BH[b]	-17.155	7.0×10^{-18}	0	
RbBr	W[a]	-21.984 ± 0.692	1.04×10^{-22}	-1.79 ± 0.59	< 0
	BH[b]	-17.051	8.9×10^{-18}	0	
RbI	W[a]	-22.043 ± 0.896	9.06×10^{-23}	-1.78 ± 0.72	< 0
	BH[b]	-17.208	6.2×10^{-18}	0	
CsCl	W[a]	-23.140 ± 0.745	7.23×10^{-24}	-3.07 ± 0.56	< 0
	BH[b]	-17.337	4.6×10^{-18}	0	
CsBr	W[a]	-22.780 ± 0.482	1.66×10^{-23}	-2.62 ± 0.39	< 0
	BH[b]	-17.387	4.1×10^{-18}	0	
CsI	W[a]	-22.345 ± 1.063	4.51×10^{-23}	-2.00 ± 0.83	< 0
	BH[b]	-17.678	2.1×10^{-18}	0	

[a] This work.
[b] Calculated from data contained in Ref. 13.

TABLE XXVI
Arrhenius Parameters for Atomic Recombination:
$Ar + M^0 + X^0 \xrightarrow{k} Ar + MX$

Species	n	$\log_{10} A$	A [cm^6 K^3/sec]	E_a [eV]	ΔH [eV]
RbCl	-3	-24.945 ± 0.568	1.14×10^{-25}	-2.72 ± 0.44	<0
RbBr	-3	-24.476 ± 0.822	3.34×10^{-25}	-2.18 ± 0.70	<0
RbI	-3	-24.288 ± 1.152	5.16×10^{-25}	-2.22 ± 0.89	<0
CsCl	-3	-24.469 ± 1.070	3.39×10^{-25}	-2.59 ± 0.76	<0
CsBr	-3	-23.894 ± 0.713	1.28×10^{-24}	-1.95 ± 0.56	<0
CsI	-3	-24.376 ± 0.209	4.21×10^{-25}	-1.71 ± 0.16	<0

activation energies do encompass the electron affinities of Br^0 and I^0, but for Cl^-, this is not so. This deviation may be one symptom that the model for the process is not precisely correct; we shall discuss this later in this section. The reverse processes show moderate decreases in rate coefficients as T increases. Such behavior is common for strongly exothermic processes involving charged particles or radicals.

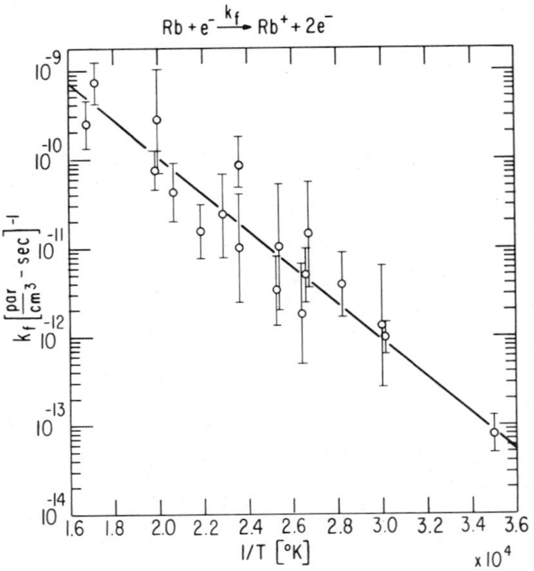

Fig. 6. Arrhenius plot of data and weighted least-squares fit for the reaction $Rb^0 + e \rightarrow Rb^+ + 2e$.

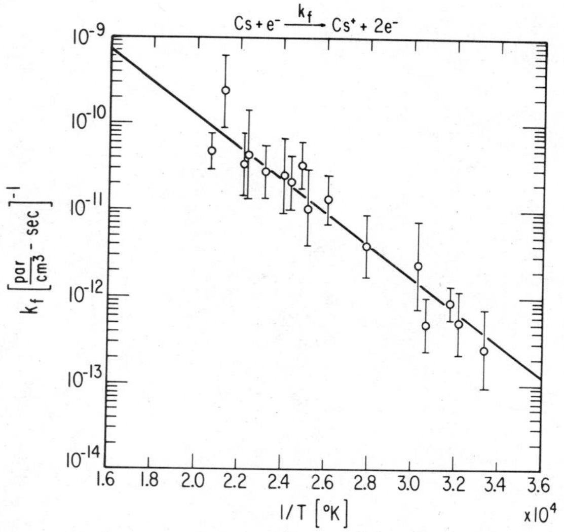

Fig. 7. Arrhenius plot of data and weighted least-squares fit for the reaction $Cs^0 + e \rightarrow Cs^+ + 2e$.

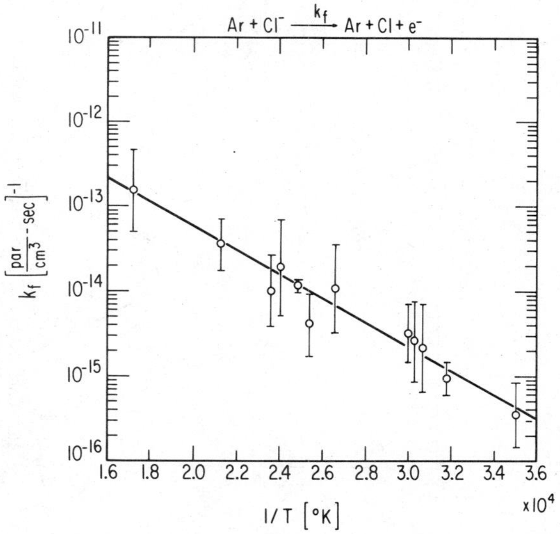

Fig. 8. Arrhenius plot of data and weighted least-squares fit for the reaction $Ar + Cl^- \rightarrow Ar + Cl^0 + e$.

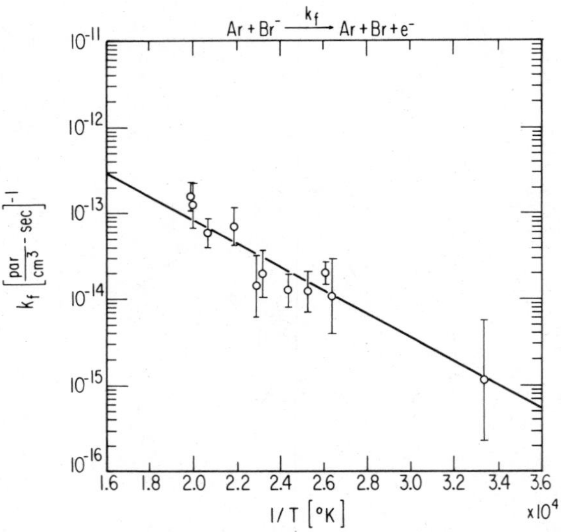

Fig. 9. Arrhenius plot of data and weighted least-squares fit for the reaction $Ar + Br^- \rightarrow Ar + Br^0 + e$.

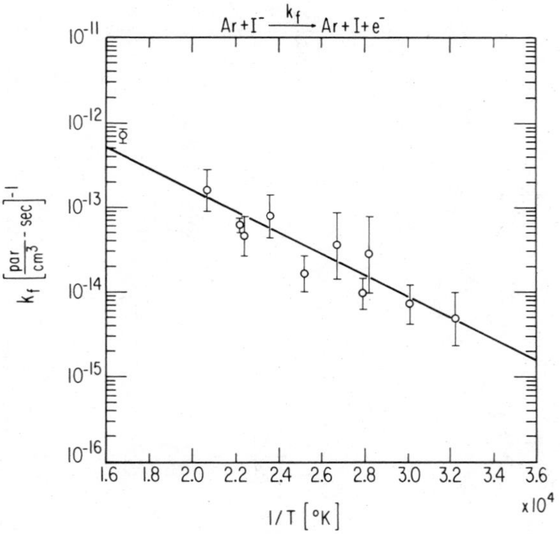

Fig. 10. Arrhenius plot of data and weighted least-squares fit for the reaction $Ar + I^- \rightarrow Ar + I^0 + e$.

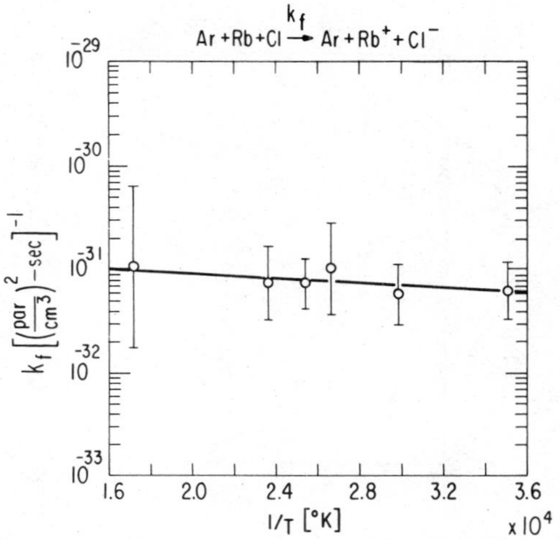

Fig. 11. Arrhenius plot of data and weighted least-squares fit for the reaction $Ar + Rb^0 + Cl^0 \rightarrow Ar + Rb^+ + Cl^-$.

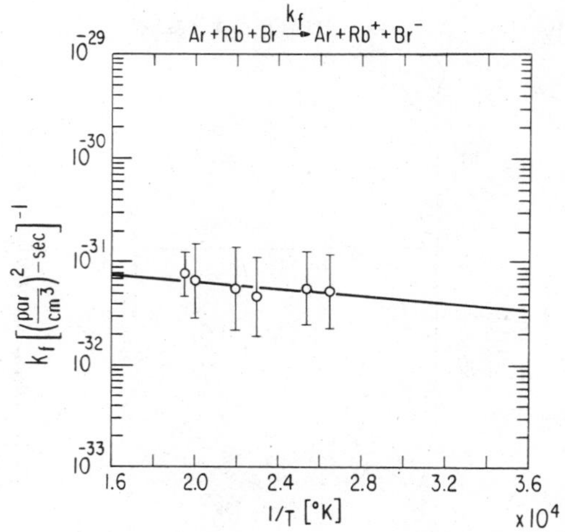

Fig. 12. Arrhenius plot of data and weighted least-squares fit for the reaction $Ar + Rb^0 + Br^0 \rightarrow Ar + Rb^+ + Br^-$.

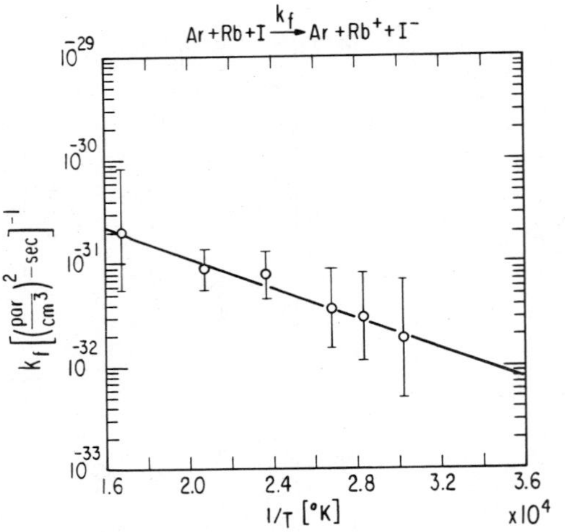

Fig. 13. Arrhenius plot of data and weighted least-squares fit for the reaction $Ar + Rb^0 + I^0 \rightarrow Ar + Rb^+ + I^-$.

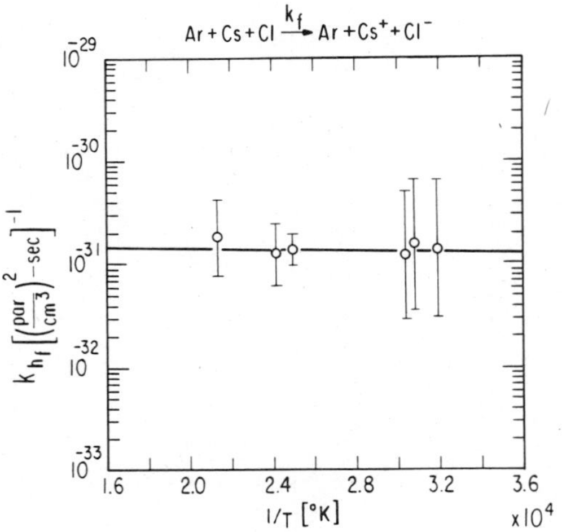

Fig. 14. Arrhenius plot of data and weighted least-squares fit for the reaction $Ar + Cs^0 + Cl^0 \rightarrow Ar + Cs^+ + Cl^-$.

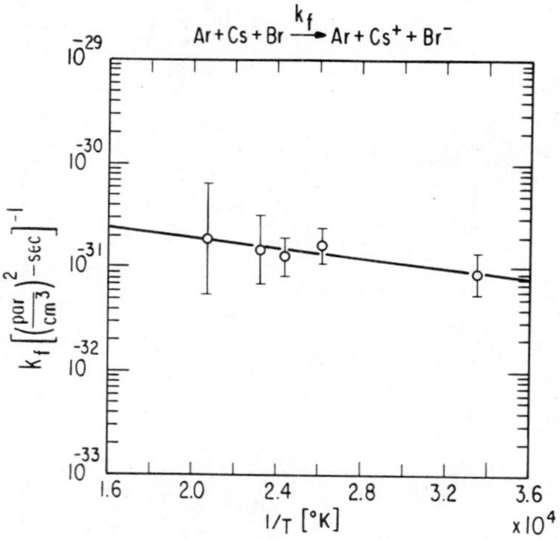

Fig. 15. Arrhenius plot of data and weighted least-squares fit for the reaction $Ar + Cs^0 + Br^0 \rightarrow Ar + Cs^+ + Br^-$.

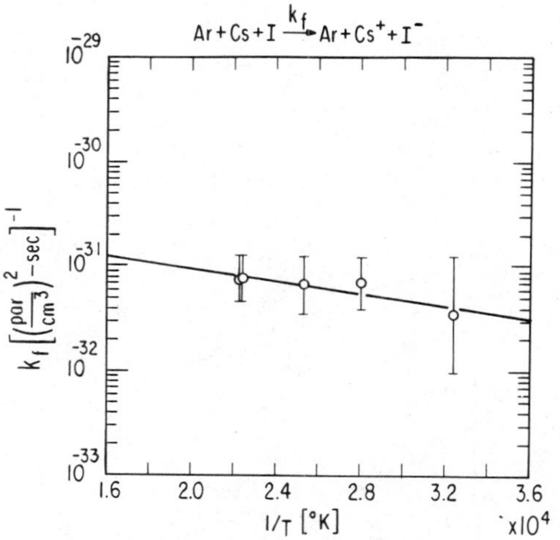

Fig. 16. Arrhenius plot of data and weighted least-squares fit for the reaction $Ar + Cs^0 + I^0 \rightarrow Ar + Cs^+ + I^-$.

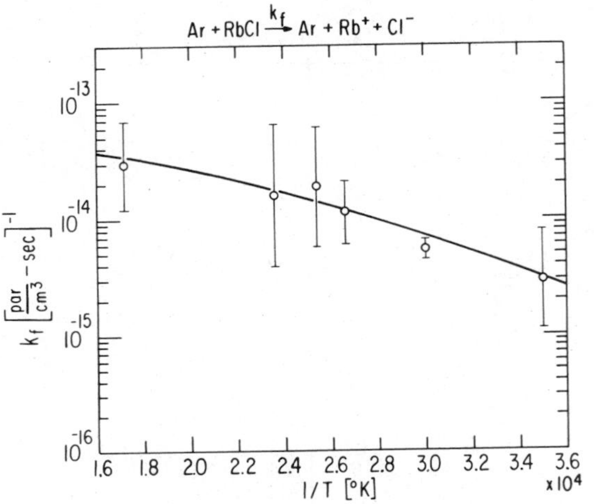

Fig. 17. Arrhenius plot of data and weighted least-squares fit for the reaction $Ar + RbCl \rightarrow Ar + Rb^+ + Cl^-$.

The ion-pair formation, ion–ion neutralization process may be addressed as a bimolecular reaction, $M^0 + X^0 \rightleftarrows M^+ + X^-$, or as a termolecular reaction, $Ar + M^0 + X^0 \rightleftarrows Ar + M^+ + X^-$. In our schema, the two are indistinguishable because we cannot vary the concentration of the argon carrier gas enough, independently of the temperature, to bring the process out of the range in which it is quasibimolecular. From other evidence discussed be-

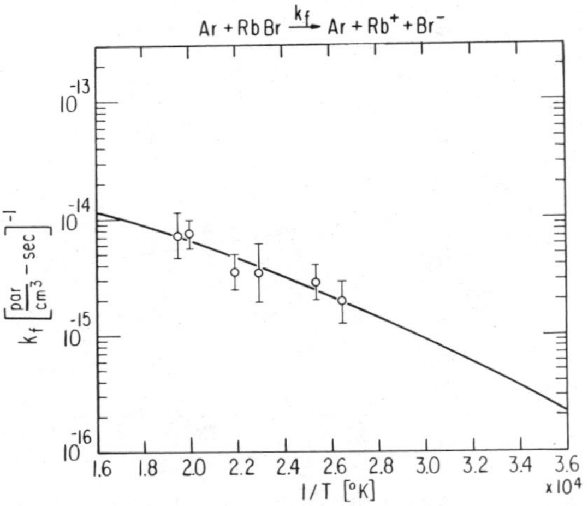

Fig. 18. Arrhenius plot of data and weighted least-squares fit for the reaction $Ar + RbBr \rightarrow Ar + Rb^+ + Br^-$.

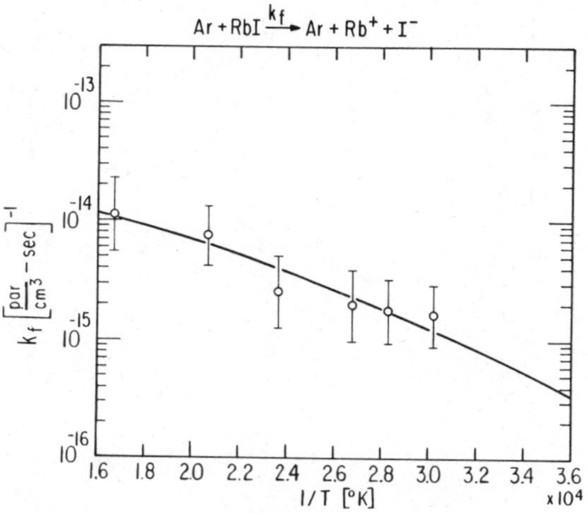

Fig. 19. Arrhenius plot of data and weighted least-squares fit for the reaction $Ar + RbI \rightarrow Ar + Rb^+ + I^-$.

low, the termolecular process is likely to be dominant under our conditions and gives good agreement with other results, from flames for example. Hence we have only presented the Arrhenius plots, Figs. 11–16, based on the termolecular interpretation. The activation energies for these reactions are small and are entirely consistent with the heats of reaction.

The two collisional-dissociation reactions, $Ar + MX \rightleftarrows Ar + M^+ + X^-$ and $Ar + MX \rightleftarrows Ar + M^0 + X^0$, are shown in Figs. 17–22 and 23–28, respec-

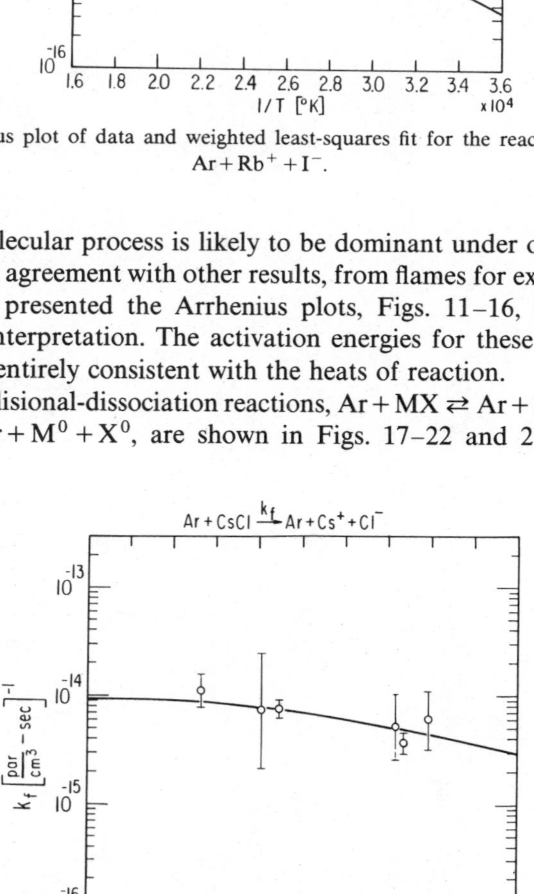

Fig. 20. Arrhenius plot of data and weighted least-squares fit for the reaction $Ar + CsCl \rightarrow Ar + Cs^+ + Cl^-$

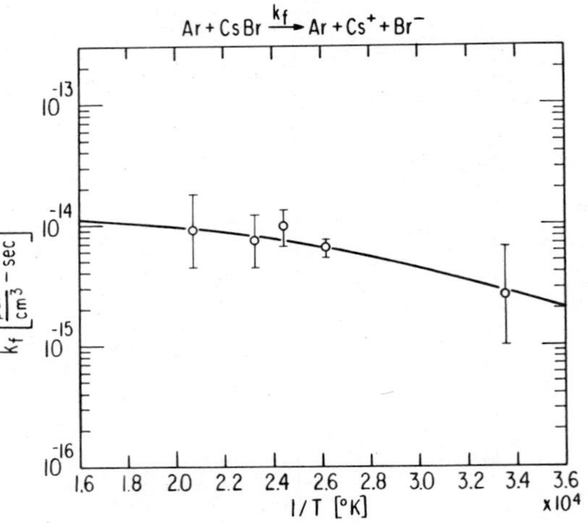

Fig. 21. Arrhenius plot of data and weighted least-squares fit for the reaction $Ar + CsBr \rightarrow Ar + Cs^+ + Br^-$.

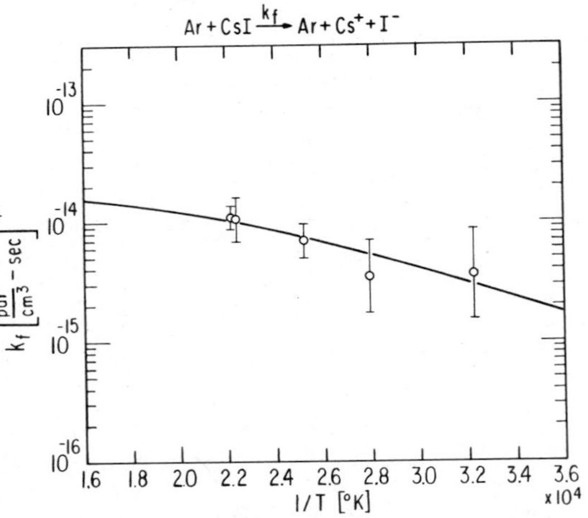

Fig. 22. Arrhenius plot of data and weighted least-squares fit for the reaction $Ar + CsI \rightarrow Ar + Cs^+ + I^-$.

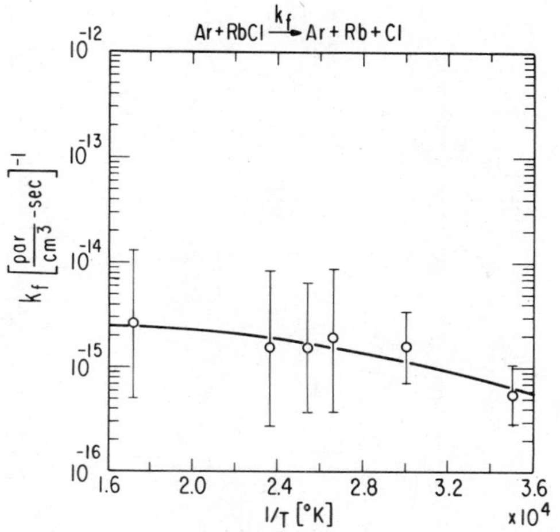

Fig. 23. Arrhenius plot of data and weighted least-squares fit for the reaction $Ar + RbCl \rightarrow Ar + Rb^0 + Cl^0$.

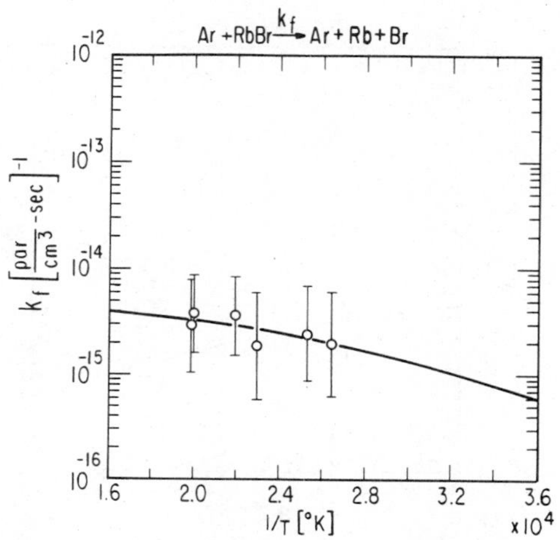

Fig. 24. Arrhenius plot of data and weighted least-squares fit for the reaction $Ar + RbBr \rightarrow Ar + Rb^0 + Br^0$.

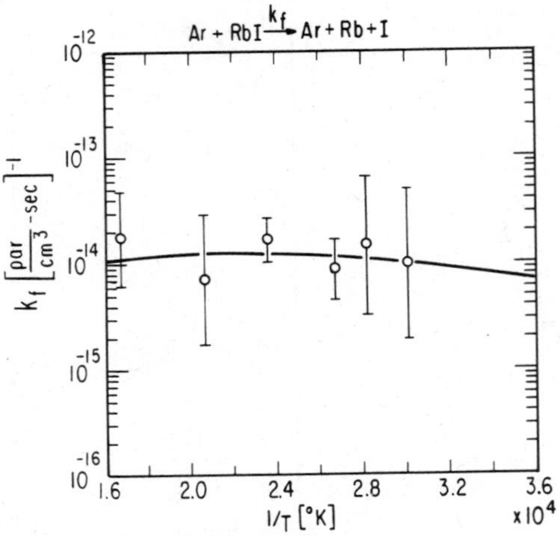

Fig. 25. Arrhenius plot of data and weighted least-squares fit for the reaction $Ar + RbI \rightarrow Ar + Rb^0 + I^0$.

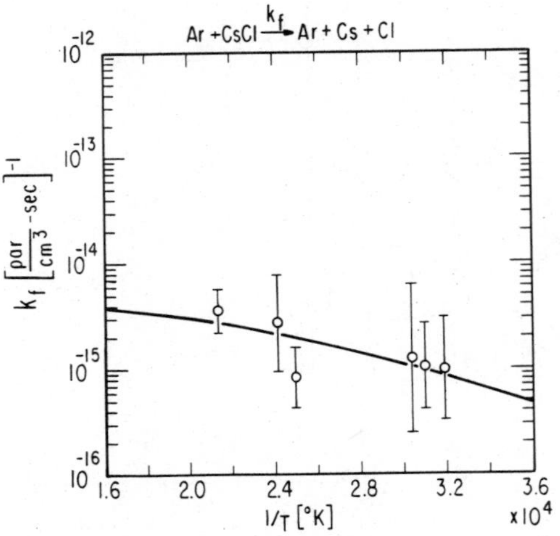

Fig. 26. Arrhenius plot of data and weighted least-squares fit for the reaction $Ar + CsCl \rightarrow Ar + Cs^0 + Cl^0$.

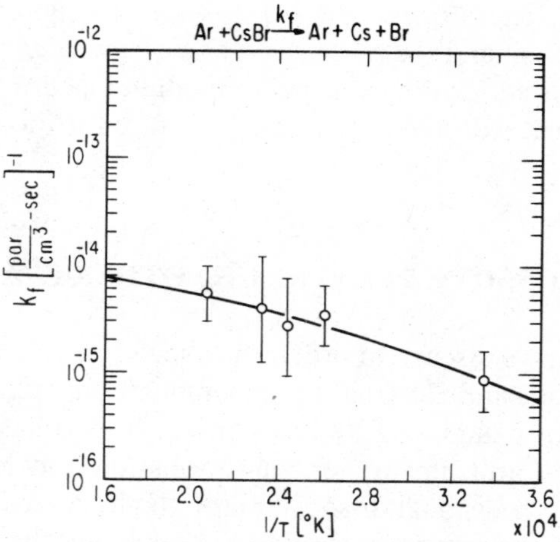

Fig. 27. Arrhenius plot of data and weighted least-squares fit for the reaction $Ar + CsBr \rightarrow Ar + Cs^0 + Br^0$.

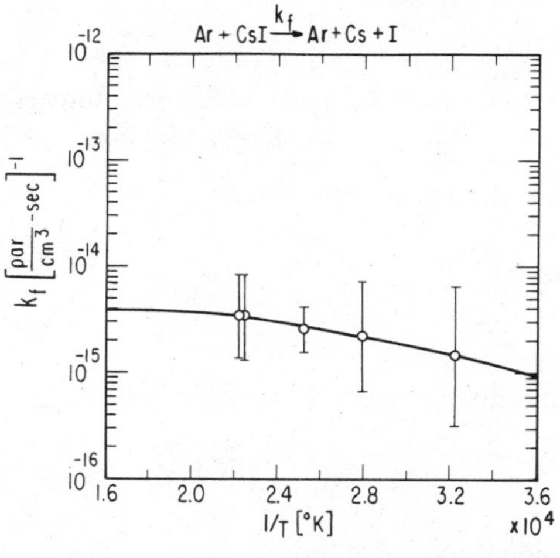

Fig. 28. Arrhenius plot of data and weighted least-squares fit for the reaction $Ar + CsI \rightarrow Ar + Cs^0 + I^0$.

tively. The activation energies for these processes are significantly lower than the heats of reaction, and the reverse reactions show large negative temperature derivatives, even for the ionic recombination reaction. These characteristics are consistent with the data in the literature for the alkali halides.

V. DISCUSSION: RELATION TO OTHER EVALUATIONS

In this section we discuss, in order, the ionization of alkali atoms by collision with electrons and electron–ion recombination; collisional detachment of electrons from halide ions by collisions with inert-gas atoms and its reverse, three-body attachment; ion-pair formation and ion–ion neutralization; and finally, collisional dissociation of alkali halides to atom pairs and to ion pairs, and their reverse processes. Throughout, the results obtained in our work are compared with relevant studies by others, not only from shock-tube and theoretical studies but from other techniques such as beam collisions and flame studies. We pay particular attention to the persistent discrepancy between shock-tube studies directed to collisional detachment and colliding-beam studies of what has been supposed to be the same process; to the apparent discrepancies between ion–ion recombination in flames and the same process in shocks; and to the relative rates of collisional dissociation of alkali halides to atoms and to ions, and what this means for the branching ratio of the dissociation and for the interpretation of how alkali halides conform to the noncrossing rule.

A. Ionization of Alkalis by Electron Impact and Electron–Ion Recombination

We begin by examining the reactions

$$e + \mathrm{M}^0 \underset{k_{5r}}{\overset{k_{5f}}{\rightleftarrows}} 2e + \mathrm{M}^+ \qquad \text{(R5)}$$

The effects of the radiative recombination process

$$e + \mathrm{M}^+ \rightarrow \mathrm{M}^0 + h\nu$$

and neutral collisional recombination

$$e + \mathrm{M}^+ + \mathrm{Ar} \rightarrow \mathrm{M}^0 + \mathrm{Ar}$$

are necessarily included implicitly in our treatment of the kinetics, as discussed briefly earlier and at length in Ref. 18.

Table XV shows the Arrhenius parameters of R5 for Rb and Cs, as obtained in the experiments of Ref. 11, the theoretical calculations of Kunc and Zgorzelski[46] (for Cs), the flame experiments of Ashton and Hayhurst,[14] and the experiments reported here. The theoretical analysis provided rate coefficients for electron–atom collisions with the atoms in the ground state or any of the lowest 17 excited states. To compute the rate coefficient to compare with ours, we constructed

$$k(T) = \sum_{i=1}^{18} P_i k_i(T)$$

where

$$P_i = \frac{g_i \exp(-E_i/kT)}{Z(T)},$$

g_i is the electronic degeneracy of the ith level (the spin degeneracy multiplied by $2J_i + 1$), and $Z(T)$ is the partition function.

The activation energies from this work and the flame study are in good agreement with the ΔH of the reaction; the theoretical E_a for Cs is as well. The activation energies inferred by Milstein and Berry are somewhat lower[11] and, as the rate coefficients of Table XXVII show, the rate coefficients based on the Arrhenius parameters of Ref. 11 are about an order of magnitude higher than those inferred from this work. By contrast, the rate constants based on Arrhenius parameters of Ashton and Hayhurst differ by only a factor of about 2 from ours.

The Arrhenius parameters for the reverse reaction, $2e + M^+ \rightarrow e + M^0$, from the flame measurements of Ashton and Hayhurst and from the theoretical study of collisional–radiative recombination of Bates, Kingston, and McWhirter (BKM),[47] are all given in Table XVI. The BKM model for H appears to work well for Cs,[48] and more recent calculations for Cs also bear this out.[49] The parameters of Bates, Kingston, and McWhirter are strikingly close to ours; those of Ashton and Hayhurst are not directly comparable because the form assumed for the preexponential temperature dependence differs from ours. The results calculated by Kunc and Zgorzelski give an Arrhenius plot too far from linear to yield meaningful parameters for the form $k \propto Ae^{-E_a/kT}$. To put our results on a basis comparable with that of Bates, Kingston, and McWhirter, we assumed that their recombination rate $R = \alpha[M^+][e]$ gives the same result as our rate $R = k_{5r}[M^+][e]^2$. We then

TABLE XXVII
Comparison of Ionization Rate Coefficients

(a) *Rubidium*: $Rb^0 + e \xrightarrow{k} Rb^+ + 2e$

Temperature [K]	W[a]	M[b]	AH[c]
2860	$8.54 \pm 20.7 \times 10^{-14}$	3.83×10^{-12}	3.36×10^{-14}
3360	$1.02 \pm 1.12 \times 10^{-12}$	1.37×10^{-11}	4.20×10^{-13}
3550	$2.19 \pm 1.85 \times 10^{-12}$	2.02×10^{-11}	9.12×10^{-13}
3765	$4.71 \pm 3.16 \times 10^{-12}$	3.00×10^{-11}	2.00×10^{-12}
3960	$8.79 \pm 5.18 \times 10^{-12}$	4.13×10^{-11}	3.79×10^{-12}
4240	$1.95 \pm 1.13 \times 10^{-11}$	6.21×10^{-11}	8.60×10^{-12}
4370	$2.72 \pm 1.66 \times 10^{-11}$	7.37×10^{-11}	1.21×10^{-11}
4560	$4.29 \pm 2.91 \times 10^{-11}$	9.31×10^{-11}	1.94×10^{-11}
5000	$1.08 \pm 0.98 \times 10^{-10}$	1.49×10^{-10}	5.03×10^{-11}
5135	$1.38 \pm 1.37 \times 10^{-10}$	1.70×10^{-10}	6.53×10^{-11}
5810	$4.07 \pm 5.79 \times 10^{-10}$	2.96×10^{-10}	2.01×10^{-10}
5945	$4.91 \pm 7.43 \times 10^{-10}$	3.25×10^{-10}	2.44×10^{-10}

(b) *Cesium*: $Cs^0 + e \xrightarrow{k} Cs^+ + 2e$

Temperature (K)	W[a]	KZ[d]	M[b]	AH[c]
2990	$3.59 \pm 2.81 \times 10^{-13}$	3.16×10^{-11}	1.38×10^{-11}	1.11×10^{-13}
3145	$7.35 \pm 4.54 \times 10^{-13}$	6.76×10^{-11}	2.27×10^{-11}	2.44×10^{-13}
3300	$1.41 \pm 0.69 \times 10^{-12}$	1.35×10^{-10}	3.56×10^{-11}	4.99×10^{-13}
3590	$4.07 \pm 1.41 \times 10^{-12}$	4.16×10^{-10}	7.48×10^{-11}	1.61×10^{-12}
3835	$8.81 \pm 2.68 \times 10^{-12}$	9.46×10^{-10}	1.28×10^{-10}	3.79×10^{-12}
3975	$1.31 \pm 0.41 \times 10^{-11}$	1.44×10^{-9}	1.69×10^{-10}	5.90×10^{-12}
4090	$1.79 \pm 0.52 \times 10^{-11}$	2.00×10^{-9}	2.10×10^{-10}	8.29×10^{-12}
4155	$2.11 \pm 0.72 \times 10^{-11}$	2.39×10^{-9}	2.36×10^{-10}	9.97×10^{-12}
4310	$3.07 \pm 1.17 \times 10^{-11}$	3.56×10^{-9}	3.06×10^{-10}	1.51×10^{-11}
4500	$4.70 \pm 2.07 \times 10^{-11}$	5.59×10^{-9}	4.12×10^{-10}	2.43×10^{-11}
4695	$7.01 \pm 3.56 \times 10^{-11}$	8.56×10^{-9}	5.45×10^{-10}	3.81×10^{-11}
4840	$9.25 \pm 5.18 \times 10^{-11}$	1.15×10^{-8}	6.61×10^{-10}	5.19×10^{-11}

[a] This work.
[b] Calculated from data contained in Ref. 11.
[c] Calculated from data contained in Ref. 14. See footnote *c* in Table XV.
[d] Calculated from data contained in Ref. 46.

TABLE XXVIII

Comparison of Electronic Collisional–Radiative Recombination Rate Constants

(a) *Rubidium*: $Rb^+ + 2e \xrightarrow{k} Rb^0 + e$

Temperature [K]	W[a]	BKM[b]	AH[c]
2860	$4.26 \pm 10.36 \times 10^{-27}$	1.43×10^{-24}	1.26×10^{-27}
3360	$3.37 \pm 3.70 \times 10^{-27}$	4.26×10^{-25}	1.07×10^{-27}
3550	$3.13 \pm 2.66 \times 10^{-27}$	2.80×10^{-25}	1.01×10^{-27}
3765	$2.91 \pm 1.95 \times 10^{-27}$	2.22×10^{-25}	9.56×10^{-28}
3960	$2.74 \pm 1.62 \times 10^{-27}$	1.46×10^{-25}	9.09×10^{-28}
4240	$2.54 \pm 1.48 \times 10^{-27}$	1.08×10^{-25}	8.49×10^{-28}
4370	$2.47 \pm 1.51 \times 10^{-27}$	8.71×10^{-26}	8.24×10^{-28}
4560	$2.36 \pm 1.61 \times 10^{-27}$	7.07×10^{-26}	7.89×10^{-28}
5000	$2.16 \pm 1.97 \times 10^{-27}$	5.04×10^{-26}	7.20×10^{-28}
5135	$2.11 \pm 2.09 \times 10^{-27}$	4.35×10^{-26}	7.01×10^{-28}
5810	$1.91 \pm 2.72 \times 10^{-27}$	3.46×10^{-26}	6.20×10^{-28}
5945	$1.87 \pm 2.84 \times 10^{-27}$	2.83×10^{-26}	6.06×10^{-28}

(b) *Cesium*: $Cs^+ + 2e \xrightarrow{k} Cs^0 + e$

Temperature [K]	W[a]	KZ[d]	BKM[b]	AH[c]
2990	$2.77 \pm 2.15 \times 10^{-27}$	6.4×10^{-29}	8.59×10^{-25}	1.44×10^{-27}
3145	$2.53 \pm 1.55 \times 10^{-27}$	6.6×10^{-29}	6.40×10^{-25}	1.37×10^{-27}
3300	$2.34 \pm 1.14 \times 10^{-27}$	7.1×10^{-29}	4.56×10^{-25}	1.30×10^{-27}
3590	$2.06 \pm 0.71 \times 10^{-27}$	9.8×10^{-29}	2.55×10^{-25}	1.20×10^{-27}
3835	$1.87 \pm 0.57 \times 10^{-27}$	1.27×10^{-29}	1.70×10^{-25}	1.12×10^{-27}
3975	$1.78 \pm 0.55 \times 10^{-27}$	1.48×10^{-28}	1.39×10^{-25}	1.08×10^{-27}
4090	$1.72 \pm 0.56 \times 10^{-27}$	1.80×10^{-28}	1.20×10^{-25}	1.05×10^{-27}
4155	$1.68 \pm 0.57 \times 10^{-27}$	1.92×10^{-28}	1.21×10^{-25}	1.03×10^{-27}
4310	$1.61 \pm 0.61 \times 10^{-27}$	2.32×10^{-28}	9.18×10^{-26}	9.98×10^{-28}
4500	$1.52 \pm 0.67 \times 10^{-27}$	2.94×10^{-28}	7.48×10^{-26}	9.56×10^{-28}
4695	$1.45 \pm 0.73 \times 10^{-27}$	3.7×10^{-28}	7.57×10^{-26}	9.16×10^{-28}
4840	$1.40 \pm 0.78 \times 10^{-27}$	4.1×10^{-28}	5.50×10^{-26}	8.88×10^{-28}

[a] This work.
[b] Calculated from data contained in Ref. 47.
[c] Calculated from data contained in Ref. 14. See Footnote *c* in Table XVI.
[d] Calculated from data contained in Ref. 46.

fitted their data with the equation

$$\log \alpha = B_0 + \frac{B_1}{T} + B_2 \log[e]$$

Some of the calculated rate coefficients based on the Arrhenius parameters obtained this way, as well as others based on the parameters of Ashton and Hayhurst, are given in Table XXVIII. The rate constants from Bates, Kingston, and McWhirter's recombination rates are larger than ours because of an approximation we made: we supposed that the equilibrium concentration of electrons could be used to convert from their α to our k_r. The equilibrium concentrations of electrons, under conditions such as those of our experiments, is always larger than the actual concentration, so the calculated rate coefficients are too large. However, as we go to high temperatures, the actual electron concentration approaches the equilibrium value, and indeed the two sets of rate coefficients are considerably closer at the high-temperature end of the range than at lower temperatures.

The rate coefficients based on the calculations of Kunc and Zgorzelski are all two orders of magnitude larger than ours for the forward reaction of R5 and two orders smaller for the recombination reaction. Furthermore their results show a temperature dependence opposite to all the others for recombination. This may be a consequence of their calculations taking in only a limited number of states.

To conclude the discussion of collisional recombination, we address the question of the role of heavy third particles. Gousset, Sayer, and Berlande[50] have measured the rate coefficient for

$$Cs^+ + e + He \rightarrow Cs^0 + He$$

at 625 K and obtained a value of approximately 4×10^{-29} cm^6 sec^{-1}. This is in excellent agreement with calculated values.[51-53] The rate coefficient for this process should be slightly smaller at the higher temperatures of our experiments. We infer, therefore, that under our conditions the electron-mediated process is indeed the more important and that perhaps it makes a nonnegligible contribution to the recombination rate in the experiments of Ashton and Hayhurst.[14]

B. Collisional Detachment and Three-Body Attachment

Next we address the reactions

$$Ar + X^- \underset{k_{2r}}{\overset{k_{2f}}{\rightleftarrows}} Ar + X^0 + e \qquad \text{(R2)}$$

and the closely related electron-induced detachment

$$e + X^- \rightleftarrows 2e + X^0 \tag{R3}$$

and its reverse. Reaction R2 has been studied previously in shock tubes,[2,6,11] and the Arrhenius parameters of these studies are in reasonably good agreement, as Table XVII shows. Mandl[6] assumes he can identify the electron affinity with the activation energy for all three halides. The activation energies obtained in Ref. 11 for Br^- and I^- are distinctly lower than the corresponding electron affinities, and the data of Fig. 1 of Ref. 6 indicate that a better fit of data and parameters could be obtained if the activation energies were taken less than the electron affinities for bromine and iodine. Our activation energies, also in Table XVII, are distinctly lower than the electron affinities for all three halogens, and we believe our results lie within the stated error bounds, with at least 90% confidence. This result is already suggestive that the interpretation of the "collisional dissociation" might require something more sophisticated than simply reaction R2.

Table XXIX translates the Arrhenius parameters of Refs. 2, 6, and 11 into some representative rate coefficients and compares them with results of this study. All four agree within about half an order of magnitude, suggesting that the four sets of results are consistent and therefore reliable. However, all four are in striking disagreement with the beam-collision studies.[54-56] The collision studies yielded threshold energies of 7.3 ± 0.2 and 7.7 ± 0.2 eV respectively, for collisional detachment of electrons from Cl^- by Ar and from Br^- by Ar, roughly twice the electron affinities. Moreover the measured cross sections imply rate coefficients 2.5 to 3 orders of magnitude smaller than the shock-tube values. One other experimental collision study shows a detachment threshold that may be higher than the corresponding affinity: Wynn, Martin, and Bailey[57] report a threshold of 1.75 ± 0.54 eV for $Ar + O^- \rightleftarrows Ar + O^0 + e$, while the electron affinity of the oxygen atom is 1.462 eV.

A lively discussion ensued regarding the discrepancy,[32,48] which remains unresolved. One of us has discussed some of the possibilities,[58] but no answer has yet emerged. The situation, put candidly, is this. The simple collisional processes based on single collisions in beams gives much lower values of the rate coefficients than the shock-tube experiments with their complex conditions. In these circumstances, the complex system is always the one under suspicion, because it has the possibility for fast, complex processes that can overwhelm any simple, slow process. That the effective activation energy in the complex system is *lower* than the thermodynamic heat of reaction is, in our view, a definitive demonstration that some process more elaborate than reaction R2 is responsible for the apparent collisional detachment that occurs in the shock-tube experiments.

TABLE XXIX
Comparison of Collisional Detachment Rate Constants

(a) *Chlorine*: $Ar + Cl^- \xrightarrow{k} Ar + Cl^0 + e$

Temperature [K]	W[a]	Ma[b]	LTW[c]	Mi[d]
2860	$4.21 \pm 3.22 \times 10^{-16}$	5.01×10^{-17}	1.50×10^{-16}	1.86×10^{-16}
3260	$1.71 \pm 0.73 \times 10^{-15}$	3.04×10^{-16}	8.98×10^{-16}	1.24×10^{-15}
3360	$2.31 \pm 0.86 \times 10^{-15}$	4.46×10^{-16}	1.31×10^{-15}	1.86×10^{-15}
3765	$6.59 \pm 1.90 \times 10^{-15}$	1.71×10^{-15}	5.01×10^{-15}	7.66×10^{-15}
4010	$1.12 \pm 0.35 \times 10^{-14}$	3.39×10^{-15}	9.86×10^{-15}	1.57×10^{-14}
4230	$1.71 \pm 0.63 \times 10^{-14}$	5.84×10^{-15}	1.70×10^{-14}	2.78×10^{-14}
4695	$3.69 \pm 1.89 \times 10^{-14}$	1.56×10^{-14}	4.51×10^{-14}	7.84×10^{-14}
5810	$1.41 \pm 1.23 \times 10^{-13}$	8.69×10^{-14}	2.49×10^{-13}	4.78×10^{-13}

(b) *Bromine*: $Ar + Br^- \xrightarrow{k} Ar + Br^0 + e$

Temperature [K]	W[a]	Ma[b]	LTW[c]	Mi[d]
2990	$1.21 \pm 1.61 \times 10^{-15}$	2.02×10^{-16}	7.03×10^{-16}	1.61×10^{-15}
3795	$1.11 \pm 0.49 \times 10^{-14}$	3.21×10^{-15}	1.14×10^{-14}	1.35×10^{-14}
3960	$1.57 \pm 0.59 \times 10^{-14}$	4.92×10^{-15}	1.75×10^{-14}	1.87×10^{-14}
4370	$3.30 \pm 1.16 \times 10^{-14}$	1.24×10^{-14}	4.43×10^{-14}	3.80×10^{-14}
4560	$4.44 \pm 1.72 \times 10^{-14}$	1.80×10^{-14}	6.43×10^{-14}	5.05×10^{-14}
4840	$6.61 \pm 3.11 \times 10^{-14}$	2.95×10^{-14}	1.06×10^{-13}	7.39×10^{-14}
5000	$8.13 \pm 4.28 \times 10^{-14}$	3.82×10^{-14}	1.37×10^{-13}	9.01×10^{-14}
5135	$9.58 \pm 5.52 \times 10^{-14}$	4.69×10^{-14}	1.69×10^{-13}	1.05×10^{-13}

(c) *Iodine*: $Ar + I^- \xrightarrow{k} Ar + I^0 + e$

Temperature [K]	W[a]	Ma[b]	LTW[c]	Mi[d]
3105	$4.52 \pm 5.23 \times 10^{-15}$	7.56×10^{-16}	1.30×10^{-14}	5.93×10^{-16}
3320	$8.27 \pm 6.88 \times 10^{-15}$	1.59×10^{-15}	2.72×10^{-14}	1.21×10^{-15}
3550	$1.46 \pm 0.88 \times 10^{-14}$	3.17×10^{-15}	5.43×10^{-14}	2.35×10^{-15}
3740	$2.20 \pm 1.08 \times 10^{-14}$	5.27×10^{-15}	9.03×10^{-14}	3.83×10^{-15}
4240	$5.50 \pm 2.50 \times 10^{-14}$	1.61×10^{-14}	2.77×10^{-13}	1.12×10^{-14}
4500	$8.16 \pm 4.38 \times 10^{-14}$	2.62×10^{-14}	4.49×10^{-13}	1.79×10^{-14}
4840	$1.28 \pm 0.88 \times 10^{-13}$	4.56×10^{-14}	7.81×10^{-13}	3.05×10^{-14}
5945	$3.91 \pm 4.95 \times 10^{-13}$	1.78×10^{-13}	3.06×10^{-12}	1.13×10^{-13}

[a] This work.
[b] Calculated from data contained in Ref. 6.
[c] Calculated from data contained in Ref. 2.
[d] Calculated from data contained in Ref. 11.

Several processes might be considered as alternatives to R2. One that seems faintly plausible is

$$e + X^- \rightleftarrows 2e + X^0, \qquad \text{(R3)}$$

for the following reasons. The cross section for the analogue of this process for $X^0 = H$ is of order 4×10^{-15} cm^2 at an electron energy of 10 eV and falls to about 5×10^{-19} cm^2 for an electron energy of 1.6 eV. The corresponding rate coefficient would be about 2×10^{-11} cm^3/sec at 1.6 eV.[59-61] The temperatures of our experiments correspond to $\frac{1}{6}$ to about $\frac{1}{4}$ of this energy. We are trying to account for rate coefficients of order 3×10^{-13} cm^3/sec and smaller when Ar is the collision partner. Its concentration is typically 10^{19} cm^{-3}; the electron concentrations are of order 10^{15}–10^{16} cm^{-3}. Hence if R3 is to have a rate that makes it a possible alternative to R2, the rate coefficient for R3 must be of order 10^{-13} cm^3/sec at the lowest of our temperatures and of order 10^{-9} cm^3/sec at the highest. Chibisov has argued[62] that the rate coefficients for the reverse processes of R3 and R2 are in the ratio $k_{3r}/k_{2r} \sim 100$ for hydrogen. This would require the forward rate coefficients to be in the same ratio. The effects of polarization and exchange that have so far been neglected in theoretical estimates of cross sections for R3 might conceivably enhance the cross sections enough at energies within about an eV above threshold to make reaction R3 a nonnegligible contributor to the transformation of X^- into $X^0 + e$.

Nevertheless the argument at this crude level does not give us a great deal of confidence in attributing to R3 the major reason why detachment of electrons from halide ions is so much faster in hot, shocked alkali halides than in beam collisions.

Another possible mechanism (among those not instantly rejectable) that has not been investigated would invoke a two-step process: ion–ion neutralization leaving the neutral alkali atom in a high excited state, as M*, followed by ionization of M* by either an electron or an argon atom. Cross sections for ion–ion neutralization are in some cases as large as 10^{-13} cm^2, and $[M^+] \cong [e]$ at high temperatures, but the relative thermal speed of the ions is less than 0.01 of thermal electron speed, so the effective rate coefficient for the first step of this process is already a hundredfold slower than the rate coefficient of R3. Hence this process is not a good candidate to explain the discrepancy.

We make no pretense that this study reconciles the disagreement over the rate coefficient and mechanism of detachment. However, this work does show, through the E_a values lower than the corresponding electron affinities, that the shock-tube process is almost certainly something more complex than the simple two-particle collision of R2.

C. Ion-Pair Formation and Neutralization

The processes of ion-pair formation and neutralization may be interpreted as two-body reactions

$$M^0 + X^0 \rightleftarrows M^+ + X^-$$

or as three-body reactions

$$Ar + M^0 + X^0 \rightleftarrows Ar + M^+ + X^- \qquad (R8)$$

The only data with which we can compare our rate coefficients are those of Burdett and Hayhurst[13] for the two-body process. These comparisons are made for the two sets of Arrhenius parameters in Table XIX and for selected values of the rate coefficients themselves in Table XXX. Our activation energies are almost all lower than those of Burdett and Hayhurst, which are uniformly close to the ΔH's of reaction. Our rate coefficients are ten to a hundredfold larger than those of Burdett and Hayhurst. The Burdett–Hayhurst values are in good agreement with theoretical values for the rate coefficients for ion-pair formation by collision of H with Cs.[63]

We attribute at least part of the difference to a significant contribution to our system from the three-body process $Ar + M^0 + X^0 \rightleftarrows Ar + M^+ + X^-$. The particle density in the work of Burdett and Hayhurst was about $2.8–4.0 \times 10^{18}$; in our work it was $2.3–3.5 \times 10^{19}$. Moreover our temperatures were two to three times theirs, so the collision frequencies were higher by an additional factor of 1.4 to 1.75. Hence the total collision frequency in our work is about 15 times larger than in the flame experiments. The temperatures of our experiments were high enough that significant fractions of the alkalis were in excited states. This probably further enhances the rates of both the two-body and three-body processes.

The Arrhenius parameters of the two-body neutralization process calculated from data of Burdett and Hayhurst,[13] from Ref. 11, and from our work are compared in Table XXI. Both the former report no significant temperature dependence, so their E_a's are uniformly zero. Ours are also zero within their uncertainties, except possibly that of RbI. Table XXXI gives typical rate coefficients for all three sets of parameters. Those of Ref. 11 and these reported here are rather close. As with the rate coefficients for ion-pair production, the rate coefficients from Ref. 13 are markedly lower than those from the shock-tube experiments, here one or two orders of magnitude. Bates[64,65] has shown how in slow neutralization processes, the termolecular contribution becomes more important as the temperature increases and eventually dominates the process if the pressure is high enough. This is precisely how we interpret the difference between our system and that of Burdett and

TABLE XXX

Comparison of Two-Body Ion-Pair Formation Rate Constants

(a) *The Rubidium Halides*: $Rb^0 + X^0 \xrightarrow{k} Rb^+ + X^-$

Species	Temperature [K]	W[a]	BH[b]
Cl	2860	$1.75 \pm 0.98 \times 10^{-12}$	$< 1.03 \times 10^{-14}$
	3360	$1.99 \pm 0.64 \times 10^{-12}$	$< 1.45 \times 10^{-14}$
	3765	$2.16 \pm 0.56 \times 10^{-12}$	$< 1.78 \times 10^{-14}$
	4230	$2.33 \pm 0.68 \times 10^{-12}$	$< 2.15 \times 10^{-14}$
	5810	$2.75 \pm 1.59 \times 10^{-12}$	$< 3.27 \times 10^{-14}$
Br	3795	$1.51 \pm 0.16 \times 10^{-12}$	9.30×10^{-15}
	3960	$1.59 \pm 0.14 \times 10^{-12}$	1.03×10^{-14}
	4370	$1.75 \pm 0.12 \times 10^{-12}$	1.29×10^{-14}
	4560	$1.83 \pm 0.14 \times 10^{-12}$	1.41×10^{-14}
	5135	$2.03 \pm 0.24 \times 10^{-12}$	1.77×10^{-14}
I	3320	$6.39 \pm 0.84 \times 10^{-13}$	3.94×10^{-14}
	3740	$1.11 \pm 0.10 \times 10^{-12}$	6.09×10^{-14}
	4240	$1.86 \pm 0.15 \times 10^{-12}$	9.13×10^{-14}
	4840	$3.01 \pm 0.34 \times 10^{-12}$	1.33×10^{-13}
	5945	$5.64 \pm 1.02 \times 10^{-12}$	2.18×10^{-13}

(b) *The Cesium Halides*: $Cs^0 + X^0 \xrightarrow{k} Cs^+ + X^-$

Species	Temperature [K]	W[a]	BH[b]
Cl	3145	$3.47 \pm 0.45 \times 10^{-12}$	$< 3.71 \times 10^{-14}$
	3300	$3.60 \pm 0.38 \times 10^{-12}$	$< 3.88 \times 10^{-14}$
	4010	$4.08 \pm 0.30 \times 10^{-12}$	$< 4.59 \times 10^{-14}$
	4155	$4.16 \pm 0.34 \times 10^{-12}$	$< 4.72 \times 10^{-14}$
	4695	$4.44 \pm 0.55 \times 10^{-12}$	$< 5.14 \times 10^{-14}$
Br	2990	$2.86 \pm 0.60 \times 10^{-12}$	$< 3.35 \times 10^{-15}$
	3835	$3.86 \pm 0.35 \times 10^{-12}$	$< 5.21 \times 10^{-15}$
	4090	$4.12 \pm 0.35 \times 10^{-12}$	$< 5.75 \times 10^{-15}$
	4310	$4.33 \pm 0.38 \times 10^{-12}$	$< 6.20 \times 10^{-15}$
	4840	$4.80 \pm 0.56 \times 10^{-12}$	$< 7.22 \times 10^{-15}$
I	3105	$1.03 \pm 0.58 \times 10^{-12}$	5.62×10^{-15}
	3590	$1.43 \pm 0.40 \times 10^{-12}$	8.50×10^{-15}
	3975	$1.76 \pm 0.39 \times 10^{-12}$	1.10×10^{-14}
	4475	$2.18 \pm 0.69 \times 10^{-12}$	1.43×10^{-14}
	4500	$2.20 \pm 0.71 \times 10^{-12}$	1.45×10^{-14}

[a] This work.

[b] Calculated from data contained in Ref. 13.

TABLE XXXI
Comparison of Ion–Ion Mutual Neutralization Rate Constants

(a) *The Rubidium Halides*: $Rb^+ + X^- \xrightarrow{k} Rb^0 + X^0$

Species	Temperature [K]	W[a]	BH[b]	M[c]
Cl	2860	$1.39 \pm 1.10 \times 10^{-10}$	$<1.0 \times 10^{-12}$	2.0×10^{-10}
	3360	$1.28 \pm 0.56 \times 10^{-10}$	$<1.0 \times 10^{-12}$	2.0×10^{-10}
	3765	$1.21 \pm 0.43 \times 10^{-10}$	$<1.0 \times 10^{-12}$	2.0×10^{-10}
	4230	$1.16 \pm 0.46 \times 10^{-10}$	$<1.0 \times 10^{-12}$	2.0×10^{-10}
	5810	$1.05 \pm 0.86 \times 10^{-10}$	$<1.0 \times 10^{-12}$	2.0×10^{-10}
Br	3795	$1.43 \pm 0.18 \times 10^{-10}$	1.5×10^{-12}	1.5×10^{-10}
	3960	$1.40 \pm 0.15 \times 10^{-10}$	1.5×10^{-12}	1.5×10^{-10}
	4370	$1.34 \pm 0.11 \times 10^{-10}$	1.5×10^{-12}	1.5×10^{-10}
	4560	$1.31 \pm 0.12 \times 10^{-10}$	1.5×10^{-12}	1.5×10^{-10}
	5135	$1.25 \pm 0.18 \times 10^{-10}$	1.5×10^{-12}	1.5×10^{-10}
I	3320	$2.10 \pm 0.35 \times 10^{-10}$	1.8×10^{-11}	—
	3740	$2.62 \pm 0.29 \times 10^{-10}$	1.8×10^{-11}	—
	4240	$3.22 \pm 0.34 \times 10^{-10}$	1.8×10^{-11}	—
	4840	$3.89 \pm 0.57 \times 10^{-10}$	1.8×10^{-11}	—
	5945	$5.00 \pm 1.17 \times 10^{-10}$	1.8×10^{-11}	—

(b) *The Cesium Halides*: $Cs^+ + X^- \xrightarrow{k} Cs^0 + X^0$

Species	Temperature [K]	W[a]	BH[b]	M[c]
Cl	3145	$8.72 \pm 1.52 \times 10^{-11}$	$<1.0 \times 10^{-12}$	3.0×10^{-10}
	3300	$8.89 \pm 1.25 \times 10^{-11}$	$<1.0 \times 10^{-12}$	3.0×10^{-10}
	4010	$9.53 \pm 0.95 \times 10^{-11}$	$<1.0 \times 10^{-12}$	3.0×10^{-10}
	4155	$9.63 \pm 1.07 \times 10^{-11}$	$<1.0 \times 10^{-12}$	3.0×10^{-10}
	4695	$9.98 \pm 1.66 \times 10^{-11}$	$<1.0 \times 10^{-12}$	3.0×10^{-10}
Br	2990	$1.56 \pm 0.62 \times 10^{-10}$	$<2.1 \times 10^{-13}$	2.0×10^{-10}
	3835	$1.65 \pm 0.27 \times 10^{-10}$	$<2.1 \times 10^{-13}$	2.0×10^{-10}
	4090	$1.67 \pm 0.25 \times 10^{-10}$	$<2.1 \times 10^{-13}$	2.0×10^{-10}
	4310	$1.68 \pm 0.27 \times 10^{-10}$	$<2.1 \times 10^{-13}$	2.0×10^{-10}
	4840	$1.71 \pm 0.37 \times 10^{-10}$	$<2.1 \times 10^{-13}$	2.0×10^{-10}
I	3105	$1.56 \pm 0.70 \times 10^{-10}$	1.5×10^{-12}	2.5×10^{-10}
	3590	$1.58 \pm 0.35 \times 10^{-10}$	1.5×10^{-12}	2.5×10^{-10}
	3975	$1.59 \pm 0.29 \times 10^{-10}$	1.5×10^{-12}	2.5×10^{-10}
	4475	$1.60 \pm 0.41 \times 10^{-10}$	1.5×10^{-12}	2.5×10^{-10}
	4500	$1.60 \pm 0.42 \times 10^{-10}$	1.5×10^{-12}	2.5×10^{-10}

[a] This work.
[b] Calculated from data contained in Ref. 13.
[c] Calculated from data contained in Ref. 11.

Hayhurst for ion–ion neutralization and therefore inevitably for ion-pair formation as well.

As we shall see in the next section, our best estimate for the rate coefficient for

$$\mathrm{Ar+M^+ + X^- \to Ar + MX}$$

is about 1×10^{-30} cm^6/sec, with an uncertainty of about one power of ten. This is the same magnitude as the rate coefficients of Table IX; the two processes differ only in whether the final state of MX is vibrationally bound or not, so their rate coefficients should not differ a great deal so long as the density of available bound states is comparable to that of the available free states.

D. Ionic and Atomic Dissociation, Branching Ratios, and the Reverse Recombination Reactions

Finally we turn to the processes that initially motivated this study, the competing collisional dissociation processes

$$\mathrm{Ar+MX} \xrightarrow{k_{1f}} \mathrm{Ar+M^+ + X^-} \tag{R1}$$

and

$$\mathrm{Ar+MX} \xrightarrow{k_{7f}} \mathrm{Ar+M^0 + X^0} \tag{R7}$$

and their reverse reactions,

$$\mathrm{Ar+M^+ + X^-} \xrightarrow{k_{1r}} \mathrm{Ar+MX}$$

and

$$\mathrm{Ar+M^0 + X^0} \xrightarrow{k_{7r}} \mathrm{Ar+MX}$$

The Arrhenius parameters for these reactions from our work, from Ref. 11, and calculated from data on Ref. 13 are given in Tables XXIII–XXVI. The shock-tube studies[11] dealt directly with both the forward reactions of dissociation; Burdett and Hayhurst, in their flame studies,[13] considered the forward collisional dissociation to ions and the reverse recombination process of the ions to give the neutral bound molecule. Our results and those of Ref.

TABLE XXXII
Comparison of Rate Constants for Collisional Dissociation into Ions

(a) *The Rubidium Halides*: $Ar + RbX \xrightarrow{k} Ar + Rb^+ + X^-$

Species	Temperature [K]	W[a]	BH[b]	M[c]
Cl	2860	$3.04 \pm 1.80 \times 10^{-15}$	9.54×10^{-15}	1.24×10^{-15}
	3360	$7.36 \pm 2.44 \times 10^{-15}$	1.08×10^{-13}	1.81×10^{-15}
	3765	$1.20 \pm 0.35 \times 10^{-14}$	4.55×10^{-13}	2.17×10^{-15}
	4230	$1.81 \pm 0.66 \times 10^{-14}$	1.62×10^{-12}	2.44×10^{-15}
	5810	$3.56 \pm 2.67 \times 10^{-14}$	2.14×10^{-11}	2.58×10^{-15}
Br	3795	$1.99 \pm 0.63 \times 10^{-15}$	8.53×10^{-13}	4.88×10^{-15}
	3960	$2.49 \pm 0.63 \times 10^{-15}$	1.35×10^{-12}	5.45×10^{-15}
	4370	$3.97 \pm 0.80 \times 10^{-15}$	3.59×10^{-12}	6.77×10^{-15}
	4560	$4.73 \pm 1.06 \times 10^{-15}$	5.26×10^{-12}	7.31×10^{-15}
	5135	$7.23 \pm 2.65 \times 10^{-15}$	1.36×10^{-11}	8.62×10^{-15}
I	3320	$1.22 \pm 0.95 \times 10^{-15}$	1.85×10^{-13}	—
	3740	$2.34 \pm 1.10 \times 10^{-15}$	7.37×10^{-10}	—
	4240	$4.10 \pm 1.67 \times 10^{-15}$	2.54×10^{-12}	—
	4840	$6.51 \pm 3.76 \times 10^{-15}$	7.55×10^{-12}	—
	5945	$1.07 \pm 1.11 \times 10^{-14}$	2.83×10^{-11}	—

(b) *The Cesium Halides*: $Ar + CsX \xrightarrow{k} Ar + Cs^+ + X^-$

Species	Temperature [K]	W[a]	BH[b]	M[c]
Cl	3145	$4.56 \pm 2.34 \times 10^{-15}$	4.01×10^{-14}	1.12×10^{-14}
	3300	$5.18 \pm 2.08 \times 10^{-15}$	7.80×10^{-14}	1.36×10^{-14}
	4010	$7.60 \pm 2.05 \times 10^{-15}$	7.93×10^{-13}	2.49×10^{-14}
	4155	$7.98 \pm 2.42 \times 10^{-15}$	1.14×10^{-12}	2.72×10^{-14}
	4695	$9.02 \pm 4.32 \times 10^{-15}$	3.50×10^{-12}	3.44×10^{-14}
Br	2990	$2.88 \pm 1.36 \times 10^{-15}$	3.39×10^{-14}	1.87×10^{-15}
	3835	$6.37 \pm 1.20 \times 10^{-15}$	7.18×10^{-13}	4.70×10^{-15}
	4090	$7.34 \pm 1.24 \times 10^{-15}$	1.36×10^{-12}	5.66×10^{-15}
	4310	$8.10 \pm 1.39 \times 10^{-15}$	2.21×10^{-12}	6.39×10^{-15}
	4840	$9.58 \pm 2.14 \times 10^{-15}$	5.69×10^{-12}	7.93×10^{-15}
I	3105	$3.01 \pm 3.26 \times 10^{-15}$	8.48×10^{-14}	—
	3590	$5.43 \pm 2.78 \times 10^{-15}$	4.30×10^{-13}	—
	3975	$7.51 \pm 2.99 \times 10^{-15}$	1.13×10^{-12}	—
	4475	$1.01 \pm 0.55 \times 10^{-14}$	2.95×10^{-12}	—
	4500	$1.02 \pm 0.57 \times 10^{-14}$	3.07×10^{-12}	—

[a] This work.
[b] Calculated from data contained in Ref. 13.
[c] Calculated from data contained in Ref. 11.

TABLE XXXIII
Comparison of Rate Constants for Collisional Dissociation into Atoms

(a) *The Rubidium Halides*: $Ar + RbX \xrightarrow{k} Ar + Rb^0 + X^0$

Species	Temperature [K]	W[a]	M[b]
Cl	2860	$6.53 \pm 5.15 \times 10^{-16}$	3.21×10^{-15}
	3360	$1.14 \pm 0.50 \times 10^{-15}$	2.84×10^{-15}
	3765	$1.52 \pm 0.51 \times 10^{-15}$	2.50×10^{-15}
	4230	$1.89 \pm 0.64 \times 10^{-15}$	2.13×10^{-15}
	5810	$2.48 \pm 1.61 \times 10^{-15}$	1.21×10^{-15}
Br	3795	$2.02 \pm 0.85 \times 10^{-15}$	4.40×10^{-15}
	3960	$2.25 \pm 0.75 \times 10^{-15}$	4.53×10^{-15}
	4370	$2.77 \pm 0.70 \times 10^{-15}$	4.70×10^{-15}
	4560	$2.99 \pm 0.81 \times 10^{-15}$	4.73×10^{-15}
	5135	$3.50 \pm 1.54 \times 10^{-15}$	4.63×10^{-15}
I	3320	$1.01 \pm 1.06 \times 10^{-14}$	—
	3740	$1.16 \pm 0.78 \times 10^{-14}$	—
	4240	$1.25 \pm 0.80 \times 10^{-14}$	—
	4840	$1.27 \pm 1.14 \times 10^{-14}$	—
	5945	$1.16 \pm 1.83 \times 10^{-14}$	—

(b) *The Cesium Halides*: $Ar + CsX \xrightarrow{k} Ar + Cs^0 + X^0$

Species	Temperature [K]	W[a]	M[b]
Cl	3145	$8.97 \pm 6.15 \times 10^{-16}$	1.49×10^{-15}
	3260	$1.04 \pm 0.60 \times 10^{-15}$	1.73×10^{-15}
	3300	$1.09 \pm 0.59 \times 10^{-15}$	1.81×10^{-15}
	4010	$2.05 \pm 1.13 \times 10^{-15}$	3.41×10^{-15}
	4155	$2.24 \pm 1.41 \times 10^{-15}$	3.73×10^{-15}
	4695	$2.88 \pm 2.81 \times 10^{-15}$	4.79×10^{-15}
Br	2990	$9.00 \pm 7.70 \times 10^{-16}$	1.12×10^{-15}
	3835	$2.75 \pm 0.87 \times 10^{-15}$	3.00×10^{-15}
	4090	$3.40 \pm 1.04 \times 10^{-15}$	3.61×10^{-15}
	4310	$3.96 \pm 1.32 \times 10^{-15}$	4.11×10^{-15}
	4840	$5.23 \pm 2.43 \times 10^{-15}$	5.20×10^{-15}
I	3105	$1.44 \pm 0.28 \times 10^{-15}$	—
	3590	$2.18 \pm 0.22 \times 10^{-15}$	—
	3975	$2.70 \pm 0.22 \times 10^{-15}$	—
	4475	$3.24 \pm 0.38 \times 10^{-15}$	—
	4500	$3.27 \pm 0.39 \times 10^{-15}$	—

[a] This work.
[b] Calculated from data contained in Ref. 11.

11 are quite similar and yield activation energies significantly lower than the heats of reaction. The activation energies from the flame studies are essentially equal to the heats of reaction. The difference in the temperature ranges of the two kinds of experiments is so large, however, that these two apparently divergent results are not necessarily in conflict, as the following discussion and graphs will indicate.

Typical rate coefficients for the three studies are compared in Tables XXXII and XXXIII. Agreement between the two shock studies is good. The rate coefficients based on Arrhenius parameters of the flame experiments, computed at roughly half the temperature of the shock studies, are far larger than those observed at these temperatures. How this difference can arise becomes apparent when we look at Arrhenius plots for all the studies put on common axes. These combined plots are shown in Figs. 29–38 for all six salts undergoing reaction R1, dissociation to ions. The combined data define approximate curves, when the entire temperature range is included. Note that this range spans almost a *factor of three*, surely one of the wider ranges over which kinetic measurements of gaseous reaction have been made.

The uncertainty limits shown in these Arrhenius plots were estimated as follows. Burdett and Hayhurst[13] give limits of approximately a factor of 2, but earlier results from flame studies[14] quoted considerably larger uncertainties, of order a factor of 5. We chose to use the larger limit, to try to be

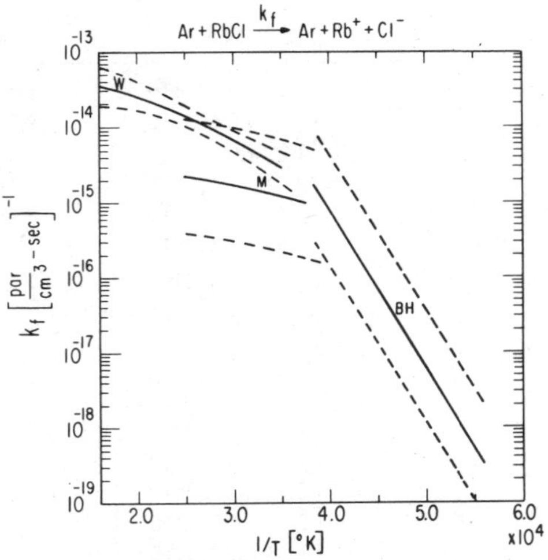

Fig. 29. Arrhenius plot of the best fit to the data of various investigators with estimates of experimental error in the temperature range of their experiments for the reaction $Ar + RbCl \rightarrow Ar + Rb^+ + Cl^-$. BH refers to the work of Burdett and Hayhurst referenced in the text; M refers to the work of Milstein also referenced in the text. W refers to the results of this study.

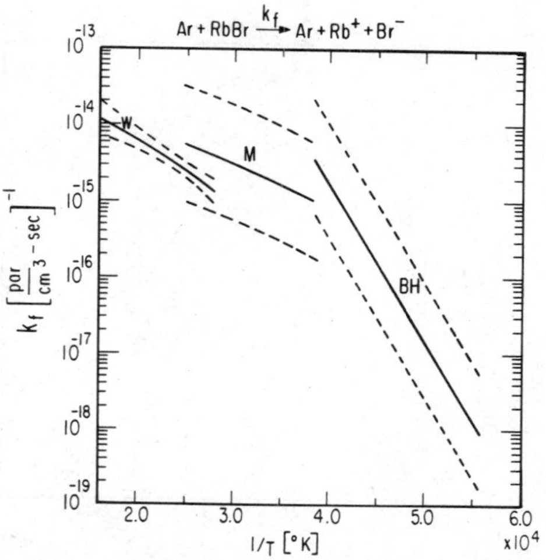

Fig. 30. Arrhenius plot of the best fit to the data of various investigators with estimates of experimental error in the temperature range of their experiments for the reaction $Ar + RbBr \rightarrow Ar + Rb^+ + Br^-$. See caption of Fig. 29 for identifications.

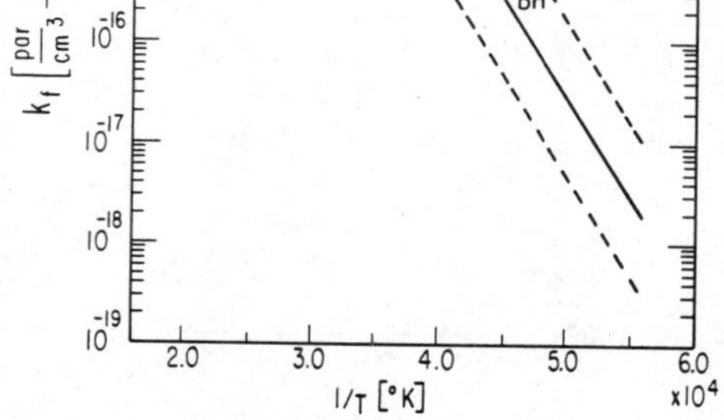

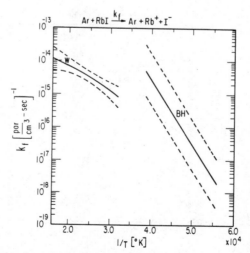

Fig. 31. Arrhenius plot of the best fit to the data of various investigators with estimates of experimental error in the temperature range of their experiments for the reaction $Ar + RbI \rightarrow Ar + Rb^+ + I^-$. See caption of Fig. 29 for identifications.

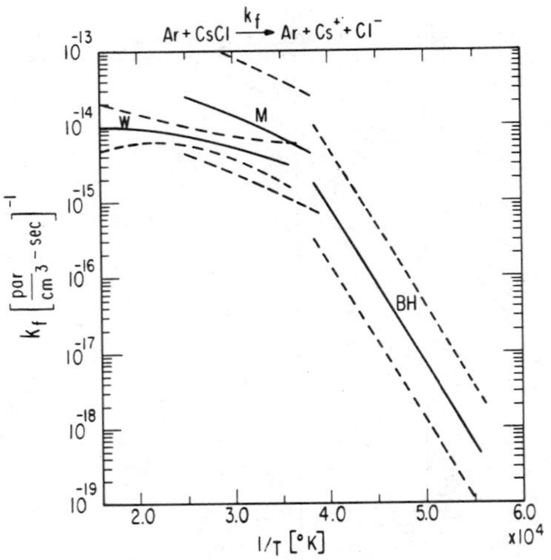

Fig. 32. Arrhenius plot of the best fit to the data of various investigators with estimates of experimental error in the temperature range of their experiments for the reaction $Ar + CsCl \rightarrow Ar + Cs^+ + Cl^-$.

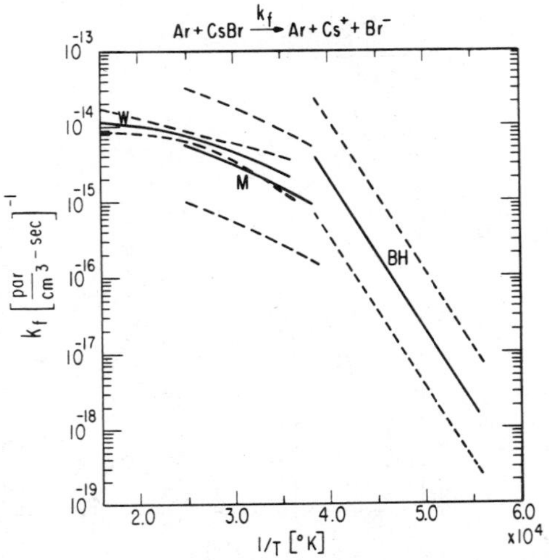

Fig. 33. Arrhenius plot of the best fit to the data of various investigators with estimates of experimental error in the temperature range of their experiments for the reaction $Ar + CsBr \rightarrow Ar + Cs^+ + Br^-$.

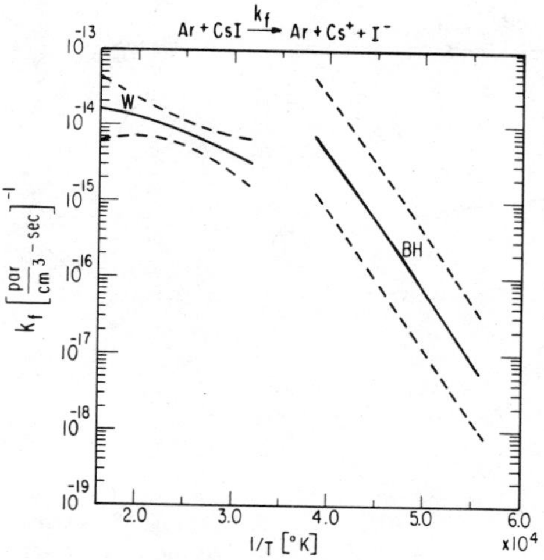

Fig. 34. Arrhenius plot of the best fit to the data of various investigators with estimates of experimental error in the temperature range of their experiments for the reaction $Ar + CsI \rightarrow Ar + Cs^+ + I^-$.

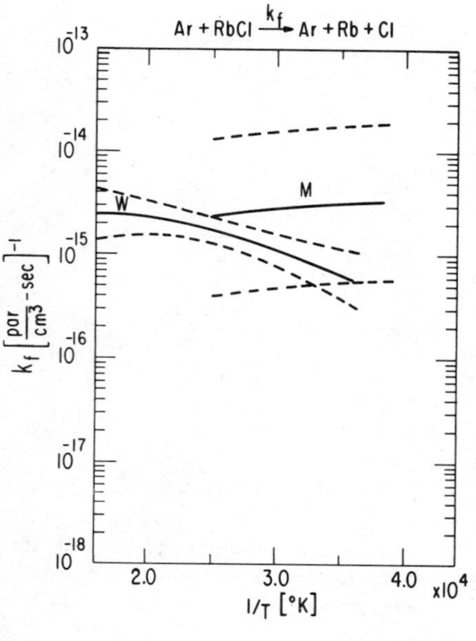

Fig. 35. Arrhenius plot of the best fit to the data of various investigators with estimates of experimental error in the temperature range of their experiments for the reaction $Ar + RbCl \rightarrow Ar + Rb^0 + Cl^0$.

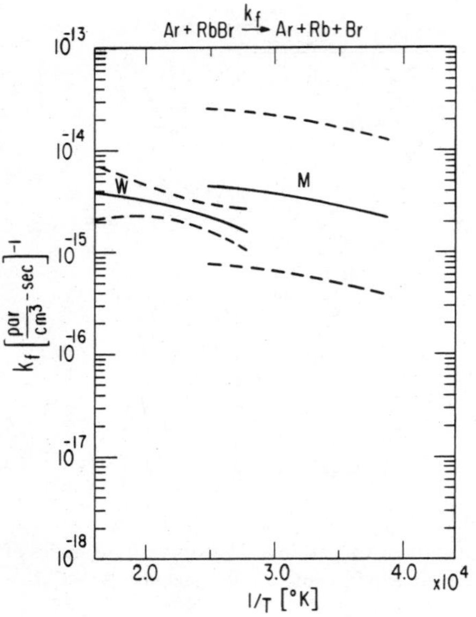

Fig. 36. Arrhenius plot of the best fit to the data of various investigators with estimates of experimental error in the temperature range of their experiments for the reaction $Ar + RbBr \rightarrow Ar + Rb^0 + Br^0$.

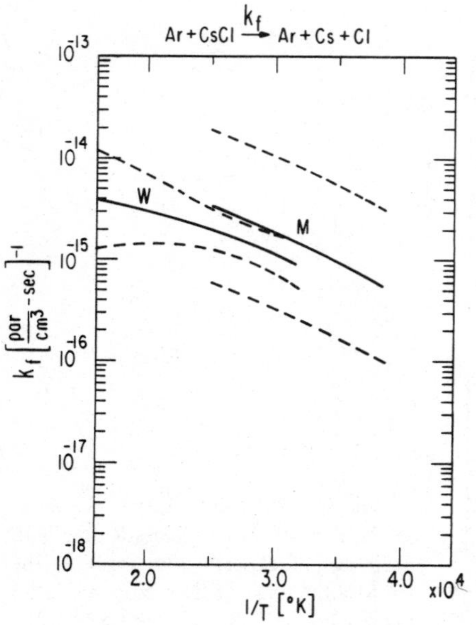

Fig. 37. Arrhenius plot of the best fit to the data of various investigators with estimates of experimental error in the temperature range of their experiments for the reaction $Ar + CsCl \rightarrow Ar + Cs^0 + Cl^0$.

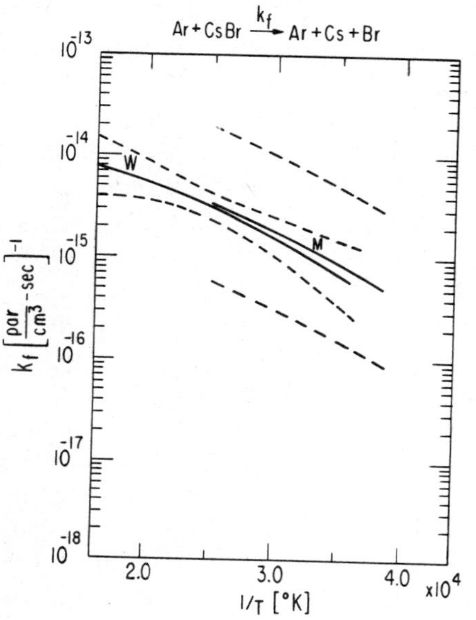

Fig. 38. Arrhenius plot of the best fit to the data of various investigators with estimates of experimental error in the temperature range of their experiments for the reaction $Ar + CsBr \rightarrow Ar + Cs^0 + Br^0$.

consistent with the 90% confidence limits associated with the uncertainties indicated for this work, based on the least-squares fits. The uncertainties in the Arrhenius curves of Ref. 11 are about a factor of 5. With these uncertainties, as indicated in Figs. 29–38, it is easy to picture consistent, smooth, curves for all the processes. The ionic dissociation processes show sharp downslopes in the low-temperature portion of their graphs; no data are available for the corresponding low-temperature region of the dissociation to atoms.

The branching ratio b for dissociation to atoms vs. dissociation to ions is given in a complex system such as ours by

$$b = \frac{k_{7f}}{k_{1f}}$$

The values of b are given in Table XXXIV. It is interesting that although our rate coefficients are generally in fairly good agreement with those of Ref. 11, our results yield far less extreme values of b. For all six of the salts discussed here except RbI, the branching ratio b is 1 or less. If dissociation to atoms is restricted to the one ${}^1\Sigma^+$ channel of the ground electronic state, then a b of 1 corresponds to equal probability for the two channels. If all eight channels corresponding to $M^0({}^2S)+X^0({}^2P_{3/2})$ are available, then a b of 8 corresponds to the atomic and ionic channels being statistically equal. If all

12 channels corresponding to $M^0(^2S)+X^0(^2P_{3/2,1/2})$ are available in the primary dissociation process, then a value of b between 8 and 12, specifically a value of $8+4\exp(-\Delta/kT)$, where Δ is the spin–orbit splitting of the $^2P_{3/2}$ and $^2P_{1/2}$ states, corresponds to statistical equality of the atomic and ionic dissociation. Without more information about the mechanism of the primary dissociation step, we cannot decide among these options. We conclude from our experiments on rate coefficients that the dissociation process occurring in Ar in the temperature range of about 3000–5000 K, for the six salts described here, yields an excess of ions in most cases, but not necessarily for RbBr at the low end of the temperature or for RbI under any of our conditions.

In some of his early experiments, Mandl[5] found rate coefficients for dissociation of CsF to ions that fall between the high values one would obtain from the Arrhenius parameters derived from the flame experiments[13] and the lower values reported here. Dissociation to ions is expected to dominate for CsF,[10] so a high rate coefficient is not surprising, but the absolute magnitude of the CsF results may now seem worth reexamining. The temperature dependence of Mandl's rate coefficient implies that the activation energy for dissociation of CsF is less than the dissociation energy, just as we find. In fact the values of Mandl's rate coefficients and of ours are approximately parallel at high temperatures and appear to converge at lower temperatures.

TABLE XXXIV
Branching Ratios
$b =$ [atom pairs]/[ion pairs] $= k_{7f}/k_{1f}$, at Upper and Lower Limits of the Temperature Ranges of This Study

Salt	T [K]	b
RbCl	2860	0.21
	5810	0.07
RbBr	3795	1.01
	5135	0.48
RbI	3320	8.3
	5945	1.08
CsCl	3145	0.20
	4695	0.32
CsBr	2990	0.31
	4840	0.54
CsI	3105	0.48
	4500	0.32

To get some insight into the factors affecting the rate coefficients, we can write orderings for the processes R1 and R7 as we observe them, as they would be if the dissociation energies or heats of dissociation determined the order, and as they would be if the model of Ewing, Milstein, and Berry[10] were followed in which the branching ratio is determined by the effective width of the adiabatic dissociation channel relative to the vibrational spacing of the diabatic ionic potential. The observed orderings are:

Ionic products	Rate ↑	Atomic products
RbCl		RbI
CsCl ≅ CsBr ≅ CsI		CsBr ≅ CsI
RbBr		RbBr ≅ CsCl
RbI		RbCl

Because of the experimental uncertainties, the placement of nearest neighbors might be reversed, but the overall top-to-bottom order is significant.

If the heats of reaction governed the rates, we would get this altogether different order:

Ionic products	Rate ↑	Atomic products
CsI		RbI
RbI ≅ CsBr		CsI
CsCl ≅ RbBr		RbBr
RbCl		CsBr
		RbCl
		CsCl

This left column is the order of the rate coefficients of Burdett and Hayhurst, consistent with their activation energies being essentially equal to their heats of reaction. The orderings based on the kinetic argument of Ref. 10 are

Ionic products	Rate ↑	Atomic products
RbCl, CsCl, CsBr		RbI
RbBr		CsI
CsI		RbBr
RbI		RbCl, CsCl, CsBr

TABLE XXXV
Comparison of Ionic Recombination Rate Constants

(a) *The Rubidium Halides*: $Ar + Rb^+ + X^- \xrightarrow{k} Ar + RbX$

Species	Temperature [K]	W[a]	BH[b]
Cl	2860	$1.09 \pm 0.63 \times 10^{-28}$	2.99×10^{-28}
	3360	$1.48 \pm 0.48 \times 10^{-29}$	1.85×10^{-28}
	3765	$4.13 \pm 1.17 \times 10^{-30}$	1.31×10^{-28}
	4230	$1.24 \pm 0.44 \times 10^{-30}$	9.25×10^{-29}
	5810	$7.35 \pm 5.31 \times 10^{-32}$	3.57×10^{-29}
Br	3795	$4.53 \pm 1.46 \times 10^{-31}$	1.63×10^{-28}
	3960	$3.18 \pm 0.81 \times 10^{-31}$	1.43×10^{-28}
	4370	$1.44 \pm 0.29 \times 10^{-31}$	1.07×10^{-28}
	4560	$1.04 \pm 0.24 \times 10^{-31}$	9.39×10^{-29}
	5135	$4.38 \pm 1.63 \times 10^{-32}$	6.57×10^{-29}
I	3320	$1.24 \pm 0.98 \times 10^{-30}$	1.69×10^{-28}
	3740	$4.31 \pm 2.06 \times 10^{-31}$	1.19×10^{-28}
	4240	$1.54 \pm 0.64 \times 10^{-31}$	8.13×10^{-29}
	4840	$5.68 \pm 3.34 \times 10^{-32}$	5.47×10^{-29}
	5945	$1.39 \pm 1.46 \times 10^{-32}$	2.95×10^{-29}

(b) *The Cesium Halides*: $Ar + Cs^+ + X^- \xrightarrow{k} Ar + CsX$

Species	Temperature [K]	W[a]	BH[b]
Cl	3145	$1.95 \pm 1.01 \times 10^{-29}$	1.48×10^{-28}
	3300	$9.91 \pm 4.00 \times 10^{-31}$	1.28×10^{-28}
	4010	$8.15 \pm 2.21 \times 10^{-31}$	7.13×10^{-29}
	4155	$5.37 \pm 1.64 \times 10^{-31}$	6.41×10^{-29}
	4695	$1.39 \pm 0.67 \times 10^{-31}$	4.44×10^{-29}
Br	2990	$1.63 \pm 0.96 \times 10^{-29}$	1.53×10^{-25}
	3835	$8.20 \pm 1.89 \times 10^{-31}$	7.27×10^{-29}
	4090	$4.12 \pm 0.85 \times 10^{-31}$	5.99×10^{-29}
	4310	$2.41 \pm 0.50 \times 10^{-31}$	5.12×10^{-29}
	4840	$7.85 \pm 2.15 \times 10^{-32}$	3.62×10^{-29}
I	3105	$2.62 \pm 2.90 \times 10^{-30}$	7.02×10^{-29}
	3590	$6.19 \pm 3.23 \times 10^{-31}$	4.54×10^{-29}
	3975	$2.44 \pm 0.99 \times 10^{-31}$	3.34×10^{-29}
	4475	$8.93 \pm 4.96 \times 10^{-32}$	2.34×10^{-29}
	4500	$8.53 \pm 4.84 \times 10^{-32}$	2.30×10^{-29}

[a] This work.
[b] Calculated from data contained in Ref. 13.

This pattern is clearly considerably closer to the one we observed than is the pattern based on the heats of reaction.

The colliding-beam experiments of Tully et al.[66,67] yield an ordering very different from that predicted by the model of Ewing, Milstein, and Berry.[10] This result is not unexpected, because the latter model is based on a ladder-climbing picture of the dissociation process, in which each molecule accumulates the energy to dissociate by a series of collisions and may, on its way up to the ionic limit, find itself within the adiabatic channel and thereby be able to dissociate adiabatically to neutral atoms. The colliding-beam experiments of course involve dissociation by single energetic collisions, and therefore by an altogether different process.

The Arrhenius parameters for ionic and atomic recombination are given in Tables XXV and XXVI. Some comparisons of rate coefficients themselves for the recombination of ions are given in Table XXXV. All the recombination processes from our studies show rate coefficients that decline with increasing temperature. This is merely a reflection of how the equilibrium at high temperatures favors the dissociated species. The recombination rate coefficients of Burdett and Hayhurst show virtually no temperature dependence. The flame experiments, done at lower temperatures, correspond to conditions under which bound molecules are the dominant form of M^0 and X^0, so the position of equilibrium is essentially unaffected by temperature and the recombination rate coefficient is related to the dissociation rate coefficient through the exponential of the latter, which is essentially $\exp(-\Delta H/kT)$. Under our higher-temperature conditions, this is no longer the case, the activation barrier is less than ΔH, and the equilibrium shifts noticeably as T increases, thus making the rate coefficient for recombination drop accordingly.

Acknowledgments

We would like to thank Tetsushi Noguchi for his assistance in carrying out the shock experiments; the Standard Oil Company of Indiana, and especially Glen Whitesell and Richard Palus, for their gracious permission and assistance in using their hybrid computer; and John Duncanson, Anders Lindgård, Peter Salamon, and Hung-tai Wang for many helpful suggestions and discussions.

The major part of this research was supported by a Grant from the Army Research Office, with some assistance in the final stages by a Grant from the National Science Foundation.

APPENDIX: STATISTICAL ANALYSIS OF THE DATA

The following discussion reviews the methods used for calculating χ^2 and percentage deviations of the experimental data for the atomic alkali concentrations and halide-ion concentrations at each temperature. Then the significance of the measurements is scrutinized on the basis of these measures.

The chi-square statistic χ^2 is defined in terms of the frequencies $f_{c,k}$ and $f_{o,k}$ at which points are, respectively, calculated and observed to fall in each kth interval, out of K possible intervals:

$$\chi^2 = \sum_{k=1}^{K} \frac{(f_{c,k} - f_{o,k})^2}{f_{c,k}} \tag{A.1}$$

To apply this, we assumed that the data points are distributed normally about the theoretical curve, that the mean of the data falls on the theoretical curve, and that at each value of the independent variables the standard deviation of the distribution is the same. We choose a division of the normal distribution into K parts, each representing the same probability (K intervals of equal area). The value of K and the number of data points N must satisfy[68,69]

$$\frac{N}{K} \geq 2.5 \tag{A.2}$$

Our average number of data points per run was 13, taken in equally spaced intervals along the film, so K was taken to be 5, and the expected frequency $f_{c,K} = 0.2$. The observed frequency $f_{o,k} = n_k/N$, where n_k is the number of data points in interval k. With our choice of K, then

$$\chi^2 = \frac{5\Sigma n_k^2}{N^2} - 1 \tag{A.3}$$

The χ^2 statistic was calculated for each of the measured species of each run, as follows. The approximate mean $\overline{m}$ and standard deviation s were computed from the value of the concentration $\bar{y}_i$ given by the analog curve for the ith point and the measured concentration y_i at the same point along the time (equivalent to film-distance) axis:

$$\overline{m} = \frac{1}{N} \sum_{i=1}^{N} (y_i - \bar{y}_i) \tag{A.4}$$

and

$$s = \left[\frac{1}{N-1} \sum_{i=1}^{N} (y_i - \bar{y}_i)^2 \right]^{1/2} \tag{A.5}$$

(The true unobservable mean μ and standard deviation σ of the distribution are approximated in the usual way by $\overline{m}$ and s.)

From standard tables of χ^2 for normal distributions, we can determine the confidence level P at which we are justified in supposing the data are

TABLE XXXVI
Summary of Statistics for Analog Fit to Experimental Data

Alkali Halide	Temperature	Species	N	s [$\times 10^{-2}$ m.u.]	χ^2	P ($\nu = 2$)	%Dev
RbCl	2860	M^0	12	4.63	0.875	65%	14.3
		X^-	15	4.08	0.311	86%	8.3
	3360	M^0	12	2.79	1.083	58%	17.6
		X^-	13	2.79	2.639	27%	18.7
	3765	M^0	12	6.48	0.736	69%	22.2
		X^-	13	6.86	0.627	73%	21.6
	3935	M^0	13	6.79	1.574	46%	26.2
		X^-	16	12.01	0.328	85%	62.6
	4230	M^0	13	6.57	0.154	93%	21.4
		X^-	13	4.20	0.391	82%	73.2
	5810	M^0	12	1.53	0.389	82%	16.1
		X^-	13	8.39	2.639	27%	99.4
RbBr	3795	M^0	18	4.74	0.080	96%	27.8
		X^-	17	7.56	0.159	92%	13.9
	3960	M^0	13	4.67	0.686	71%	19.6
		X^-	13	7.02	1.396	50%	11.4
	4370	M^0	13	7.85	0.982	61%	34.1
		X^-	16	3.78	0.523	77%	21.0
	4560	M^0	13	4.54	0.272	87%	23.9
		X^-	12	1.43	1.569	46%	7.8
	5000	M^0	12	2.92	0.181	91%	10.6
		X^-	13	5.39	0.627	73%	28.9
	5135	M^0	11	4.72	0.446	80%	32.3
		X^-	12	5.71	1.014	60%	24.7
RbI	3320	M^0	13	6.34	0.391	82%	17.7
		X^-	12	5.20	1.569	46%	18.4
	3550	M^0	12	5.75	0.319	85%	10.3
		X^-	11	1.40	0.529	77%	11.9
	3740	M^0	17	4.70	0.540	76%	16.1
		X^-	13	2.64	0.864	65%	16.7
	4240	M^0	13	6.03	0.272	87%	11.7
		X^-	13	1.59	1.101	58%	56.6
	4840	M^0	12	4.67	0.944	62%	18.1
		X^-	13	5.91	0.272	87%	83.7
	5945	M^0	11	1.51	1.273	53%	17.7
		X^-	13	2.88	3.290	19%	40.9
CsCl	3145	M^0	16	5.63	0.250	88%	12.4
		X^-	13	10.56	0.450	80%	25.8
	3260	M^0	11	3.01	1.521	47%	20.4
		X^-	15	7.45	0.400	82%	16.7
	3300	M^0	0	—	—	—	—
		X^-	16	4.83	0.445	80%	33.9
	4010	M^0	11	1.38	0.612	74%	7.6
		X^-	16	1.00	0.133	94%	39.5

TABLE XXXVI (*Continued*)

Alkali Halide	Temperature	Species	N	s [$\times 10^{-2}$ m.u.]	χ^2	P ($\nu = 2$)	%Dev
	4155	M^0	0	—	—	—	—
		X^-	12	4.09	0.042	98%	31.1
	4695	M^0	0	—	—	—	—
		X^-	13	3.97	0.686	71%	105.1
CsBr	2990	M^0	16	6.59	0.484	79%	13.4
		X^-	13	7.01	0.627	73%	15.7
	3835	M^0	13	2.73	0.331	85%	16.6
		X^-	13	1.98	0.154	93%	3.7
	4090	M^0	17	2.93	0.194	91%	7.3
		X^-	12	3.22	0.736	69%	11.9
	4310	M^0	16	4.40	0.406	82%	9.6
		X^-	12	2.02	1.431	49%	13.3
	4840	M^0	13	4.19	0.154	93%	13.1
		X^-	12	1.43	0.042	98%	29.3
CsI	3105	M^0	13	11.68	1.041	59%	100.6
		X^-	13	6.80	1.219	54%	26.5
	3590	M^0	12	8.19	0.389	82%	29.9
		X^-	13	4.77	2.225	33%	34.5
	3975	M^0	13	3.13	0.095	95%	10.6
		X^-	13	1.23	0.450	80%	12.9
	4475	M^0	12	6.09	0.319	85%	38.1
		X^-	13	8.44	1.041	59%	52.1
	4500	M^0	13	4.94	1.278	53%	18.5
		X^-	13	2.30	0.371	82%	12.3

indeed normally distributed about the theoretical (analog) curve with a mean equal to the value of the theoretical curve. For this step of our analysis we note that we are using two free parameters, the mean and standard distribution of the observed concentrations, so the number of degrees of freedom, ν, which is $K-1$ less the number of free parameters, is 2. The values of N, s, χ^2, and P are collected in Table XXXVI. The standard deviations s are given in machine units rather than absolute concentrations and should only be taken as indicative of the variability in the scatter of the data. The conversion factors for the alkali concentrations are given in Table XXXVII below.

Table XXXVI also contains another statistic, the average percentage deviation, %Dev, defined as

$$\%\text{Dev} = \frac{1}{N} \sum_{i=1}^{N} \frac{|y_i - \bar{y}_i|}{\bar{y}_i} \times 100. \tag{A.6}$$

The percentage deviation and confidence levels are generally adequate; the exceptions occur for the halide ions at the very highest temperatures. The quality of fits of the data reported here is comparable with but slightly less than that of Ref. 11. Measurements at the lower temperatures could be made a little more precisely and, the results suggest, perhaps with a little higher accuracy.

The reliability of the data can be judged by combining the means and standard deviations to perform a t test as to whether the means of the observed differences differ significantly from the assumed (analog) distribution. At the 95% confidence level, 75% of the means, and at the 90% confidence level, 90% of the means, do not differ significantly from the values given by the analog fits to the data.

Thus the reason for some low values of the confidence levels P appears to lie in the occasional deviation of the data points from a normal distribution about the mean, rather than in these being significant differences. This deviation is quite visible in a few of the cases with high χ^2 values, where the analog curve clearly deviates in a small systematic way from the data, over part of the duration of the experiment. Such deviation can be seen in Fig. 4, although that example gives a χ^2 of only 0.523 and P of 77% for the Br^- ion.

The occasional large values of %Dev arise primarily from cases in which the denominator $\bar{y}_i$ is the value of a halide concentration at a high temperature and hence is a small number. Secondarily, this statistic is sensitive to large deviations in one or two data points. The regions of the steep initial portions of the analog curves are particularly vulnerable to such errors. Because Eq. (A.6) reflects only deviations in the ordinate, a data point may be very close to the curve yet make a large contribution to the %Dev. (We did not construct a statistic based on the perpendicular distance from each point to the analog curve).

Another approach to quantifying the quality of the fit could be carried out for the metal-atom concentrations for those of the runs for which two or three principal-series lines could be used to measure those concentrations independently. To compare these data with the standard deviations of the data points about the analog curves, we calculated means of the experimental points and standard deviations of the experimental points from their means. From these standard deviations we evaluated rms deviations

$$s(av) = \frac{1}{N'} \sum_{j=1}^{N'} s_j^2 \tag{A.7}$$

where N' is the number of data points for a run in which $[M^0]$ was determined for more than one line. Figure 2 showed the data points, their mean, and their standard deviations for one such run. Table XXXVII displays a

TABLE XXXVII
Comparison of Estimates to the Standard Deviation of the Data About the Experimental Mean or the Analog Curve

Sample		Calculated About Experimental Mean			Calculated About Analog Curve		Conversion Factor
Alkali Halide	Temperature	Number of Lines	N'	s(av) [$\times 10^{-2}$ m.u.]	N	s [$\times 10^{-2}$ m.u.]	(cm^{-3}/m.u.)
RbCl	2860	2	9	7.83	12	4.63	1×10^{16}
RbCl	3360	2	12	5.58	12	2.79	1×10^{16}
RbCl	3765	2	12	8.25	12	6.48	5×10^{15}
RbCl	3935	2	13	5.84	13	6.79	1×10^{16}
RbCl	4230	2	11	9.37	13	6.57	5×10^{15}
RbBr	3795	2	12	6.98	18	4.74	5×10^{16}
RbBr	3960	2	12	5.24	13	4.67	2×10^{16}
RbI	3320	2	13	4.62	13	6.34	2×10^{16}
RbI	3550	3	12	12.26	12	5.75	2×10^{16}
RbI	3740	3	12	6.66	17	4.70	2×10^{16}
CsBr	2990	2	10	7.07	16	6.59	1×10^{16}
CsBr	3835	2	11	5.10	13	2.73	2×10^{16}
CsBr	4090	2	16	9.57	17	2.93	2×10^{16}
CsI	3105	3	13	9.37	13	11.68	1×10^{16}
CsI	3590	2	11	7.18	12	8.19	1×10^{16}
CsI	3975	2	11	5.04	13	3.13	2×10^{16}

comparison of the standard deviations $s(av)$ and s based on the analog curve, as defined in Eq. (A.5). Again, the standard deviations are given in machine units; the conversion factors to concentration units are included in the table. The standard deviations of the points about the analog curves are comparable to and generally less than those about the experimental means.

To recapitulate, the analog curves are good representations of the time evolution of the concentrations of the observed species at the 90% confidence level, and lie within the estimated experimental uncertainties of the concentration profiles.

REFERENCES

1. V. R. Hartig, H. A. Olschewski, J. Troe, and H. G. Wagner, *Ber. Bunsenges. Physik. Chem.* **72**, 1016 (1968).
2. K. Luther, J. Troe, and H. G. Wagner, *Ber. Bunsenges. Physik. Chem.* **76**, 53 (1972).
3. A. Mandl, E. W. Evans, and B. Kivel, *Chem. Phys. Lett.* **5**, 307 (1970).
4. A. Mandl, B. Kivel, and E. W. Evans, *J. Chem. Phys.* **53**, 2363 (1970).
5. A. Mandl, *J. Chem. Phys.* **55**, 2918 (1971).
6. A. Mandl, *J. Chem. Phys.* **64**, 903 (1976).

7. A. Mandl, in *Alkali Halide Vapors: Structure, Spectra and Reaction Dynamics*, P. Davidovits and D. L. McFadden (eds.), Academic, New York, 1979, Chapter 12, pp. 389 ff.
8. R. S. Berry, T. Cernoch, M. Coplan, and J. J. Ewing, *J. Chem. Phys.* **49**, 127 (1968).
9. M. Coplan, T. Cernoch, and R. S. Berry, *Phys. Fluids* **1969**, Suppl. 1, I-118 (1969).
10. J. J. Ewing, R. Milstein, and R. S. Berry, *J. Chem. Phys.* **54**, 1752 (1971).
11. R. Milstein and R. S. Berry, J. Chem. Phys. **80**, 6025 (1984). see also R. Milstein, Doctoral Dissertation, Univ. of Chicago, 1972.
12. A. N. Hayhurst and T. M. Sugden, *Trans. Farad. Soc.* **63**, 1375 (1967).
13. N. A. Burdett and A. N. Hayhurst, *Phil. Trans. Roy. Soc. London* **290**, 299 (1979).
14. A. F. Ashton and A. N. Hayhurst, *Combustion and Flame* **21**, 69 (1973).
15. N. A. Burdett and A. N. Hayhurst, *Chem. Phys. Lett.* **48**, 95 (1977).
16. P. D. Gait and R. S. Berry, *J. Chem. Phys.* **66**, 2387, 2764 (1977).
17. P. D. Gait and R. S. Berry, in *Electronic Transition Lasers*, Vol. II, L. E. Wilson, S. N. Suchard, and J. I. Steinfeld (eds.), MIT, Cambridge, Mass., 1977, p. 211.
18. J. N. Weber, Doctoral Dissertation, Univ. of Chicago, 1981.
19. C. G. James and T. M. Sugden, *Nature* **171**, 428 (1953).
20. Eastman Kodak Co., *Kodak Plates and Films for Scientific Photography*, Eastman Kodak, Rochester, N.Y., 1973.
21. T. H. James and G. C. Higgins, *Fundamentals of Photographic Theory*, Morgan and Morgan, Hastings-on-Hudson, N.Y., 1968, Chapter 4.
22. G. V. Marr and D. M. Creek, *Proc. Roy. Soc. (London)* **A304**, 245 (1968).
23. G. McGinn, *J. Chem. Phys.* **50**, 1404 (1969).
24. J. C. Weisheit, *Phys. Rev. A* **5**, 1621 (1972).
25. R. J. Exton, *J. Quant. Spectros. Rad. Transfer* **16**, 309 (1976).
26. G. Pichler, *J. Quant. Spectros. Rad. Transfer* **16**, 147 (1976).
27. D. W. Norcross, *Phys. Rev. A* **7**, 606 (1973).
28. H. P. Popp, *Phys. Reports* **16C**, 170 (1975).
29. H. W. Drawin and P. Felenbok, *Data for Plasmas in Local Thermodynamic Equilibrium*, Gauthier-Villars, Paris, 1965.
30. N. Davidson, *Statistical Mechanics*, McGraw-Hill, New York, 1962, Chapter 8.
31. K. S. Pitzer, *Quantum Chemistry*, Prentice-Hall, New York, 1953, p. 210 and Appendix 14.
32. J. E. Mayer and M. G. Mayer, *Statistical Mechanics*, Wiley, New York, 1940, Chapter 7.
33. C. E. Moore, *Ionization Potentials and Ionization Limits Derived from the Analysis of Optical Spectra*, NSRDS-NBS Publication 34, U.S. Government Printing Office, 1970.
34. H. Hotop and W. C. Lineberger, *J. Phys. Chem. Ref. Data* **4**, 539 (1975).
35. K. P. Huber and G. Herzberg, *Constants of Diatomic Molecules*, Van Nostrand Reinhold, New York, 1979.
36. A. Mandl, *J. Chem. Phys.* **54**, 4129 (1971).
37. A. Mandl, *J. Chem. Phys.* **57**, 5617 (1972).
38. A. Mandl, *J. Chem. Phys.* **59**, 3423 (1973).
39. A. Mandl, *J. Chem. Phys.* **65**, 2483 (1976).
40. H. S. W. Massey, E. H. S. Burhop, and H. B. Gilbody, *Electronic and Ionic Impact Phenomena*, 2nd ed., Vol. 2, Clarendon, Oxford, 1969, Chapters 12, 13.

41. R. A. Willoughby (ed.), *Stiff Differential Systems*, Plenum, New York, 1974.

42. A. Jones, in *Reaction Kinetics*, Vol. 1, P. G. Ashmore (ed.), Chemical Society, London, 1975, Chapter 7.

43. A. S. Jackson, *Analog Computation*, McGraw-Hill, New York, 1960.

44. N. R. Drapper and H. Smith, *Applied Regression Analysis*, Wiley, New York, 1966.

45. R. J. Cvetanović, R. P. Overend, and G. Paraskevopoulos, *Int. J. Chem. Kinetics Symp.* **1**, 249 (1975).

46. J. Kunc and M. Zgorzelski, *Atomic Data and Nuclear Data Tables* **19**, 1 (1977).

47. D. R. Bates, A. E. Kingston, and R. W. P. McWhirter, *Proc. Roy. Soc. (London)* **A267**, 297 (1962).

48. H. S. W. Massey, E. H. S. Burhop, and H. B. Gilbody, *Electronic and Ionic Impact Phenomena*, 2nd ed., Vol. 4, Clarendon, Oxford, 1974, p. 2237.

49. B. P. Curry, *Phys. Rev. A* **1**, 166 (1970).

50. G. Gousset, B. Sayer, and J. Berlande, *Phys. Rev. A* **16**, 1070 (1977).

51. J. J. Thompson, *Phil. Mag.* **47**, 337 (1924).

52. L. P. Pitaevski, *Sov. Phys. JETP* **15**, 919 (1962).

53. D. R. Bates and S. P. Khare, *Proc. Phys. Soc.* **85**, 231 (1965).

54. R. L. Champion and L. D. Doverspike, *Phys. Rev. A* **13**, 609 (1976).

55. R. L. Champion and L. D. Doverspike, *J. Chem. Phys.* **65**, 2482 (1976).

56. B. T. Smith, W. R. Edwards, L. D. Doverspike, and R. L. Champion, *Phys. Rev. A* **18**, 945 (1978).

57. M. J. Wynn, J. D. Martin, and T. L. Bailey, *J. Chem. Phys.* **52**, 191 (1970).

58. R. S. Berry, *J. Chim. Phys.* **77**, 759 (1980).

59. F. H. M. Faisal and A. K. Bhatia, *Phys. Rev. A* **5**, 2144 (1972).

60. B. Peart, D. S. Walton, and K. T. Dolder, *J. Phys. B* **3**, 1346 (1970).

61. B. Peart, R. Grey, and K. T. Dolder, *J. Phys. B* **9**, L369, L373 (1976).

62. M. I. Chibisov, *Sov. Phys. JETP* **22**, 593 (1966).

63. R. K. Janev and Z. M. Radulović, *Phys. Rev. A* **17**, 889 (1978).

64. D. R. Bates, *Adv. Atom. Molec. Phys.* **15**, 235 (1979).

65. D. R. Bates, *Proc. Roy. Soc. (London)* **A369**, 327 (1980).

66. F. P. Tully, N. H. Cheung, H. Haberland, and Y. T. Lee, *J. Chem. Phys.* **73**, 4460 (1980).

67. F. P. Tully, Y. T. Lee, and R. S. Berry, *Chem. Phys. Lett.* **9**, 80 (1971).

68. A. H. Bowker and G. T. Lieberman, *Engineering Statistics*, 2nd ed., Prentice-Hall, Englewood Cliffs, N.J., 1972, pp. 458–461.

69. S. L. Meyer, *Data Analysis for Scientists and Engineers*, Wiley, New York, 1975, Chapters 26 and 32.

THE VIRIAL THEOREM

GUILHEM MARC[‡] AND W. G. McMILLAN

Department of Chemistry and Biochemistry
University of California, Los Angeles
Los Angeles, California 90024

CONTENTS

[‡]Based on the Dissertation of Guilhem Marc for the degree of Doctor of Philosophy in Chemistry, University of California, Los Angeles, 1983.

I. INTRODUCTION

A. Motivation

The virial theorem (VT) is a corollary of the equation of motion of the (linear) momentum, that is, of Newton's second law and its quantum-mechanical counterpart. It is a very fundamental and general relation that can be placed on almost the same plane as the conservation laws for momentum, angular momentum, and energy. As with these conservation laws, the range of validity of the VT includes all domains of physics: classical mechanics, quantum mechanics, and statistical mechanics, including both the special and general relativistic regimes. The VT is thus a remarkable theoretical tool that has found numerous applications in physical problems ranging from molecules to galaxies, from nuclear physics to cosmology. This wide range of applicability is reflected in the large number of research articles that have appeared in the literature since the VT was first discovered by Clausius[‡] in 1870.

Despite its great utility, the VT has not received a correspondingly wide prominence and recognition, either at the elementary level or even in graduate textbooks. Typically, the only exposure of most students to the word "virial" is in the virial equation of state and the associated virial coefficients, mention being made only rarely of the VT as a basis for this equation. In part, this obscurity of the VT results from the lack of any comprehensive treatment; the few review articles[1–5] that exist are very brief and of limited scope.

To help rectify this obscurity, this work comprises a review of the various formulations, generalizations, extensions, and applications of the VT. It is addressed to graduate students of physics and physical chemistry. Although there are obvious difficulties in encompassing such a multidisciplinary subject, the treatment is fairly comprehensive and self-contained. The VT cannot be appreciated in isolation from the diverse contexts in which it is embedded; consequently, we have recalled, wherever needed, all of the fundamental results and background information necessary to the comprehension of each subject treated.

We have also attempted to trace the historical development of the VT. The text is sprinkled with historical notes, mainly concerning the early history of the VT, to which are attached the names of many pioneers of modern science—Clausius, Maxwell, Rayleigh, Poincaré, Eddington, to cite only a few of the earliest.

[‡]Attributions for various developments are given in the order in which they appear in the body of the text, with references listed at the end.

Following this Introduction, the work is split into two main sections: II. General Theory and III. Applications, the contents of which are summarized below.

B. Summary

Section II, General Theory, presents various alternative derivations of the VT within the domains of classical mechanics (Section II.A) and quantum mechanics (Section II.B), while stressing the parallel between the two formalisms. Section II concludes with a few relativistic generalizations (Section II.C).

Starting with Clausius' original derivation, the scalar product of Newton's Second Law with the position vector $\vec{r}_a$ of particle a yields the equation of motion for the position·momentum product:

$$\frac{d}{dt}(\vec{r}_a \cdot \vec{p}_a) = \vec{r}_a \cdot \vec{F}_a + 2T_a \tag{1.1}$$

Here $\vec{p}_a$ is the linear momentum of particle a, $\vec{F}_a$ is the total force on particle a, and T_a its kinetic energy.

Upon averaging over time with the assumption that the quantity $\vec{r}_a \cdot \vec{p}_a$ remains bounded, the VT for particle a becomes

$$2\overline{T}_a = -\overline{\vec{r}_a \cdot \vec{F}_a} \tag{1.2}$$

Summing over all particles, the VT for the whole system is then

$$2\overline{T} = 2\sum_a \overline{T}_a = -\overline{\sum_a \vec{r}_a \cdot \vec{F}_a} \tag{1.3}$$

that is, twice the average total kinetic energy is equal to the virial of the forces for the total system.

For a system of particles enclosed in volume V and exerting a uniform pressure P on the wall of the container, the contribution to the virial of the forces exerted by the wall is shown to be:

$$-\left(\overline{\sum_a \vec{r}_a \cdot \vec{F}_a}\right)_{\text{wall}} = 3PV \tag{1.4}$$

This is the basis of the applications of the VT in determining the equations of state for gases, plasmas and solids (without shear strength) under pressure (Section III.B).

The virial of internal forces that derive from a potential-energy function U that is homogeneous in the particle coordinates turns out to be proportional to the potential energy itself. The VT then takes a form, particularly useful in applications, that expresses the balance between the average kinetic and potential energies of the system:

$$3PV = 2\overline{T} - \gamma\overline{U} \tag{1.5}$$

where γ is the degree of homogeneity of the potential-energy function. That this simple form requires the potential energy to be a homogeneous function of the coordinates may appear as a serious limitation. But we note that the electrostatic potential that governs all Coulomb interactions of charges—and thus the whole of chemistry—is homogeneous of degree $\gamma = -1$, as is also the gravitational potential that governs the motion of stars, star clusters, and galaxies. Since together the Coulomb and gravitational potentials encompass a very large body of chemistry, physics, and astronomy, the virial theorem in the form

$$3PV = 2\overline{T} + \overline{U} \tag{1.6}$$

has an enormous range of applicability. Further, the harmonic-oscillator model, which applies to all vibrations of small amplitude, has a potential-energy function that is homogeneous of degree $\gamma = +2$.

By including mixed terms in the position•momentum product, one obtains a more general tensor relation whose antisymmetric part is the equation of motion for the angular momentum, and whose symmetric part is the equation of motion for the moment of inertia. The trace of this symmetric part is simply the scalar VT.

Generalizing further, the VT is one of many exact relations, collectively included by Hirschfelder under the designation *hypervirial theorem* (HVT). These relations are obtained by taking the moment of any fundamental equation of dynamics with any arbitrary dynamical quantity. The applications of the HVT are reviewed in Section III.E, and the method of moments is systematically applied to the Fokker–Planck equation in Section III.F.

Still in the classical domain, the VT is rederived in the Hamiltonian formalism, then in the Lagrangian formalism using the principle of least action. It is also shown that a variational principle leads directly to the VT with the help of a coordinate scale transformation. In elaboration of this result, the relations between invariance and conservation laws are examined. It emerges that the position•momentum product is the generator of the contact transformation that performs the scaling.

In Section II.B, the development of the quantum-mechanical VT follows closely its classical antecedent. However, it proves beneficial here to invert the order to take advantage of the facility of scaling. To this end the discussion starts with the position•momentum product operator as the generator of the unitary transformation that scales the coordinates. Then, the quantum-mechanical VT is obtained successively from the variational principle, from the Heisenberg equation of motion, and from the Schrödinger equation. The quantum-mechanical tensor generalizations and the HVT are also given.

For enclosed systems, the quantum-mechanical VT and HVT require an extra term, a surface integral, that can be written as a generalized current density.

At this stage, we introduce the energy derivative that leads to the Hellmann–Feynman theorem (HFT), which is often used in conjunction with the VT and HVT. A general form of the VT for systems subject to external constraints is derived, combining the VT and HFT; this applies, for example, to molecules, particles in a box, and the electron gas.

Section II concludes with three relativistic generalizations of the VT for, successively:

1. A single non-quantum-mechanical relativistic particle.
2. A system of non-quantum-mechanical relativistic particles in interaction with an electromagnetic field.
3. A Dirac particle in an electromagnetic field.

The enormous scope of applications of the VT to physical problems has been remarked earlier. Section III, which deals with these applications, is divided into six subsections. Each subsection is self-contained and can be read independently of the others.

In Section III.A, which treats unconstrained systems involving only inverse-square forces, the virial relation,

$$\overline{E} = -\overline{T} = \tfrac{1}{2}\overline{U} < 0 \tag{1.7}$$

is used to explain the stability of atoms—and matter generally—as the result of the balance between the attractive potential energy of the electrons and their kinetic energy that tends to resist collapse. The same relation can be used also to predict an order of magnitude for the mass of stars. Since the mean kinetic energy is proportional to the mean temperature (equipartition theorem), the loss of energy of a star by radiation raises its mean temperature (negative heat capacity!). The VT removes the mystery: the loss of energy through radiation is accompanied by a contraction that makes the potential energy more negative and the mean kinetic energy—therefore the mean temperature—more positive. Finally, in Section III.A a simple and accurate model based on the VT is shown to predict that the cohesive energy (per atom) of monovalent metals should nearly equal the average kinetic energy of the valence electrons in the Fermi sea.

In Section III.B, the VT in the form

$$3PV = 2\overline{T} - \sum_{a} \overline{\vec{r}_a \cdot \vec{\nabla}_a U} \tag{1.8}$$

provides a way to calculate the equation of state of matter under various conditions. In particular, the VT constitutes a theoretical foundation for the virial equation of state that derives its name from the theorem. Contrary to the statements of some researchers in the early 1950s, the quantum-mechanical pressure calculated from the VT is the same as that calculated from statistical mechanics. The equations of state for three- and two-dimensional plasmas are also derived with, respectively, screened Coulomb and logarithmic potentials. The utility of the VT as a pedagogical tool is illustrated here by deriving statistical-mechanical results usually obtained by more complicated means. Many applications of the VT to the solid state are also reviewed in Section III.B. A model based on the VT that gives an equation of state for many solids under pressure is selected for its simplicity, generality, and excellent agreement with experiment.

Since the VT is an exact relation, there is no *a priori* reason that an approximate model will satisfy this requirement, although this is a necessary (but not sufficient) condition for the model to be accurate. We see in Section III.C that the two main approximations of quantum mechanics for many-particle systems—namely, the self-consistent-field (SCF) and the statistical model—do indeed satisfy the VT. The SCF method is first outlined, including the Hartree and Hartree–Fock methods, the Slater local-exchange approximation, and the X_α method. The results of many recent applications involving the VT in this context are sketched. Next we review the statistical model of Thomas and Fermi, improved by the Dirac exchange correction. Finally, in Section III.C, we prove directly that the VT is satisfied in both the TF and TFD models.

Section III.D deals with the application of the VT to molecules. This includes a discussion of the Born–Oppenheimer approximation, the chemical bond, and a molecular partitioning technique. We present in this section a new application in which the VT for the Schrödinger equation governing the nuclear motion in the Born–Oppenheimer approximation is derived and applied, with the help of the HFT, using the reduced mass as parameter, to the isotope effect in diatomic molecules for an arbitrary potential function.

Section III.E is devoted to the HVT. Collectively, the HVTs constitute a wide class of generalizations of the VT and have found many applications in recent years, mainly as constraints to improve variational wavefunctions and to calculate average values without the need for integral computations. We review these main applications, then focus on what we consider to be the most promising: a recent perturbation technique where the perturbed wavefunction is not needed to calculate exactly the energy series and certain other expectation values. So far, this technique has been applied only to one-dimensional problems: the harmonic oscillator and the hydrogen atom with a radial perturbation. Section III.E concludes with the derivation of a sum rule useful in radiation theory. It is based on the HVT and a formula derived by Hausdorff.

In Section III.F, the method of moments is applied systematically to the Fokker–Planck equation to obtain time-dependent generalizations of the virial and fluctuation–dissipation theorems, with a disgression on Langevin's equation and Einstein's relation between the diffusion coefficient and the mobility.

References to a number of other works on the VT not discussed here are included in the Appendix. These references are classified under the headings: Mathematics; Relativistic Quantum Mechanics; Scattering; Surfaces; Low Temperatures; and Astrophysics, General Relativity, and Magnetohydrodynamics.

II. GENERAL THEORY

A. The Classical Virial Theorem

"The mean *vis viva* of the system is equal to its *virial.*"—Clausius.

1. The Virial Theorem from Newton's Second Law

In the above quotation, Rudolf Clausius[6] summarized the essence of his virial theorem (VT), first published in 1870. Today the "mean *vis viva*" is better known as the average kinetic energy, although the term "*virial*" is still retained and lends its name to this powerful and very general theorem. The VT applies to any system of particles whose positions and momenta remain bounded. Its proof, essentially the original one given by Clausius, runs as follows.

Let $\vec{F}_a$ be the total force acting on particle a, which has mass m_a, position $\vec{r}_a$, and momentum $\vec{p}_a \equiv m_a\dot{\vec{r}}_a$. Newton's second law reads

$$\vec{F}_a = m_a\ddot{\vec{r}}_a = \dot{\vec{p}}_a \tag{2.1}$$

where the superior dots indicate time derivatives. The scalar product of this equation by the position vector $\vec{r}_a$ contains the kinetic energy T_a of particle a:

$$\vec{r}_a\cdot\vec{F}_a = \vec{r}_a\cdot\dot{\vec{p}}_a = \frac{d}{dt}(\vec{r}_a\cdot\vec{p}_a) - m_a\dot{\vec{r}}_a^2$$

or

$$\frac{d}{dt}(\vec{r}_a \cdot \vec{p}_a) = \vec{r}_a \cdot \vec{F}_a + 2T_a \tag{2.2}$$

This equation may be called the time-dependent or dynamic virial theorem, and is simply the equation of motion for $\vec{r}_a \cdot \vec{p}_a$.

It proves fruitful to invoke the time average, defined for any function f by

$$\bar{f} = \frac{1}{\tau}\int_0^\tau f\,dt \tag{2.3}$$

where τ is an arbitrary time interval (but usually assumed to be long compared with the characteristic time for the system).

Applying time averaging, denoted by the vinculum, to Eq. (2.2) yields

$$\frac{1}{\tau}[\vec{r}_a \cdot \vec{p}_a]_0^\tau = \overline{\vec{r}_a \cdot \vec{F}_a} + 2\overline{T}_a \tag{2.4}$$

The integrated term on the left-hand side vanishes if the motion is periodic (i.e., if all dynamical quantities—such as $\vec{r}_a \cdot \vec{p}_a$—repeat after a certain time) and τ is chosen to be any multiple of the period of the motion. It vanishes also for nonperiodic motion provided that $\vec{r}_a \cdot \vec{p}_a$ remains within finite limits, since τ can be taken sufficiently long to make this term as small as desired. We thus obtain the VT for particle a as

$$-2\overline{T}_a = \overline{\vec{r}_a \cdot \vec{F}_a} \tag{2.5}$$

or, by summing over all the particles, the VT for the whole system becomes

$$-2\overline{T} = \sum_a \overline{\vec{r}_a \cdot \vec{F}_a} \tag{2.6}$$

where $T\ (\equiv \Sigma T_a)$ is the total kinetic energy of the system. The term on the right-hand side of Eq. (2.6) will be referred to as the virial,‡ from the Latin *vis* (plural: *vires*) meaning force.

Milne[7] (1925) remarked that frictional forces proportional to the velocity make no contribution to the virial, since the scalar product of $\vec{r}_a$ with the velocity $-\dot{\vec{r}}_a$ yields $-\vec{r}_a \cdot \dot{\vec{r}}_a = -d(\vec{r}^2/2)/dt$, which vanishes upon averaging. (But, of course, any motion is damped out by friction in the long-time aver-

‡Note that Clausius' definition of the virial was half this value.

age unless energy is continually added to the system). More generally, when the virial of any force can be written as a total time derivative of a bounded quantity, this force makes no contribution to the virial.

Although it is evident that a form of the VT can be obtained for a single particle [Eq. (2.5)], or even for a single Cartesian coordinate of a single particle, such forms are useful only in special circumstances, since the total force on the particle generally includes force contributions from the other particles in the system. However, it is clearly useful in Eq. (2.6) to be able to separate the summation over all particles into sums over subsets of particles such as, for example, the atoms in a molecule or the stars in a cluster. This separation is discussed in the following subsection.

2. *Separability of the Virial*

To display the issues in the separability of the virial, we consider a natural grouping ("set") of particles (e.g., a molecule, star cluster, etc.), possibly within an ensemble. It is convenient to partition the total force $\vec{F}_a$ acting on particle a according to whether the source is external to the set ($\vec{F}_a^e$), or internal ($\vec{f}_{ab}$) due to the other particles b within the set:

$$\vec{F}_a = \vec{F}_a^e + \sum_b \vec{f}_{ab} \tag{2.7}$$

The position vector $\vec{R}$ of the center of mass of the set is defined as the mass-average position

$$\vec{R} = \frac{\sum m_a \vec{r}_a}{M} \tag{2.8}$$

where M ($\equiv \Sigma m_a$) is the total mass. Correspondingly, the internal coordinate $\vec{\rho}_a$ relative to the center of mass is given by

$$\vec{\rho}_a = \vec{r}_a - \vec{R} \tag{2.9}$$

so that

$$\sum m_a \vec{\rho}_a = 0 \tag{2.10}$$

In these terms, Newton's second law (2.1) applied to the set takes the form

$$\vec{F} = \sum_a \vec{F}_a = \sum_a \vec{F}_a^e + \sum_a \sum_b \vec{f}_{ab} = \sum m_a \left(\ddot{\vec{\rho}}_a + \ddot{\vec{R}} \right)$$

Noting that the double sum of the internal forces vanishes owing to Newton's third law (i.e., $\vec{f}_{ba} = -\vec{f}_{ab}$) and that from Eq. (2.10)

$$\sum m_a \ddot{\vec{\rho}}_a = 0$$

we obtain

$$\vec{F} = \sum \vec{F}_a^e = M\ddot{\vec{R}} \tag{2.11}$$

This has exactly the form of Newton's second law applied to the motion of the center of mass, provided we regard the "total force" $\vec{F}$—being the resultant of all the (external) forces acting on all particles of the set—as though it acted at the center of mass. Taking the scalar product of Eq. (2.11) with $\vec{R}$ immediately leads to the dynamic VT for the center of mass:

$$\frac{d}{dt}(M\vec{R}\cdot\dot{\vec{R}}) = M\dot{\vec{R}}^2 + \vec{R}\cdot\vec{F} \tag{2.12}$$

Returning to Eq. (2.2) and summing over all particles a yields the full dynamic VT for the set:

$$\frac{d}{dt}\left(\sum m_a \vec{r}_a\cdot\dot{\vec{r}}_a\right) = \sum m_a \dot{\vec{r}}_a^2 + \sum \vec{r}_a\cdot\vec{F}_a \tag{2.13}$$

Upon transforming to the relative coordinates of Eq. (2.9), the total kinetic energy is readily separated into two terms respectively dependent upon the (mutually exclusive sets of) center-of-mass coordinates $\vec{R}$ and internal coordinates $\vec{\rho}_a$:

$$\sum m_a\left(\dot{\vec{R}} + \dot{\vec{\rho}}_a\right)^2 = M\dot{\vec{R}}^2 + \sum m_a \dot{\vec{\rho}}_a^2 \tag{2.14}$$

the expected mixed terms $\dot{\vec{\rho}}_a\cdot\dot{\vec{R}}$ vanishing because of Eq. (2.10). However, it is not possible in general to achieve such complete separability for the virial of the forces:

$$\sum_a (\vec{R} + \vec{\rho}_a)\cdot\left(\vec{F}_a^e + \sum_b \vec{f}_{ab}\right) = \vec{R}\cdot\vec{F} + \sum_{a>b}\sum \vec{\rho}_{ab}\cdot\vec{f}_{ab} + \sum \vec{\rho}_a\cdot\vec{F}_a^e \tag{2.15}$$

wherein

$$\vec{r}_a - \vec{r}_b \equiv \vec{r}_{ab} = \vec{\rho}_{ab} \equiv \vec{\rho}_a - \vec{\rho}_b \tag{2.16}$$

In relative coordinates the full dynamic VT (2.13) becomes

$$-\left[\frac{d}{dt}(M\vec{R}\cdot\dot{\vec{R}})-(M\dot{\vec{R}}^2+\vec{R}\cdot\vec{F})\right]$$

$$=\frac{d}{dt}\left(\sum m_a\vec{\rho}_a\cdot\dot{\vec{\rho}}_a\right)-\sum m_a\dot{\vec{\rho}}_a^2-\sum_{a>b}\sum\vec{\rho}_{ab}\cdot\vec{f}_{ab}-\sum\vec{\rho}_a\cdot\vec{F}_a^e \tag{2.17}$$

Since the left-hand side of this equation has already been shown to vanish [Eq. (2.12)], we obtain

$$\sum m_a\dot{\vec{\rho}}_a^2=\frac{d}{dt}\left(\sum m_a\vec{\rho}_a\cdot\dot{\vec{\rho}}_a\right)-\sum_{a>b}\sum\vec{\rho}_{ab}\cdot\vec{f}_{ab}-\sum\vec{\rho}_a\cdot\vec{F}_a^e \tag{2.18}$$

Evidently the first two terms on the right of Eq. (2.15) involve respectively only external or only internal coordinates and forces, but the last term involves the scalar product of internal coordinates with external forces. This same last term is carried over into Eq. (2.18), and appears to thwart the desired separability. Nevertheless, there are important circumstances in which the mixed term vanishes:

1. Where there are (virtually) no external forces $\vec{F}_a^e$ acting on the set, as for an isolated molecule in a field-free space, the total solar system, and so on.
2. Where the external forces $\vec{F}_a^e$ are proportional to the respective masses m_a, as in a gravitational field that is uniform over the whole set.

With varying degrees of approximation, case 1 obviously extends to situations in which interactions of members of the set with external agencies (e.g., collisions with the wall or other sets) are relatively rare and thus negligible. For example, in air at STP, a molecule spends only an average time fraction $\sim 10^{-3}$ in collision with other molecules. Further, in some circumstances a force originating from a particle external to the set can be treated as an internal force by enlarging the set to include the particle. Finally, if there is no correlation in the direction of the particle relative coordinates $\vec{\rho}_a$ and the directions of the external forces $\vec{F}_a^e$, there will likely be at least a partial cancellation in the sum between terms of opposite signs, a cancellation greatly enhanced by taking the time average.

For these various reasons, the last term in Eq. (2.18) is generally neglected in the averaging over time, which then leads to the internal VT

$$\overline{\sum m_a\dot{\vec{\rho}}_a^2}=\overline{-\sum_{a>b}\sum\vec{\rho}_{ab}\cdot\vec{f}_{ab}} \tag{2.19}$$

A similar averaging over time of Eq. (2.12) leads to the VT for the center of mass:

$$\overline{M\dot{\vec{R}}^2} = -\overline{\vec{R}\cdot\vec{F}} \tag{2.20}$$

However, it is only when the term $\Sigma\vec{\rho}_a\cdot\vec{F}_a^e$ vanishes that the VT is separately applicable to the internal and center-of-mass motions.

3. *Virial of the Wall Forces*

Amongst the variety of forces arising from sources external to the set(s) of particles described above, those due to interactions with the wall of an enclosure are of special interest. If P is the local pressure in a fluid system —here for definiteness epitomized by an ensemble of molecules—the (element of) force exerted on a wall element of surface $d\vec{S}$, oriented outward, has the time average $P\,d\vec{S}$; to this corresponds the equal but oppositely directed reaction force $d\vec{F}^w$ exerted by the wall on the ensemble:

$$d\vec{F}^w = -P\,d\vec{S} \tag{2.21}$$

This time-average force acts only on those molecules undergoing collision with the surface element $d\vec{S}$, during which the center-of-mass position $\vec{R}$ coincides with the position of $d\vec{S}$. Thus in the VT for the center of mass, Eq. (2.20), the contribution of the wall forces to the virial, summed over all molecules s, is

$$\overline{\sum\vec{R}_s\cdot\vec{F}_s^w} = -\oint d\vec{S}\cdot\vec{R}P = -\int_V d^3\vec{R}\,\mathrm{div}(P\vec{R}) \tag{2.22}$$

wherein Gauss's theorem has been applied to the surface integral and the resulting three-dimensional integral extends over the total volume V within the enclosure.

Denoting the resultant of all other (nonwall) external forces on molecule s by $\vec{F}_s^e$, the center-of-mass VT becomes

$$\overline{\sum M_s\dot{\vec{R}}_s^2} = -\overline{\sum\vec{R}_s\cdot\vec{F}_s^e} + \int_V d^3\vec{R}\,(\vec{R}\cdot\mathrm{grad}\,P) + 3\int_V d^3\vec{R}\,(P) \tag{2.23}$$

where the last term on the right has utilized the three-dimensional vector identity $\mathrm{div}\,\vec{R} = 3$.

Of the several situations in which the volume integrals can be readily evaluated, we consider first the case of uniform pressure.

a. Uniform Pressure. Noting that grad $P = 0$ for a pressure that is uniform throughout the volume V, Eq. (2.23) immediately reduces to

$$2\overline{T} = -\overline{\sum \vec{R}_s \cdot \vec{F}_s^e} + 3PV \tag{2.24}$$

where T is now the total translational kinetic energy (of the centers of mass) of the molecules. As will be seen in Section III.B, this form of the VT provides a basis for calculating the equation of state for any fluid—gases, liquids, plasmas, and (the electron gas in) metals—from a knowledge of the center-of-mass kinetic energy and the virial of the intermolecular forces $\vec{F}_s^e$.

In particular, Eq. (2.24) constitutes a theoretical foundation for the virial equation of state of a gas, as could be expected from the similarity of names. Here, we merely remark in passing that an ideal gas has, by definition, no intermolecular forces; using in Eq. (2.24) the equipartition theorem applied to the three-dimensional center-of-mass (translational) coordinates,

$$2\overline{T} = 3Nk\Theta \tag{2.25}$$

for N particles at a temperature $k\Theta$ (in energy units), we obtain the well-known equation of state for the ideal gas:

$$PV = Nk\Theta \tag{2.26}$$

If the system is composed of different fluid phases, labeled with the superscript α, respectively of volume V^α and uniform pressure P^α, the virial of the wall forces becomes a sum over phases α:

$$2\overline{T} = -\overline{\sum_s \vec{R}_s \cdot \vec{F}_s^e} + 3\sum_\alpha P^\alpha V^\alpha \tag{2.27}$$

b. Nonuniform Pressure. Examples of nonuniform pressure arise in many diverse physical problems, such as a planetary atmosphere or the electron atmosphere in the field of the nucleus within the statistical model of the atom. Here we illustrate the main principles by considering the planetary atmosphere. Since the pressure varies with altitude above the surface, the gradient of the pressure in Eq. (2.23) cannot be set equal to zero. Nevertheless, if (as in this example) the gradient of pressure arises from an external body force $\vec{g}$ per unit mass, that is,

$$\frac{\vec{F}_s^e}{M_s} = \vec{g} = \frac{\vec{\nabla} P}{\rho} \tag{2.28}$$

where ρ is the density of fluid (air), the term in the virial due to this external force density cancels exactly the integral over the pressure gradient in Eq. (2.23); we are left with a net contribution

$$2\overline{T} = 3\int d^3\vec{R}(P) \tag{2.29}$$

Depending on the circumstances, this integral takes a variety of useful forms. For a thin stratum of air essentially all at the same altitude h, the pressure and temperature are nearly uniform throughout and the integral is $3PV$, leading again to the ideal-gas equation of state, Eq. (2.26). If the volume extends over such a large altitude range that the pressure varies significantly, the volume can be partitioned into a number of thin strata or "phases" α, so that the integral becomes $3\Sigma P^{\alpha}V^{\alpha}$ [cf. Eq. (2.27)]. Alternatively, the integral can be set equal to $3\overline{P}V$, where $\overline{P}$ is the volume-average pressure:

$$\overline{P} = \frac{1}{V}\int d^3\vec{R}(P) \tag{2.30}$$

For an isothermal atmosphere the pressure decreases exponentially with altitude according to the "barometric formula,"

$$P = P_0 e^{-h/H} \tag{2.31}$$

Here the scale height,

$$H = \overline{h} = \frac{k\Theta}{mg} \tag{2.32}$$

is the density-weighted mean altitude $\overline{h}$ of the distribution, and corresponds to the thickness that the atmospheric blanket would have if its molecular number density had throughout the value n_0 at the planetary surface. Using P from Eq. (2.31), the integral (2.29) becomes $3HP_0$ per unit of surface area. But since $P_0 = n_0 mgH = n_0 mg\overline{h}$, we obtain

$$\frac{\overline{T}}{Hn_0} = \tfrac{3}{2}mg\overline{h} \tag{2.33}$$

This result, that per particle the mean kinetic energy is proportional to the mean potential energy, is a special case of a more general theorem to be discussed in the next section.

4. *Virial for a Homogeneous Potential Function*

The virial takes a particularly simple form when the forces derive from a potential-energy function U,

$$\vec{F}_a = -\vec{\nabla}_a U \tag{2.34}$$

that is a homogeneous function of the coordinates. Such a function has the property that multiplication of all its variables $\vec{r}_a$ by the same scale factor λ merely multiplies the whole function by the factor λ^γ, where γ is the degree of homogeneity:

$$U(\lambda\vec{r}_a, \lambda\vec{r}_b, \ldots) = \lambda^\gamma U(\vec{r}_a, \vec{r}_b, \ldots) \tag{2.35}$$

Application of Euler's theorem on homogeneous functions[8] shows that the virial is then simply proportional to the potential energy:

$$-\sum \vec{r}_a \cdot \vec{F}_a = \sum \vec{r}_a \cdot \vec{\nabla}_a U = \gamma U \tag{2.36}$$

Accordingly, in the absence of any other forces, the VT takes a most useful form:

$$2\overline{T} = \gamma \overline{U} \tag{2.37}$$

This relation gives the balance between kinetic and potential energies in the system; the total energy $E = \overline{T} + \overline{U}$ can thus be expressed in terms of either $\overline{T}$ or $\overline{U}$:

$$E = \frac{\gamma + 2}{2}\overline{U} = \frac{\gamma + 2}{\gamma}\overline{T} \tag{2.38}$$

Where the total force $\vec{F}_a$ acting upon particle a is a superposition of interparticle forces $\vec{f}_{ab}$ [cf. Eq. (2.7)] that are themselves each derivable from a two-body potential $u_{ab}(\vec{r}_{ab})$,

$$\vec{f}_{ab} = -\vec{\nabla}_{\vec{r}_{ab}} u_{ab}(\vec{r}_{ab}) \tag{2.39}$$

the total potential is the sum of all pair potentials:

$$U\{\vec{r}_a\} = \sum_{a>b}\sum u_{ab}(\vec{r}_{ab}) \tag{2.40}$$

Thus if the pair potentials are themselves homogeneous functions of the interparticle coordinates[‡] $\vec{r}_{ab}$ in Eq. (2.16), all of a common degree γ, then so is the total potential energy.

In the simplest case the pair potentials do not depend upon direction, but are functions of only the interparticle distance:

$$r_{ab} = |\vec{r}_a - \vec{r}_b| = \rho_{ab} \tag{2.41}$$

This applies to Coulombic and Newtonian forces, both involving inverse-square force laws and both of paramount importance. Newtonian gravitational forces account for the motion of planets and stars in the nonrelativistic regime; Coulombic forces describe the attraction and repulsion of charged particles, and govern most atomic and molecular phenomena. The corresponding potential-energy functions are both homogeneous of degree $\gamma = -1$, and lead to the VT in the form

$$2\overline{T} = -\overline{U} = -2E \tag{2.42}$$

Here U can represent either the gravitational potential energy of a set of massive bodies $m_a, m_a, \ldots,$

$$U = -\sum_{a>b}\sum \frac{Gm_a m_b}{r_{ab}} \tag{2.43}$$

where G is the universal gravitational constant, or the electrostatic potential energy between charges $q_a, q_b, \ldots$ obeying Coulomb's law:

$$U = \sum_{a>b}\sum \frac{q_a q_b}{r_{ab}} \tag{2.44}$$

Since in order to have a stable (bound) system the total energy E must be negative, the potential energy must also be negative and outweigh the kinetic energy, which is always positive; indeed, the VT [Eq. (2.42)] explains the stability of atoms and molecules, held together by Coulomb forces (cf. Sections III.C, III.D), and the stability of stars, held together by gravitation. Armed with this VT, one can obtain from simple qualitative arguments some remarkable insight into the main factors that determine the stability of matter, and thus understand some paradoxical facts such as the negative heat capacity of stars. This will be discussed in Section III.A.

Apart from magnetic and other relativistic effects, essentially all of chemistry is a consequence of Coulomb's law (plus quantum mechanics, of course).

[‡]Although there are evidently $N(N-1)/2$ interparticle vectors $\vec{r}_a - \vec{r}_b$, clearly these are obtainable from only the N independent vectors $\vec{r}_a$.

Nevertheless, potential functions for other types of forces often have great utility. These include such intermolecular interactions as the attractive London dispersion potential, the potentials between various multipoles (e.g., dipole–dipole), various power or exponential functions representing repulsion, and so on; and such interatomic interactions as the quasiharmonic potentials governing intramolecular vibrations.

If the potential energy is a sum of terms U_γ of differing degrees of homogeneity γ, the VT (including the center-of-mass motion) becomes

$$2\overline{T} = \sum_\gamma \gamma \overline{U}_\gamma + 3PV \tag{2.45}$$

A familiar example is the Lennard–Jones intermolecular "6–12 potential,"

$$u(r) = -\varepsilon\left[2\left(\frac{r_e}{r}\right)^6 - \left(\frac{r_e}{r}\right)^{12}\right] \tag{2.46}$$

where ε is the well depth and r_e the equilibrium separation. For this potential, evidently

$$3PV = 2\overline{T} - 12\varepsilon\left[\overline{\left(\frac{r_e}{r}\right)^6} - \overline{\left(\frac{r_e}{r}\right)^{12}}\right] \tag{2.47}$$

Another example is the interaction potential of two point dipoles $\vec{\mu}$ and $\vec{\mu}'$:

$$u(\vec{r}) = \vec{\mu}\cdot(3\vec{r}_1\vec{r}_1 - \mathbf{1})\cdot\vec{\mu}'/r^3 \tag{2.48}$$

wherein $\vec{r}_1$ is the unit vector along r and $\mathbf{1}$ stands for the unit dyad.

That the potential-energy functions in these examples appear to bear little resemblance to the Coulomb potentials from which they ultimately derive is evidently a consequence of the grouping together ("lumping") of subsets of charged particles. One example of such grouping (into molecules) was encountered earlier in deriving the VT for the center-of-mass motion (Section 2). Besides the clustering of electrons and nuclei into atoms and of atoms into molecules, the most evident physical manifestations occur in the electrical asymmetries that give rise to permanent electric dipole moments in certain molecules (e.g., HCl), electric quadrupole moments in other molecules (e.g., CO_2), and so on.

In considering this grouping into subsets of particles, one might well ask how far down such a hierarchical grouping should extend. The issue is illustrated in Table I, using examples ranging from chemistry into nuclear physics.

TABLE I
Possible Hierarchies and Groupings

Level	Example
Ensemble of systems	Gibbs grand canonical ensemble
System of sets	Gas of molecules
Set	Molecule
Subset	Atom in molecule
Particle	Electron or nucleus within atom
Subnucleus	Nucleon in nucleus
Subnucleon	Quark
⋮	⋮

For most of chemistry it suffices to stop at the electron–nucleus level, treating the nucleus as a charged entity without further elaborating its composition and structure. This is permissible because the nuclear radius is much smaller than typical electron-orbit radii, and the non-Coulombic nuclear forces are of such short range that their effect on the electrons can for most purposes be neglected. The same argument applies *a fortiori* to the subnuclear and subnucleon forces and particles. But clearly such an approximation would be inadequate for applications of the VT to the nucleus itself, as well as to interactions between electrons and nuclei that involve the nuclear structure: K capture, muonic atoms, internal conversion, nuclear multipole radiation, nuclear magnetic resonance, nuclear quadrupole resonance, etc.

For most problems of interest in the chemical physics of the VT, we are thus left with electrons and nuclei as the basic building blocks. Nevertheless, just as in the separation into center-of-mass and internal parts discussed in Section II.A.2, the VT must be able to treat certain natural groupings or subsets of these charged particles—atoms, molecules, ions—as single entities without the necessity of treating each constituent particle separately. Fortunately, the familiar multipole expansion provides a fundamental and convenient way to describe the Coulomb potential for such charge groupings and their interactions. And for the treatment of intramolecular vibrations, the Born–Oppenheimer approximation provides a quantum-mechanical basis for grouping electrons and nuclei into atoms within a molecule. Also, as Murdoch[9] has recently emphasized, the VT is essential in justifying the traditional treatment of molecular fragments as group entities, for example, in organic chemical reactions. The applications of the VT to the chemical bond will be discussed in detail in Section III.D.

To sum up the results so far obtained, the VT for a system of particles exerting (uniform) pressure P in volume V is

$$2\overline{T} - \sum \overline{\vec{r}_a \cdot \nabla_a U} = 3PV \tag{2.49}$$

If, in addition, the potential-energy function U is homogeneous of degree γ,

$$2\bar{T} - \gamma\bar{U} = 3PV \tag{2.50}$$

These scalar equations will be generalized in the next subsection.

5. *Tensor Generalizations*

In 1870, almost simultaneously with Clausius but quite independently, Maxwell,[10] published the first tensor form of the VT. This work related to such static engineering applications as the stability of rigid-frame structures‡ (e.g., bridges), for which the kinetic energy vanishes and time-averaging is unnecessary. Maxwell's ideas were elaborated some decades later by Rayleigh.[12]

In the derivation below we follow a modification of the tensor formulation due to Parker.[13] We consider a system of particles subject to forces which, as in Eq. (2.7), we partition into those ($\vec{F}_a^e$) arising from external causes, and those ($\vec{f}_{ab}$) resulting from the interaction with other particles b. Newton's second law thus requires that

$$\vec{F}_a^e + \sum_b \vec{f}_{ab} = m_a \ddot{\vec{r}}_a \tag{2.51}$$

Taking the dyadic product of $\vec{r}_a$ with this equation yields

$$\vec{r}_a\vec{F}_a^e + \sum_b \vec{r}_a\vec{f}_{ab} = m_a\vec{r}_a\ddot{\vec{r}}_a = \frac{d}{dt}\left(m_a\vec{r}_a\dot{\vec{r}}_a\right) - m_a\dot{\vec{r}}_a\dot{\vec{r}}_a \tag{2.52}$$

If we now sum over all particles of the system, the second term may be written

$$\sum_a \sum_b \vec{r}_a\vec{f}_{ab} = \sum\sum_{a>b} (\vec{\rho}_a - \vec{\rho}_b)\vec{f}_{ab} = \sum\sum_{a>b} \vec{\rho}_{ab}\vec{f}_{ab} \tag{2.53}$$

in which Newton's third law has been used to equate $\vec{f}_{ba} = -\vec{f}_{ab}$. Then de-

‡In a later article[11] (1874), Maxwell wrote: "As an example of the generality of this theorem—Clausius's virial theorem—we may mention that in any framed structure consisting of struts and ties, the sum of the products of the pressure in each strut into its length, exceeds the sum of the products of the tension of each tie into its length, by the product of the weight of the whole structure into the height of its centre of gravity above the foundations."

fining the tensors (or dyads)

$$\begin{aligned} \mathbf{J} &\equiv \sum m_a \vec{r}_a \dot{\vec{r}}_a \\ 2\mathbf{T} &\equiv \sum m_a \dot{\vec{r}}_a \dot{\vec{r}}_a \\ \mathbf{\Psi} &\equiv \sum \vec{r}_a \vec{F}_a^e \\ \mathbf{\Phi} &\equiv \sum_{a>b}\sum \vec{\rho}_{ab} \vec{f}_{ab} \end{aligned} \tag{2.54}$$

we obtain from Eqs. (2.52) and (2.53) the tensor virial equation

$$\dot{\mathbf{J}} - \mathbf{\Psi} = 2\mathbf{T} + \mathbf{\Phi} \tag{2.55}$$

The two tensors on the right may be related to their more familiar scalar counterparts through the process of contraction, by which the diagonal sum of a dyad yields the scalar product:

$$\text{Trace}\,\vec{A}\vec{B} = \vec{A}\cdot\vec{B} \tag{2.56}$$

Thus,

$$T = \text{Trace}\,\mathbf{T} = \tfrac{1}{2}\sum m_a \dot{\vec{r}}_a^2 \tag{2.57}$$

which we recognize as the total kinetic energy of the system. Accordingly, $\mathbf{T}$ is called the kinetic-energy tensor.

If the interparticle forces $\vec{f}_{ab}$ are conservative, they are derivable in the usual way from a potential-energy function U:

$$\vec{f}_{ab} = -\vec{\nabla} U \tag{2.58}$$

wherein the differentiation contained in the gradient is taken with respect to the interparticle vector $\vec{\rho}_{ab}$. Thus, if U is a homogeneous function of degree γ in the interparticle distances, contraction of $\mathbf{\Phi}$ followed by application Euler's theorem yields

$$\Phi = \text{Trace}\,\mathbf{\Phi} = -\sum_{a>b}\sum \vec{\rho}_{ab}\cdot\vec{\nabla} U = -\gamma U \tag{2.59}$$

Accordingly, $\mathbf{\Phi}$ is called the potential tensor.

The quantity $\mathbf{\Psi}$ is known as the (tensor) virial of the external forces. The external forces of interest are again of two principal types: those derivable from a homogeneous potential function as above, and those exerted on the system by the walls of an enclosure. In the latter case, contracting $\boldsymbol{\Psi}$ followed by averaging over time yields the (scalar) virial for the forces [cf. Eq. (2.22)]:

$$\Psi = \text{Trace}\,\boldsymbol{\Psi} = \sum \vec{r}_a \cdot \vec{F}_a^w = \oint d\vec{S} \cdot \vec{r} P \Rightarrow -3PV \tag{2.60}$$

wherein the last form results, as in Eq. (2.24), for a uniform pressure.

Returning to the tensor virial equation (2.55), an average over a time interval τ gives

$$\frac{1}{\tau}\int_0^\tau \dot{\mathbf{J}}\,dt = \frac{\mathbf{J}(\tau)-\mathbf{J}(0)}{\tau} = 2\overline{\mathbf{T}} + \overline{\boldsymbol{\Phi}} + \overline{\boldsymbol{\Psi}} \tag{2.61}$$

The time averages on the right are, of course, finite quantities. But if both $\vec{r}_a$ and $\dot{\vec{r}}_a$ (which occur in the definition of $\mathbf{J}$) are bounded, as in an enclosed system, the difference $\mathbf{J}(\tau)-\mathbf{J}(0)$ is also bounded, so that when divided by the time interval τ—which may be taken as large as we please—the ratio goes to zero; thus,

$$-\overline{\boldsymbol{\Psi}} = 2\overline{\mathbf{T}} + \overline{\boldsymbol{\Phi}} \tag{2.62}$$

In contracted form, using Eqs. (2.57), (2.59), and (2.60), Eq. (2.62) reduces to Clausius's scalar virial theorem, which—again for inverse-square interparticle forces—takes the form

$$3PV = 2\overline{T} + \overline{U} \tag{2.63}$$

So far the tensor $\mathbf{J}$ has been limited to the role of vanishing from Eq. (2.61). To return to its physical interpretation, we perform a separation into its symmetric (S) and antisymmetric (A) parts,

$$\mathbf{J} \equiv \mathbf{J}_S + \mathbf{J}_A \tag{2.64}$$

respectively defined by

$$2\mathbf{J}_S = \mathbf{J} + \tilde{\mathbf{J}} \tag{2.65}$$

and

$$2\mathbf{J}_A = \mathbf{J} - \tilde{\mathbf{J}} \tag{2.66}$$

where $\tilde{\mathbf{J}}$ is the transpose of $\mathbf{J}$. Twice the symmetric part is thus the time derivative of the tensor moment of inertia $\mathbf{I}$ about the position of the center of mass‡ (c.m.), defined by

$$\mathbf{I} \equiv \sum m_a \vec{r}_a \vec{r}_a \tag{2.67}$$

Twice the antisymmetrical part is just the tensor form of the angular momentum:

$$\mathbf{P} \equiv \mathbf{J} - \tilde{\mathbf{J}} \tag{2.68}$$

for which the (x, y) or $(1,2)$ element, for example, is

$$\mathbf{P}_{xy} = \sum m_a (x_a \dot{y}_a - \dot{x}_a y_a) \tag{2.69}$$

that is, just the z component of the vector product $\Sigma(\vec{r}_a \times \vec{p}_a)$.

If we similarly define the symmetric (S) and antisymmetric (A) parts of the total tensor virial $(\boldsymbol{\Phi} + \boldsymbol{\Psi})$ to be

$$(\boldsymbol{\Phi} + \boldsymbol{\Psi})_S \equiv 2\boldsymbol{\sigma} \tag{2.70}$$

$$(\boldsymbol{\Phi} + \boldsymbol{\Psi})_A \equiv 2\boldsymbol{\tau} \tag{2.71}$$

and note that the kinetic-energy tensor is symmetric, we may finally separate Eq. (2.55) into its corresponding symmetric and antisymmetric parts:

$$\frac{\ddot{\mathbf{I}}}{2} = 2\mathbf{T} + \boldsymbol{\sigma} \tag{2.72}$$

and

$$\dot{\mathbf{P}} = \boldsymbol{\tau} \tag{2.73}$$

Eq. (2.72) is seen to be the equation of motion of the (c.m.) moment of inertia, while (2.73) is evidently the tensor generalization of the theorem that the time rate of change of the angular momentum is equal to the torque. The tensor VT in the form of the equation of motion of the moment of inertia was noted early by Poincaré[15] (1911) and Eddington[16] (1916), who were the first to apply the VT to problems in astronomy (see the Appendix).

‡Moments of inertia may be defined about a plane, line, or point (cf. MacMillan[14]).

6. *Method of Moments*

The foregoing derivation of the tensor virial equation is an example of the method of moments, which can be used to generate a large class of what Hirschfelder[17] (1960) has called hypervirial theorems (HVT) on the ground that the VT is just one of them (cf. Sections I.A.7 and II.B.3). The tensor virial equation (2.55) was arrived at by forming the dyadic product—that is, the *moment*—of the equation of motion for the linear momentum (Newton's second law) with the position vector $\vec{r}_a$. The generalization now considered is to form the moment of any fundamental equation with any dynamical quantity. Although so far very few of the relations so generated have been found particularly useful, some applications of the quantum HVT are discussed in Section III.E. What makes the VT especially useful is the happy circumstance that, for a homogeneous potential, the virial—otherwise a quantity neither very extraordinary nor simple—is just proportional to the potential energy itself, thanks to Euler's theorem. Unlike most hypervirial theorems, the VT is a relation between two physically meaningful quantities, the kinetic and potential energies.

A further generalization of the method of moments is to take not only the first, but also the second, third,... moments of the same equation; Chandrasekhar[18] (1969) has applied this treatment to the conservation of momentum in hydrodynamics, employing the second- and third-order tensor VT in his very complete study of gravitational equilibrium of uniformly rotating bodies. We illustrate the method of moments by applying it systematically to the Fokker–Planck equation in Section III.F.

As an illustration of the method of moments in continuum mechanics, and also to accommodate the Lorentz force that—even though not homogeneous—merits our attention, we consider a continuous electrical fluid (e.g., a plasma) with electric charge density ρ and current density $\vec{j}$ subject to an electromagnetic field $\vec{\mathscr{E}}, \vec{\mathscr{B}}$ (Schwinger,[19] 1982). The total force on the system is given by

$$\vec{F} = \int d^3\vec{r}\,\vec{f} \tag{2.74}$$

where $\vec{f}$ is the Lorentz force density,

$$\vec{f} = \rho\vec{\mathscr{E}} + \frac{1}{c}\vec{j} \times \vec{\mathscr{B}} \tag{2.75}$$

Eliminating the source densities through Maxwell's equations,

$$\rho = \frac{1}{4\pi}\vec{\nabla}\cdot\vec{\mathscr{E}} \tag{2.76}$$

and

$$\vec{j} = \frac{c}{4\pi}\left(\vec{\nabla} \times \vec{\mathscr{B}} - \frac{1}{c}\frac{\partial \vec{\mathscr{E}}}{\partial t}\right) \tag{2.77}$$

a little algebra yields

$$\vec{f} = -\frac{\partial}{\partial t}\left(\frac{1}{4\pi c}\vec{\mathscr{E}} \times \vec{\mathscr{B}}\right) - \vec{\nabla}\cdot\left(\mathbf{1}\frac{\mathscr{E}^2 + \mathscr{B}^2}{8\pi} - \frac{\vec{\mathscr{E}}\vec{\mathscr{E}} + \vec{\mathscr{B}}\vec{\mathscr{B}}}{4\pi}\right) \tag{2.78}$$

where **1** is the unit dyadic.

We interpret this equation physically by identifying for the electromagnetic field the momentum density vector

$$\vec{g} = \frac{\vec{\mathscr{E}} \times \vec{\mathscr{B}}}{4\pi c} \tag{2.79}$$

and the momentum flux dyadic (or Maxwell stress tensor)

$$\mathbf{S} = \mathbf{1}\frac{\mathscr{E}^2 + \mathscr{B}^2}{8\pi} - \frac{\vec{\mathscr{E}}\vec{\mathscr{E}} + \vec{\mathscr{B}}\vec{\mathscr{B}}}{4\pi} \tag{2.80}$$

a symmetric tensor ($\mathbf{S} = \tilde{\mathbf{S}}$) whose trace is the energy density:

$$\text{Trace}\,\mathbf{S} = u = \frac{\mathscr{E}^2 + \mathscr{B}^2}{8\pi} \tag{2.81}$$

We thus obtain the conservation of momentum density in differential form:

$$\frac{\partial \vec{g}}{\partial t} + \vec{\nabla}\cdot\mathbf{S} + \vec{f} = 0 \tag{2.82}$$

The volume integral of this equation expresses for the total system the conservation of momentum for particles ($\vec{P}$) plus field ($\vec{G}$):

$$\frac{d}{dt}(\vec{P} + \vec{G}) = -\int d^3\vec{r}\,\vec{\nabla}\cdot\mathbf{S} = -\oint d\vec{S}\cdot\mathbf{S} \tag{2.83}$$

where we have defined

$$\frac{d\vec{P}}{dt} = \vec{F} = \int d^3\vec{r}\,\vec{f}$$

and

$$\frac{d\vec{G}}{dt} = \frac{d}{dt}\int d^3\vec{r}\vec{g} = \int d^3\vec{r}\,\frac{\partial \vec{g}}{\partial t} \tag{2.84}$$

(for fixed volume). The right-hand side of Eq. (2.83), where we have used Gauss's theorem, represents the influx of momentum through the surface enclosing the system.

As in the previous section, we form the dyad of the conservation-of-momentum equation (2.82) with the position vector $\vec{r}$ to get

$$\frac{\partial \vec{g}\vec{r}}{\partial t} + (\vec{\nabla}\cdot\mathbf{S})\vec{r} - \tilde{\mathbf{S}} + \vec{f}\vec{r} = 0 \tag{2.85}$$

where subtracting the transpose $\tilde{\mathbf{S}}$ compensates for allowing the del operator in the second term to operate on $\vec{r}$. Subtracting from Eq. (2.85) its conjugate (c) now yields a relation among antisymmetric tensors:

$$\frac{\partial}{\partial t}(\vec{g}\vec{r} - \vec{r}\vec{g}) + \{(\vec{\nabla}\cdot\mathbf{S})\vec{r} - [(\vec{\nabla}\cdot\mathbf{S})\vec{r}]_c\} + (\vec{f}\vec{r} - \vec{r}\vec{f}) = 0 \tag{2.86}$$

wherein the symmetry $\tilde{\mathbf{S}} = \mathbf{S}$ causes these terms to cancel. Each antisymmetric tensor is simply the tensor form—indicated by brackets [ ]—of the corresponding vector product:‡

$$\vec{g}\vec{r} - \vec{r}\vec{g} = [\vec{r}\times\vec{g}] \equiv \mathbf{P}, \text{ the tensor density of angular momentum} \tag{2.87}$$

$$\vec{f}\vec{r} - \vec{r}\vec{f} = [\vec{r}\times\vec{f}] \equiv \boldsymbol{\tau}, \text{ the tensor density of torque} \tag{2.88}$$

The two remaining terms can be written

$$(\vec{\nabla}\cdot\mathbf{S})\vec{r} - [(\vec{\nabla}\cdot\mathbf{S})\vec{r}]_c = -[\vec{\nabla}\cdot\mathbf{S}\times\vec{r}] = \vec{\nabla}\cdot(\vec{r}\times\mathbf{S}) \equiv \vec{\nabla}\cdot\mathbf{M} \tag{2.89}$$

‡The Cartesian tensor form of a three-dimensional (axial) vector $\vec{a}$ [$\equiv (A_x, A_y, A_z)$] is given by the antisymmetric matrix

$$[\vec{A}] \equiv \mathbf{A} = \begin{bmatrix} 0 & -A_z & A_y \\ A_z & 0 & -A_x \\ -A_y & A_x & 0 \end{bmatrix}$$

The scalar product of this matrix with another vector $\vec{b}$ gives the usual vector products: $\mathbf{A}b = \vec{a}\times\vec{b}$; $b\mathbf{A} = \vec{b}\times\vec{a}$. Thus $[\vec{A}] = \times\vec{a} = \vec{a}\times$.

where $\mathbf{M}$ is the third-order tensor (cf. Jackson[20]) defined by

$$\vec{r}\times\mathbf{S}\equiv\mathbf{M}, \text{ the angular-momentum flux density} \tag{2.90}$$

We thus obtain the tensor angular-momentum conservation law in differential form:

$$\frac{\partial \mathbf{P}}{\partial t}+\vec{\nabla}\cdot\mathbf{M}+\tau=0 \tag{2.91a}$$

or equivalently, in terms of the corresponding axial vectors

$$\frac{\partial \vec{P}}{\partial t}+\vec{\nabla}\cdot\mathbf{M}+\vec{\tau}=0 \tag{2.91b}$$

where $\mathbf{M}$ is the second-order tensor resulting from the product of the antisymmetrical tensor $(\vec{r}\times)$ with $\mathbf{S}$.

Finally, by contraction of Eq. (2.85) and use of Eq. (2.81), we obtain the electromagnetic VT of Schwinger:[19]

$$\frac{\partial}{\partial t}(\vec{r}\cdot\vec{g})+\vec{\nabla}\cdot(\mathbf{S}\cdot\vec{r})-u+\vec{r}\cdot\vec{f}=0 \tag{2.92}$$

This result will be generalized to the relativistic regime in Section II.C.2.

7. *Poisson Brackets and the Hypervirial Theorem*

This and the following section present alternative proofs of the VT based on the Hamiltonian and Lagrangian formalisms of classical mechanics. We begin with the Hamiltonian formalism; then, since quantum mechanics is based upon this formalism, the quantum VT can easily be derived by analogy (cf. Section II.B.3).

In terms of the generalized coordinates q_k and their conjugate momenta p_k for a classical system with hamiltonian H, Hamilton's equations of motion are

$$\begin{aligned}\dot{q}_k&=\frac{\partial H}{\partial p_k}\\ \dot{p}_k&=-\frac{\partial H}{\partial q_k}\end{aligned} \tag{2.93}$$

The equation of motion of an arbitrary dynamical variable $f(q_k, p_k, t)$ is thus

$$\frac{df}{dt}=\frac{\partial f}{\partial t}+\sum_k\left(\frac{\partial f}{\partial q_k}\dot{q}_k+\frac{\partial f}{\partial p_k}\dot{p}_k\right)$$
$$=\frac{\partial f}{\partial t}+\sum_k\left(\frac{\partial f}{\partial q_k}\frac{\partial H}{\partial p_k}-\frac{\partial f}{\partial p_k}\frac{\partial H}{\partial q_k}\right)$$

or, in Poisson bracket { } notation,

$$\frac{df}{dt}=\frac{\partial f}{\partial t}+\{f,H\} \tag{2.94}$$

For that class of dynamical variables wherein (1) f does not depend explicitly on time ($\partial f/\partial t=0$), and (2) $f(q_k, p_k)$, like the q_k and the p_k, remains bounded in the course of time, the time average of Eq. (2.94) yields a class of relation that has been called the classical hypervirial theorem [Hirschfelder[17] (1960)]:

$$\overline{\{f,H\}}=0 \tag{2.95a}$$

or more symmetrically,

$$\sum\overline{\frac{\partial f}{\partial q_k}\frac{\partial H}{\partial p_k}}=\sum\overline{\frac{\partial f}{\partial p_k}\frac{\partial H}{\partial q_k}} \tag{2.95b}$$

Taking f to be the function $G=\Sigma q_k p_k$, we can write this HVT in the form

$$\sum\overline{p_k\frac{\partial H}{\partial p_k}}=\sum\overline{q_k\frac{\partial H}{\partial q_k}} \tag{2.96}$$

The use of this form is limited, however, because in generalized coordinates the kinetic energy usually involves not only the momenta but also the coordinates; and the generalized coordinates may be dissimilar—lengths, angles, etc. For our present (nonrelativistic) applications we shall use the Hamiltonian as a function of the vector positional coordinates $\vec{r}_a$ and their corresponding momenta $\vec{p}_a$:

$$H=\sum_a\frac{\vec{p}_a^{\,2}}{2m_a}+U\{\vec{r}_a\}=T+U \tag{2.97}$$

The VT, Eq. (2.96), then reads

$$2\overline{T} = \sum \overline{\vec{r}_a \cdot \vec{\nabla}_a U} \tag{2.98}$$

As already mentioned, the most useful form of the VT arises when, in addition, U is a homogeneous function, say of degree γ, in the coordinates, which yields the simplified form,

$$2\overline{T} = \gamma \overline{U} \tag{2.99}$$

8. *Principle of Least Action*

This section addresses the question whether the most general formulation of mechanics, the Lagrangian formalism, has anything further to offer in the development of the VT. The Lagrangian L of a system is the difference between its kinetic and potential energies, the main actors in the VT. The equation of motion results from minimizing the action integral (Hamilton's principle of least action):

$$S = \int_{t_1}^{t_2} dt\, L(q_k, \dot{q}_k, t) \tag{2.100}$$

We seek that "path," defined by the set of time-dependent coordinate functions $\{q_k(t)\}$, for which any set of (small) variations δq_k causes the corresponding variation δS in the action to vanish:

$$\delta S = 0 \tag{2.101}$$

These variations are made subject to no change at the fixed time limits t_1 and t_2; that is, we require

$$\delta q_k(t_1) = 0 = \delta q_k(t_2) \quad (\text{all } k)$$

With the help of an integration by parts,

$$\begin{aligned} 0 = \delta S &= \int_1^2 dt \sum_k \left(\frac{\partial L}{\partial q_k} \delta q_k + \frac{\partial L}{\partial \dot{q}_k} \delta \dot{q}_k \right) \\ &= \sum_k \frac{\partial L}{\partial \dot{q}_k} \delta q_k \bigg|_1^2 + \int_1^2 dt \left[\frac{\partial L}{\partial q_k} - \frac{d}{dt}\left(\frac{\partial L}{\partial \dot{q}_k} \right) \right] \delta q_k \end{aligned} \tag{2.102}$$

and using the boundary condition and the arbitrariness of δq_k, we get for each generalized coordinate an equation of motion known as the Euler–

Lagrange equation:

$$\frac{d}{dt}\frac{\partial L}{\partial \dot{q}_k} - \frac{\partial L}{\partial q_k} = 0 \quad (\text{all } k) \tag{2.103}$$

The VT can as usual be derived by multiplying each of these equations by q_k:

$$\frac{d}{dt}\left(q_k \frac{\partial L}{\partial \dot{q}_k}\right) - \dot{q}_k \frac{\partial L}{\partial \dot{q}_k} - q_k \frac{\partial L}{\partial q_k} = 0 \quad (\text{all } k) \tag{2.104}$$

Since $q_k\, \partial L / \partial \dot{q}_k$ is assumed to be bounded, its time average vanishes, and leads to the VT in the form

$$\overline{\dot{q}_k \frac{\partial L}{\partial \dot{q}_k}} + \overline{q_k \frac{\partial L}{\partial q_k}} = 0 \quad (\text{all } k)$$

Summing over the generalized coordinates yields

$$\sum_k \left(\overline{\dot{q}_k \frac{\partial L}{\partial \dot{q}_k}} + \overline{q_k \frac{\partial L}{\partial q_k}} \right) = 0 \tag{2.105}$$

This VT in Lagrangian form connects immediately to the VT in Hamiltonian form, Eq. (2.97), through the definitions

$$p_k \equiv \frac{\partial L}{\partial \dot{q}_k} = \frac{\partial T}{\partial \dot{q}_k} \tag{2.106}$$

(wherein the second equality holds when the potential energy is independent of velocity); and

$$L \equiv \sum p_k \dot{q}_k - H = T - U \tag{2.107}$$

Since the kinetic energy is always bilinear (i.e., homogeneous of degree 2) in the generalized velocities $\dot{q}_k$,

$$\sum \dot{q}_k \frac{\partial T}{\partial \dot{q}_k} = \sum \dot{q}_k p_k = 2T \tag{2.108}$$

and the total energy becomes

$$H = \tfrac{1}{2} \sum \dot{q}_k p_k + U\{q_k\} \tag{2.109}$$

Equation (2.105) is equivalent to Eq. (2.96), and their derivations are quite similar. A still more perspicuous route to the VT follows.

The intimate connection between the VT and Euler's theorem on homogeneous functions, Eq. (2.35)—which is evidently a consequence of a scale transformation of all linear dimensional coordinates—raises the question whether the VT can be obtained directly from such a scale transformation. Fock and Krutov[21] (1932) answered this question affirmatively by showing that the VT results directly from Hamilton's principle by a special choice of variation involving a change of scale or dilation of the coordinates. To this end, we set in Eq. (2.102)

$$\delta q_k = \epsilon q_k(t), \qquad \delta \dot{q}_k = \epsilon \dot{q}_k(t) + \dot{\epsilon} q_k(t) \tag{2.110}$$

where $\epsilon(t)$ is a small parameter, to obtain

$$0 = \delta S = \int_1^2 dt \sum \left(\frac{\partial L}{\partial q_k} q_k + \frac{\partial L}{\partial \dot{q}_k} \dot{q}_k \right) \epsilon + \int_1^2 dt \sum \frac{\partial L}{\partial \dot{q}_k} q_k \dot{\epsilon} \tag{2.111}$$

The parameter $\epsilon(t)$ is chosen to have a constant value ϵ on the interval $t_1 < t < t_2$, but to vanish outside in order to satisfy the endpoint conditions, leading to the time derivative in terms of Dirac δ functions,

$$\dot{\epsilon}(t) = \epsilon[\delta(t - t_1) - \delta(t - t_2)] \tag{2.112}$$

The last integral in Eq. (2.111) thus has the value $\epsilon[\Sigma q_k(\partial L/\partial \dot{q}_k)]_1^2$. If this term remains bounded, its average over a long time interval $t_2 - t_1$ vanishes. The arbitrariness of the parameter ϵ then leads again to the Lagrangian form of the VT, Eq. (2.105).

To understand how a scale change of the coordinates can lead to the virial theorem, it proves illuminating to examine the relation between symmetry and the conservation laws. This involves the introduction of contact transformations (which pave the way for their quantum analogs, the unitary transformations), to which we now turn.

9. *Symmetry and the Conservation Laws*

We consider again a system of particles described by the Hamiltonian $H(q_k, p_k)$ in terms of the generalized coordinates q_k and their conjugate momenta p_k introduced in Section II.A.7. For such a system we shall show that the dynamic VT is simply the equation of motion [Eq. (2.94)] for the quantity $G = \Sigma q_k p_k$:

$$\frac{dG}{dt} = 2T - \sum q_k \frac{\partial U}{\partial q_k} \tag{2.113}$$

On the other hand, in the previous section the VT was obtained by performing an infinitesimal scale transformation of the coordinates,

$$q_k \to q_k(1+\epsilon) \tag{2.114}$$

and invoking the variational principle. We shall show presently that G is the generator of the infinitesimal contact transformation which effects the scale change of the coordinates.[22,23] In the next section, these results will be carried over isomorphically to quantum mechanics.

A contact transformation[23] (sometimes also called a canonical transformation[24]) allows the change from one set of independent coordinates and their conjugate momenta (q_k, p_k) to another set (Q_k, P_k) while leaving Hamilton's equations invariant; when the transformation is infinitesimal, the new variables differ from the old ones by infinitesimal amounts:

$$\begin{aligned} Q_k &= q_k + \delta q_k \\ P_k &= p_k + \delta p_k \end{aligned} \tag{2.115}$$

An infinitesimal contact transformation (ICT) differs infinitesimally from the identity operator and is characterized by a generating function $F(q, p)$ and an infinitesimal parameter ϵ such that

$$\mathrm{ICT}(F,\epsilon) \qquad \begin{cases} \delta q_k = \epsilon \dfrac{\partial F}{\partial p_k} \\ \delta p_k = -\epsilon \dfrac{\partial F}{\partial q_k} \end{cases} \tag{2.116a}$$

or, in Poisson-bracket notation,

$$\begin{aligned} \delta q_k &= \epsilon\{q_k, F\} \\ \delta p_k &= \epsilon\{p_k, F\} \end{aligned} \tag{2.116b}$$

More generally, for any dynamical function $u(q_k, p_k)$ we have

$$\delta u = \epsilon\{u, F\} \tag{2.117}$$

Thus the change under an ICT of a Hamiltonian that does not depend explicitly on time, such as (2.97), is given by

$$\delta H = \epsilon\{H, F\} = -\epsilon \frac{dF}{dt} \tag{2.118}$$

If F is a constant of the motion, Eq. (2.118) shows that the energy is conserved: the constants of the motion are generating functions of those ICTs that leave the Hamiltonian invariant. For example, by setting $F = p_j$ in Eq. (2.116) we get

$$\begin{aligned} \delta q_k &= \epsilon \delta_{kj} \\ \delta p_k &= 0 \end{aligned} \tag{2.119}$$

A momentum component generates an infinitesimal translation along the conjugate coordinate; the conservation of momentum for an isolated system is thus a consequence of the homogeneity of space, that is, of the translational symmetry. Similarly, an infinitesimal rotation‡ about an axis conserves the angular-momentum projection on this axis, a consequence of the isotropy of space; this projection is the generator of an infinitesimal rotation. Taking F as the Hamiltonian itself generates a translation in time ($\epsilon = \delta t$), and the conservation of energy results from the homogeneity of time; this provides the equation of motion for an arbitrary operator u,

$$\delta u = \delta t\{u, H\}$$

or, when u is explicitly time-dependent,

$$\frac{du}{dt} = \frac{\partial u}{\partial t} + \{u, H\} \tag{2.120}$$

in agreement with Eq. (2.94).

We now show that the infinitesimal unitary transformation (IUT) using $F = G = pq$ produces an infinitesimal scale transformation of the coordinates. This appears immediately by application of Eq. (2.116):

$$\begin{aligned} \delta q_k &= \epsilon\left\{q_k, \sum p_j q_j\right\} = \epsilon q_k \\ \delta p_k &= \epsilon\left\{p_k, \sum p_j q_j\right\} = -\epsilon p_k \end{aligned} \tag{2.121}$$

‡Note that the limitation specified in connection with Eq. (2.96) does not apply to the infinitesimal transformations considered here.

The ICT with generating function G and parameter ϵ scales each coordinate q_k into $q_k(1+\epsilon)$, while each momentum scales inversely from p_k into $p_k(1-\epsilon)$; also each generalized velocity $\dot{q}_k$ scales into $\dot{q}_k(1-\epsilon)$. Inserting these infinitesimal scalings into the Hamiltonian in Eq. (2.109) yields

$$H+\delta H=\tfrac{1}{2}\sum \dot{q}_k p_k(1-\epsilon)^2+U\{q_k+\epsilon q_k\} \tag{2.122}$$

Thus, keeping only terms first order in ϵ, the change δH in H is

$$\delta H=-\epsilon\sum \dot{q}_k p_k+\epsilon\sum q_k\frac{\partial U}{\partial q_k}=\epsilon\left\{H,\sum p_k q_k\right\} \tag{2.123}$$

This provides a check of consistency, since the same relation can also be obtained by setting $F=G$ in Eq. (2.118). A little Poisson-bracket algebra followed by division with ϵ then leads to the equation of motion for $G=\Sigma q_j p_j$:

$$\frac{dG}{dt}=2T-\sum q_j\frac{\partial U}{\partial q_j} \tag{2.124}$$

This is the desired result, the VT in dynamic form, Eq. (2.113). This result, and most of the previous ones, are carried over into quantum mechanics in the following sections.

B. The Quantum-Mechanical Virial Theorem

1. *Transition to Quantum Mechanics*

The quantum analogs of the classical contact transformations are the unitary (canonical) transformations. Their role is to transform from one basis set in Hilbert space to another while conserving the form of the equations of motion. Corresponding to the ICT(F,ϵ) of Section II.A.9 are the infinitesimal unitary transformations, IUT$(\hat{F},\epsilon)$, defined‡ as[25]

$$\hat{U}_\epsilon=1+\frac{i}{\hbar}\epsilon\hat{F} \tag{2.125}$$

where ϵ is an infinitesimal parameter and $\hat{F}$ is the generator of the IUT, a quantum-mechanical operator that must be Hermitian ($\hat{F}=\hat{F}^\dagger$)§ for $\hat{U}_\epsilon$ to be unitary, that is,

$$\hat{U}_\epsilon\hat{U}_\epsilon^\dagger=\hat{U}_\epsilon^\dagger\hat{U}_\epsilon=1 \tag{2.126}$$

to the first order in ϵ.

‡Quantum-mechanical operators wear a circumflex accent in this transition section, but in what follows the circumflex will be omitted unless there is danger of confusion.

§The dagger superscript symbolizes the Hermitian conjugate.

Unitary transformations preserve the norms and inner products of the vectors in Hilbert space, and the forms of operator equations. Under the unitary transformation $\hat{U}$, an operator $\hat{A}$ becomes

$$\hat{A}' = \hat{U}\hat{A}\hat{U}^{\dagger} \tag{2.127}$$

while the wavefunction ψ becomes

$$\psi' = \hat{U}\psi \tag{2.128}$$

Under the IUT($\hat{F}, \epsilon$) of Eq. (2.125), neglecting terms of second- and higher-order in ϵ, $\hat{A}$ becomes

$$\begin{aligned} \hat{A}' &= \left\{1 + \frac{i\epsilon}{\hbar}\hat{F}\right\}\hat{A}\left\{1 - \frac{i\epsilon}{\hbar}\hat{F}\right\} \\ &= \hat{A} + \frac{\epsilon}{i\hbar}[\hat{A}, \hat{F}] \end{aligned} \tag{2.129}$$

where $[\hat{A}, \hat{F}]$ is the commutator of $\hat{A}$ and $\hat{F}$. Under an IUT, the change in an operator $\hat{A}$ that does not depend explicitly on time is thus

$$\delta\hat{A} = \frac{\epsilon}{i\hbar}[\hat{A}, \hat{F}] \tag{2.130}$$

This is in agreement with Eq. (2.117), since, as Dirac first pointed out, the classical Poisson bracket translates into the corresponding quantum-mechanical commutator divided by $i\hbar$ under the condition that $\hbar$ approach zero:

$$\lim_{\hbar \to 0} \frac{[\hat{A}, \hat{B}]}{i\hbar} = \{A, B\} \tag{2.131}$$

By application of (2.130) to q_k and p_k we recover the quantum-mechanical operator equivalents of Eqs. (2.116) and (2.118):

$$\begin{aligned} \delta\hat{q}_k &= \frac{\epsilon}{i\hbar}[\hat{q}_k, \hat{F}] = \epsilon\frac{\partial\hat{F}}{\partial\hat{p}_k} \\ \delta\hat{p}_k &= \frac{\epsilon}{i\hbar}[\hat{p}_k, \hat{F}] = -\epsilon\frac{\partial\hat{F}}{\partial\hat{q}_k} \end{aligned} \tag{2.132}$$

and

$$\delta\hat{H} = \frac{\epsilon}{i\hbar}[\hat{H}, \hat{F}] \tag{2.133}$$

All of the classical results of the last section carry over into quantum-mechanical operator form. From Eq. (2.133), the generators of all IUT that commute with H are constants of the motion, that is, they leave the Hamiltonian unchanged. The generators of infinitesimal translation and rotation are respectively components of the linear momentum and angular momentum, whose conservations derive from the homogeneity and isotropy of space. Translations in time are generated by the Hamiltonian, as can be seen by setting $\epsilon = \delta t$ and $\hat{F} = \hat{H}$ in Eq. (2.130):

$$\delta\hat{A} = \frac{\delta t}{i\hbar}[\hat{A}, \hat{H}] \tag{2.134}$$

or

$$\frac{d\hat{A}}{dt} = \frac{1}{i\hbar}[\hat{A}, \hat{H}] \tag{2.135}$$

This is the equation of motion for an operator that does not depend explicitly on the time, as will be confirmed in Section II.B.3.

With this brief review we turn to the IUT generated by the "virial operator," $\hat{G} = \Sigma \hat{q}_i \hat{p}_i$, or more precisely, since generators must be Hermitian, its symmetrized[‡] form,

$$\hat{G} = \tfrac{1}{2}\sum(\hat{q}_i\hat{p}_i + \hat{p}_i\hat{q}_i) \tag{2.136}$$

Inserting this operator into Eqs. (2.133) and (2.135) yields

$$\delta\hat{H} = \frac{\epsilon}{i\hbar}[\hat{H}, \hat{G}] = -\frac{\epsilon}{i\hbar}\frac{d\hat{G}}{dt} \tag{2.137}$$

After a little commutator algebra and converting the generalized coordinates to the usual coordinates and momenta in order to have a Hamiltonian operator with kinetic energy quadratic in the momenta (cf. Section I.A.7), namely,

$$H = \sum \frac{\hat{\vec{p}}_a^2}{2m_a} + \hat{U}\{\vec{r}_a\} \tag{2.138}$$

[‡]In this particular case, the same results would obtain with the unsymmetrized generator, since from Eq. (2.137) and $[q_i p_j] = i\hbar\delta_{ij}$,

$$\hat{G} = \sum \hat{q}_i \hat{p}_i - in\hbar/2,$$

where n is the number of generalized coordinates; thus, Hermitian and nonhermitian G's differ only by a constant, and their commutators with another operator are the same.

we obtain the dynamical VT in operator form

$$\frac{d}{dt}\left(\sum \hat{\mathrm{r}}_a \cdot \hat{\mathrm{p}}_a\right) = 2\hat{T} - \sum \hat{\mathrm{r}}_a \cdot \vec{\nabla}_a \hat{U} \tag{2.139}$$

This is the exact quantum-mechanical analog of Eq. (2.124).

The quantum-mechanical average of an operator $\hat{A}$ is defined as

$$\langle \hat{A} \rangle = \langle \psi | \hat{A} | \psi \rangle = \int d^{3N}\{\vec{\mathrm{r}}\}\, \psi^* \hat{A} \psi \tag{2.140}$$

where ψ^* is the complex conjugate of ψ and the symbol $\int d^{3N}\{\vec{\mathrm{r}}\}$ indicates integration over the coordinates of all N particles and, if necessary, summation over spins. We now take the quantum-mechanical average of Eq. (2.139), adopting for the operator time derivative the natural definition (cf. Section II.B.3),

$$\frac{d\langle \hat{A} \rangle}{dt} = \left\langle \frac{d\hat{A}}{dt} \right\rangle \tag{2.141}$$

to get the scalar quantum-mechanical VT,

$$\frac{d}{dt}\left\langle \sum \hat{\mathrm{r}}_a \cdot \hat{\mathrm{p}}_a \right\rangle = 2\langle \hat{T} \rangle - \sum \left\langle \hat{\mathrm{r}}_a \cdot \vec{\nabla}_a \hat{U} \right\rangle \tag{2.142}$$

For a stationary quantum state of a system whose potential-energy function U is homogeneous of degree γ with respect to the coordinates (cf. Section III.A.3), we are left with

$$2\langle \hat{T} \rangle = \gamma \langle \hat{U} \rangle \tag{2.143}$$

This most useful relation gives the partition of the energy between its two main forms, kinetic and potential, for a quantum-mechanical system.

2. *Scaling*

We have seen in the previous section how a scale transformation leads to the VT, the operator G being the generator of the transformation. Following Fock[26] (1930), we now present a direct and perspicuous derivation of the VT, valid for a stationary state, based on scaling and the variational principle.

A stationary state is a state of definite energy E represented by a wavefunction $\psi\{\vec{\mathrm{r}}\}$ that satisfies the eigenvalue equation

$$\hat{H}\psi = E\psi \tag{2.144}$$

The Hamiltonian operator $\hat{H}$ is given by

$$\hat{H} = \hat{T} + \hat{U} = \sum \left(\frac{-\hbar^2}{2m_a} \right) \nabla_a^2 + \hat{U}\{\vec{r}_a\} \tag{2.145}$$

in which we assume that $\hat{U}$ is homogeneous of degree γ.

We now introduce a scale transformation of all N particle vector coordinates which sends each $\vec{r}_a$ into $\vec{r}_a' \equiv \lambda \vec{r}_a$, and define the new scaled wavefunction as

$$\phi_\lambda\{\vec{r}\} \equiv \lambda^{3N/2} \psi\{\lambda \vec{r}\} \tag{2.146}$$

In this definition the power of λ is chosen to maintain normalization of the new wavefunction over the same physical domain as the original:

$$\begin{aligned} \int d^{3N}\{\vec{r}\}\, \phi_\lambda^* \phi_\lambda &= \int d^{3N}\{\lambda \vec{r}\}\, \psi^*\{\lambda \vec{r}\}\, \psi\{\lambda \vec{r}\} \\ &= \int d^{3N}\{\vec{r}'\}\, \psi^*\{\vec{r}'\}\, \psi\{\vec{r}'\} = 1 \end{aligned} \tag{2.147}$$

[provided that the integration limits (e.g., 0, $\pm\infty$) are unaffected by the scale change].

Upon taking the expectation value of H with ϕ_λ, the energy evidently becomes a function of λ:

$$E_\lambda = \langle \phi_\lambda | \hat{H} | \phi_\lambda \rangle = \int d^{3N}\{\vec{r}\}\, \phi_\lambda^* \hat{T} \phi_\lambda + \int d^{3N}\{\vec{r}\}\, \phi_\lambda^* \hat{U} \phi_\lambda \tag{2.148}$$

Considering first the kinetic energy operator,

$$\hat{T} = -\hbar^2 \sum \frac{\nabla_a^2}{2m_a} = -\frac{\hbar^2}{2} \sum \frac{(\lambda \nabla_a')^2}{m_a} = \lambda^2 \hat{T}' \tag{2.149}$$

so that

$$\begin{aligned} \langle \phi_\lambda | \hat{T} | \phi_\lambda \rangle &= \int d^{3N}\{\vec{r}'\}\, \psi^*\{\vec{r}'\}\, \lambda^2 \hat{T}' \psi\{\vec{r}'\} \\ &= \lambda^2 \langle \psi | \hat{T} | \psi \rangle. \end{aligned} \tag{2.150a}$$

Similarly, for $U\{\vec{r}\}$ homogeneous of degree γ,

$$\langle \phi_\lambda | \hat{U} | \phi_\lambda \rangle = \lambda^{-\gamma} \langle \psi | \hat{U} | \psi \rangle \tag{2.150b}$$

The resulting energy

$$E_\lambda = \lambda^2\langle \hat{T} \rangle + \lambda^{-\gamma}\langle \hat{U} \rangle \tag{2.151}$$

must be a minimum for the true wavefunction ψ which corresponds to $\lambda = 1$, that is,

$$0 = \left[\frac{\partial E_\lambda}{\partial \lambda} \right]_{\lambda=1} = 2\langle \hat{T} \rangle - \gamma\langle \hat{U} \rangle \tag{2.152}$$

Thus again the quantum-mechanical VT emerges for a homogeneous potential.

We shall use the scaling device repeatedly to prove the VT for the particle in a box (Section II.B.7.b), for a molecule in the Born–Oppenheimer approximation (Section III.D.2), for the Thomas–Fermi statistical model of the atom (Section III.C.2) and for the Dirac equation (Section II.C.3). The VT in scattering theory can also be treated by this device.[27]

Scaling can also be used to improve an approximate wavefunction that does not fulfill the VT. As above, we scale the coordinates and wavefunction according to Eq. (2.146), and then choose for λ that value λ_0 that minimizes the energy:

$$0 = \left[\frac{\partial E_\lambda}{\partial \lambda} \right]_{\lambda_0} = 2\lambda_0\langle \hat{T} \rangle - \gamma\lambda_0^{-\gamma-1}\langle \hat{U} \rangle$$

for which

$$\lambda_0 = \left(\frac{\gamma\langle \hat{U} \rangle}{2\langle \hat{T} \rangle} \right)^{1/(\gamma+2)} \tag{2.153}$$

The new wavefunction ψ_{λ_0}, though still approximate, now satisfies the VT and is thus certainly improved over the initial one. Of course, this does not assure that ψ_{λ_0} is accurate. Rather, the VT is a necessary but not sufficient condition for an accurate wavefunction.[28]

3. *Commutators and the Hypervirial Theorem*

Recalling the usual definition of the quantum-mechanical average of an operator $\hat{f}$ over the (time-dependent) state Ψ,

$$\langle \hat{f} \rangle = \langle \Psi | \hat{f} | \Psi \rangle \tag{2.154}$$

and the time-dependent Schrödinger equation,

$$\hat{H}\Psi = i\hbar \frac{\partial \Psi}{\partial t} \tag{2.155}$$

one immediately obtains the Heisenberg equation of motion for $\langle \hat{f} \rangle$:

$$\frac{d\langle \hat{f} \rangle}{dt} = \left\langle \frac{\partial \hat{f}}{\partial t} \right\rangle + \frac{1}{i\hbar}\langle [\hat{f}, \hat{H}] \rangle \tag{2.156}$$

Here we note that the average brackets $\langle \ \rangle$ enclose each term on the right, but stand under the time-derivative operator on the left. Thus, with a view toward removing all such brackets—returning, so to speak, to the operator prior to averaging—we define a new entity $\hat{\dot{f}}$ whose average is the total time derivative of the average of $\hat{f}$:

$$\langle \hat{\dot{f}} \rangle = \frac{d\langle \hat{f} \rangle}{dt} \tag{2.157}$$

With this substitution in Eq. (2.156), removal of all brackets leads to

$$\hat{\dot{f}} = \frac{\partial \hat{f}}{\partial t} + \frac{1}{i\hbar}[\hat{f}, \hat{H}] \tag{2.158}$$

(plus, perhaps, other quantities whose average values vanish). We have already met a closely similar formula, Eq. (2.120), in connection with the infinitesimal canonical transformation of Section 2.A.9.

Although Eq. (2.158) is called the Heisenberg equation of motion and looks much like the equation of motion in the Heisenberg picture—in which the operators depend on time while the wavefunctions do not—the two equations of motion have somewhat different meanings.[29] In the Schrödinger picture $\hat{\dot{f}}$ defines the operator corresponding to the physical quantity $\dot{f}$ and goes over into this quantity in the classical limit. The Heisenberg (H) and Schrödinger (S) operators are related by the unitary transformation $e^{i\hat{H}t/\hbar}$:

$$\hat{f}_H = e^{i\hat{H}t/\hbar}\hat{f}_S e^{-i\hat{H}t/\hbar} \tag{2.159}$$

The total time derivative of this equation yields

$$\begin{aligned} \hat{\dot{f}} &= e^{i\hat{H}t/\hbar}\left(\frac{\partial \hat{f}_s}{\partial t} + \frac{[\hat{f}_s, \hat{H}]}{i\hbar} \right) e^{-i\hat{H}t/\hbar} \\ &= e^{i\hat{H}t/\hbar}\hat{\dot{f}}_s e^{-i\hat{H}t/\hbar} \end{aligned} \tag{2.160}$$

In any case, since

$$\Psi_S = e^{-i\hat{H}t/\hbar}\psi_H \tag{2.161}$$

with ψ_H constant in time, the quantum-mechanical average values—the only measurable quantities—are the same in both pictures, and the equation of motion is always Eq. (2.156).

Nevertheless, the use of Eq. (2.158) is often convenient, and enhances the parallel between classical and quantum mechanics. For example, one can define a velocity operator $\hat{\vec{v}}$ for a particle by use of Eq. (2.158):

$$\hat{\vec{v}} = \dot{\hat{\vec{r}}} = \frac{1}{i\hbar}[\hat{\vec{r}}, \hat{H}] = \frac{\hat{\vec{p}}}{m} \tag{2.162}$$

The simplest route to this result is through the momentum representation in which $\hat{\vec{r}} \to i\hbar\nabla\vec{p}$, using the Hamiltonian from Eq. (2.97). One notes that the momentum operator does not depend on the mass, while the velocity operator does. Similarly, we find for the acceleration operator

$$\dot{\hat{\vec{v}}} = \frac{1}{i\hbar}[\hat{\vec{v}}, \hat{H}] = \frac{1}{mi\hbar}[\hat{\vec{p}}, \hat{U}]$$

or

$$m\dot{\hat{\vec{v}}} = -\nabla\hat{U} = \dot{\hat{\vec{p}}} \tag{2.163}$$

since for any $\hat{u}$

$$[\hat{\vec{p}}, \hat{u}] = -i\hbar\vec{\nabla}\hat{u} \tag{2.164}$$

This is the analog of the second of Eqs. (2.132). Equations (2.162) and (2.163), or their equivalents with average values

$$\langle\hat{\vec{p}}\rangle = m\frac{d\langle\hat{\vec{r}}\rangle}{dt} \tag{2.165}$$

and

$$\frac{d\langle\vec{p}\rangle}{dt} = -\langle\vec{\nabla}\hat{U}\rangle \tag{2.166}$$

are often referred to as Ehrenfest's theorems.[30]

By now, it should be clear that the VT for a single particle is found by taking the time derivative of the operator $\hat{\vec{r}}_a\cdot\hat{\vec{p}}_a$:

$$\frac{d}{dt}(\hat{\vec{r}}_a\cdot\hat{\vec{p}}_a) = \dot{\hat{\vec{r}}}_a\cdot\hat{\vec{p}}_a + \hat{\vec{r}}_a\cdot\dot{\hat{\vec{p}}}_a \tag{2.167}$$

By summing, we obtain the VT for the whole system (or for any number of particles that we want to include in the sum):

$$\frac{d\left(\sum \hat{\vec{r}}_a \cdot \hat{\vec{p}}_a\right)}{dt} = 2\hat{T} - \sum \hat{\vec{r}}_a \cdot \vec{\nabla}_a \hat{U} \tag{2.168}$$

Using Eq. (2.157), this can be written in terms of average values,

$$\frac{d\left\langle \sum \hat{\vec{r}}_a \cdot \hat{\vec{p}}_a \right\rangle}{dt} = 2\langle \hat{T} \rangle - \sum \langle \hat{\vec{r}}_a \cdot \vec{\nabla}_a \hat{U} \rangle \tag{2.169}$$

A tensor generalization, whose trace will be the scalar VT [Eq. (2.169)] and whose antisymmetric part will be the equation of motion for the average value of the angular-momentum operator, results immediately (Vinti,[31] 1940; Cohen,[32] 1978); upon defining the tensor operators

$$\begin{aligned} \hat{\mathbf{G}} &= \sum \langle \hat{\vec{r}}_a \hat{\vec{p}}_a \rangle \\ \hat{\mathbf{T}} &= \sum \langle \hat{\vec{p}}_a \hat{\vec{p}}_a / 2m_a \rangle \\ \mathbf{\Phi} &= \sum \langle \hat{\vec{r}}_a \nabla_a \hat{U} \rangle \end{aligned} \tag{2.170}$$

we thus obtain the quantum-mechanical tensor VT

$$\frac{d\hat{\mathbf{G}}}{dt} = 2\hat{\mathbf{T}} + \mathbf{\Phi} \tag{2.171}$$

Returning to Eq. (2.169) and assuming that $\langle \Sigma \hat{\vec{r}}_a \cdot \hat{\vec{p}}_a \rangle$ remains bounded (not necessarily in a stationary state), one can write a virial relation with both quantum-mechanical average values and time averages:

$$2\overline{\langle \hat{T} \rangle} = \sum \overline{\langle \hat{\vec{r}}_a \cdot \vec{\nabla}_a \hat{U} \rangle} \tag{2.172}$$

Quantum-mechanical averages result from the probabilistic nature of quantum mechanics and have nothing to do with time averages. However, in a stationary state, when the wavefunction is, apart from a phase factor, time-independent and is associated with a definite energy E so that

$$H\psi = E\psi \tag{2.173}$$

the left-hand side of Eq. (2.169) drops out automatically, since all average values are time-independent; we are left with the best-known version of the quantum-mechanical VT,[33-35]

$$2\langle \hat{T} \rangle = \sum \langle \vec{r}_a \cdot \vec{\nabla}_a \hat{U} \rangle \tag{2.174}$$

To assert that Eq. (2.174) is the analog of the classical counterpart Eq. (2.98), although formally correct, is somewhat misleading, since in the classical limit the quantum-mechanical averages do not automatically become time averages. More exactly, the parallel is between the dynamical equations (2.168) and (2.2) which are related in exactly the same way in which Ehrenfests's theorem

$$\dot{\hat{\vec{\mathrm{P}}}}_a = -\vec{\nabla}_a \hat{U} \tag{2.175}$$

is related to Newton's second law

$$\dot{\vec{\mathrm{P}}}_a = -\vec{\nabla}_a U = \vec{\mathrm{F}}_a \tag{2.1}$$

If the potential is a sum of homogeneous terms U_γ of degree γ, Eq. (2.174) becomes [cf. Eq. (2.45)]

$$2\langle \hat{T} \rangle = \sum_\gamma \gamma \langle \hat{U}_\gamma \rangle \tag{2.176}$$

This relation between energy components is sometimes useful in applications (cf. Section III.C.2).

For atoms and molecules, in a good first approximation the potentials arise from Coulomb forces among nuclei and electrons, and are homogeneous of degree minus one. We shall examine in detail (Sections III.C and III.D) the many applications of the relevant VT,

$$E = \tfrac{1}{2}\langle \hat{U} \rangle = -\langle \hat{T} \rangle \tag{2.177}$$

If, to the Coulomb potential U_c, we add to the Hamiltonian the spin–orbit and spin–spin interactions U_s (Carr,[36] 1957), we can write immediately for the hydrogen-atom case

$$2\langle \hat{T} \rangle + \langle \hat{U}_c \rangle + 3\langle \hat{U}_s \rangle = 0 \tag{2.178}$$

since $\hat{U}_s$ is homogeneous of degree -3 in the coordinates.

As in the classical case (Section II.A.6), the quantum-mechanical VT is only one of Hirschfelder's hypervirial theorems: since in a stationary state all average values are independent of time, for any explicitly time-independent operator $\hat{A}$ that is inserted into the commutator of the Heisenberg equation of motion (2.156), a particular HVT is generated:

$$\langle [\hat{A}, \hat{H}] \rangle = 0 \tag{2.179}$$

The various HVTs, like the VT itself [cf. Eq. (2.153)], can be used as constraints to improve variational wavefunctions[37–39] to calculate average values without computing integrals[40] and to obtain the result of perturbation theory; Section III.E reviews these applications. Another very useful theorem, often used with the VT and HVT, is the energy-derivative or Hellmann–Feynman theorem (HFT), to be discussed in Section II.B.6.

Comparing the classical VT [Eq. (2.49)] with the quantum VT [Eq. (2.174)] just derived, one might wonder if the classical $3PV$ term has a quantum-mechanical analog. Of course it does, as will be seen in the following sections.

4. *The Virial Theorem from the Schrödinger Equation*

The quantum mechanical VT was first derived by Born, Heisenberg, and Jordan[41] (1925) in their seminal article on matrix mechanics, then somewhat later by Finkelstein[42] (1928) using the wave mechanics of Schrödinger. Fock[26] (1930) gave an elegant derivation based on scaling and the variational principle (cf. Section II.B.2). Following Finkelstein, Slater[43] (1933) presented a direct proof based on the Schrödinger equation. His application of the VT to molecules gave rise to many further developments (cf. Section III.D). The neatest proof is the one from the Heisenberg equation of motion (Section II.B.3), as given originally by Born, Heisenberg, and Jordan (cf. Refs. 33–35).

Nevertheless, none of these derivations consider enclosed systems, and no attention was given to the virial of the external wall forces, the quantum equivalent of the classical $3PV$ in Eq. (2.49). In order to display the role of the surface term and to connect the derivations based on the Heisenberg equation of motion and the Schrödinger equation, we start from the time-dependent Schrödinger equation,

$$H\Psi = i\hbar \frac{\partial \Psi}{\partial t} \tag{2.180}$$

using the Hamiltonian operator in the form

$$H = -\hbar^2 \sum \frac{\nabla_a^2}{2m_a} + U\{\vec{r}_a\} = T + U \tag{2.181}$$

Here, as always in quantum mechanics, $\vec{r}_a$ designates one of the set of N purely mathematical position vectors of which $\Psi\{\vec{r}\}$ is a function. It has nothing to do with the "vector position of particle a," first because for indistinguishable particles there is no "particle a," and second because the physical interpretation is that $|\Psi\{\vec{r}\}|^2 d\{N\}$ is the probability that the N

particles be found collectively to occupy the configuration $\{\vec{r}\}$ within $d\{\vec{r}\}$. The various "cells" (e.g., $d^3\vec{r}_a$ at $\vec{r}_a$) in the configuration space are, of course, distinguishable. Nevertheless, it is convenient and common—if careless—practice to treat the particles as though they are distinguishable and numbered, and thus to identify $\vec{r}_a$ with "particle a." Multiplying the Schrödinger equation from the left with the scalar operator $\Psi^*\vec{r}_b\cdot\vec{\nabla}_b$

$$\Psi^*(\vec{r}_b\cdot\vec{\nabla}_b)(T\Psi)+\Psi^*(\vec{r}_b\cdot\vec{\nabla}_b U)\Psi+\Psi^* U(\vec{r}_b\cdot\vec{\nabla}_b\Psi)=i\hbar\Psi^*(\vec{r}_b\cdot\vec{\nabla}_b)\frac{\partial\Psi}{\partial t} \tag{2.182}$$

Substitution of $\Psi^*U=U\Psi^*$ from the complex conjugate of (2.180), namely

$$\Psi^*U=-T\Psi^*-i\hbar\frac{\partial\Psi^*}{\partial t} \tag{2.183}$$

gives the expression

$$\vec{r}_b\cdot\left[\Psi^*\vec{\nabla}_b(T\Psi)-(\vec{\nabla}_b\Psi)(T\Psi^*)\right]+\Psi^*\vec{r}_b\cdot\vec{\nabla}_b U\Psi=i\hbar\vec{r}_b\cdot\left[\Psi^*\vec{\nabla}_b\frac{\partial\Psi}{\partial t}+\frac{\partial\Psi^*}{\partial t}\vec{\nabla}_b\Psi\right] \tag{2.184}$$

Had Ψ represented a stationary state, the RHS of Eq. (2.184) would have been zero. In the present more general case, we see that this RHS represents the time derivative of $-\Psi^*\vec{r}_b\cdot\vec{p}_b\Psi$, which makes contact with the LHS of the equation of motion (2.167). Using also the identity

$$(\vec{r}_b\cdot\vec{\nabla}_b)(T\Psi)=T(\vec{r}_b\cdot\vec{\nabla}_b\Psi)-2T_b\Psi \tag{2.185}$$

Eq. (2.184) becomes

$$\frac{d(\Psi^*\vec{r}_b\cdot\vec{p}_b\Psi)}{dt}=2\Psi^*T_b\Psi-\Psi^*\vec{r}_b\cdot\vec{\nabla}_b U\Psi+(\vec{r}_b\cdot\vec{\nabla}_b\Psi)(T\Psi^*)-\Psi^*T(\vec{r}_b\cdot\vec{\nabla}_b\Psi) \tag{2.186}$$

or, upon integration over all the configurational coordinates,

$$\frac{d\langle\vec{r}_b\cdot\vec{p}_b\rangle}{dt}=2\langle T_b\rangle-\langle\vec{r}_b\cdot\vec{\nabla}_b U\rangle+\int d\tau\left[(\vec{r}_b\cdot\vec{\nabla}_b\Psi)(T\Psi^*)-\Psi^*T(\vec{r}_b\cdot\vec{\nabla}_b\Psi)\right] \tag{2.187}$$

This is the VT for one particle, but it differs from the one obtained from the equation of motion because of the last integral. Replacing the kinetic energy T by its operator form, $-\Sigma(\hbar^2/2m)\nabla_a^2$, the last integral becomes a surface integral by use of Green's theorem:

$$\sum \frac{-\hbar^2}{2m_a}\int d\tau\,\left[\left(\vec{r}_b\cdot\vec{\nabla}_b\Psi\right)\nabla_a^2\Psi^* - \Psi^*\vec{\nabla}_a^2\left(\vec{r}_b\cdot\vec{\nabla}_b\Psi\right)\right]$$

$$= -\sum \frac{\hbar^2}{2m_a}\int d\tau^{(a)}\oint d\vec{S}_a\cdot\left[\left(\vec{r}_b\cdot\vec{\nabla}_b\Psi\right)\nabla_a\Psi^* - \Psi^*\nabla_a\left(\vec{r}_b\cdot\vec{\nabla}_b\Psi\right)\right] \quad (2.188)$$

Here $\int d\tau^{(a)}$ denotes integration over all coordinates except $\vec{r}_a$ (and, eventually, summation over all spins). As Cottrell and Paterson[44] (1951) noted, Slater's derivation[43] omitted this term on the ground that Ψ (or Ψ^*) should vanish on the bounding surface. However, it appears that the vanishing of the surface integral requires instead that $\vec{\nabla}_a\Psi$ (all a) vanish on the boundary. The vanishing of the surface integral is thus assured (as assumed in Slater's derivation) for bound states of an unconfined system where the range of integration extends over all space and Ψ is a well-behaved function which vanishes, and whose gradient vanishes, at infinity; but this is not the case for enclosed systems—for example, a particle in a box (cf. Section II.B.7.b) —where Ψ is typically a sine (or cosine) function with a nonvanishing derivative on the boundary. The vanishing of the surface integral is thus assured (as assumed in Slater's derivation) for bound states of an unconfined system where the range of integration extends over all space and Ψ is a well-behaved function which vanishes, and whose gradient vanishes, at infinity; but this is not the case for enclosed systems—for example, a particle in a box (cf. Section II.B.7.b)—where Ψ is typically a sine (or cosine) function with a nonvanishing derivative on the boundary.

The complete VT for the whole system is obtained by summing Eq. (2.187) over all b to yield

$$\frac{d\left\langle \sum \vec{r}_b\cdot\vec{p}_b\right\rangle}{dt} = 2\langle T\rangle - \sum\langle\vec{r}_b\cdot\vec{\nabla}_b U\rangle$$

$$-\sum \frac{\hbar^2}{2m_a}\int d\tau^{(a)}\oint d\vec{S}_a\cdot\left\{\left(\sum \vec{r}_b\cdot\vec{\nabla}_b\Psi\right)\vec{\nabla}_a\Psi^* - \Psi^*\vec{\nabla}_a\left(\sum\vec{r}_b\cdot\vec{\nabla}_b\Psi\right)\right\} \quad (2.189)$$

In the next section, we show that the surface term can be considered as a generalized current density, which arises when the operator kinetic energy T is not Hermitian with respect to the functions Ψ^* and $\Sigma\vec{r}_b\cdot\vec{\nabla}_b\Psi$.

5. *Generalized Current Density*

Starting from the Schrödinger equation, we have just seen that the VT requires, for an enclosed system, an extra term—a surface integral. One may wonder why no such term was present when the VT was derived from the Heisenberg equation of motion (Section II.B.3). The answer lies in the Hermiticity of H (more exactly of T) that was tacitly assumed in the derivation of Eq. (2.156). To show this, we consider an operator A, chosen for simplicity to be not explicitly time-dependent, and examine the limits of validity of the equation

$$\frac{d\langle A\rangle}{dt}=\frac{1}{i\hbar}\langle[A,H]\rangle \tag{2.190}$$

Invoking the time-dependent Schrödinger equation (2.173) and its complex conjugate, we can write

$$\begin{aligned}\frac{d\langle\Psi|A|\Psi\rangle}{dt}&=\left\langle\Psi\left|A\right|\frac{\partial\Psi}{\partial t}\right\rangle+\left\langle\frac{\partial\Psi}{\partial t}\left|A\right|\Psi\right\rangle\\&=\frac{i}{i\hbar}\{\langle\Psi|AH|\Psi\rangle-\langle H\Psi|A|\Psi\rangle\}\end{aligned} \tag{2.191}$$

Equation (2.190) follows when H is Hermitian with respect to the functions Ψ^* and $\Phi=A\Psi$, that is, by definition

$$\langle H\Psi|\Phi\rangle-\langle\Psi|H\Phi\rangle=0 \tag{2.192}$$

This is valid only if the surface integral (to come) vanishes. The Hermiticity of the potential energy U, a multiplicative operator, is assured, and the condition of Eq. (2.192) becomes

$$\begin{aligned}\langle H\Psi|\Phi\rangle-\langle\Psi|H\Phi\rangle&=\langle T\Psi|\Phi\rangle-\langle\Psi|T\Phi\rangle\\&=\sum\frac{\hbar^2}{2m_a}\{\langle\Phi|\nabla_a^2|\Psi\rangle^*-\langle\Psi|\nabla_a^2|\Phi\rangle\}\\&=-\sum\frac{\hbar^2}{2m_a}\int d\tau^{(a)}\oint d\vec{S}_a\cdot[\Phi\vec{\nabla}_a\Psi^*-\Psi^*\vec{\nabla}_a\Phi]\end{aligned} \tag{2.193}$$

by use of Green's theorem. Here is the expected surface integral; for most problems Ψ^* and $\vec{\nabla}_a\Psi^*$ vanish at the boundaries (often at infinity), but not necessarily for enclosed systems, as discussed in the previous section. By use of the last equality, the Heisenberg equation of motion (2.191) becomes

$$\frac{d\langle A\rangle}{dt} = \frac{1}{i\hbar}\langle[A, H]\rangle + \sum_a \frac{\hbar}{2im_a}\int d\tau^{(a)}\oint d\vec{S}_a\cdot[\Phi\vec{\nabla}_a\Psi^* - \Psi^*\vec{\nabla}_a\Phi] \tag{2.194}$$

Note that by choosing $\Phi = \Sigma\vec{r}_b\cdot\vec{p}_b\Psi$ we obtain immediately the VT of the previous section, Eq. (2.189), since we have

$$[\vec{r}_b\cdot\vec{p}_b, H] = [\vec{r}_b, T]\cdot\vec{p}_b + \vec{r}_b\cdot[\vec{p}_b, U] = i\hbar(2T - \vec{r}_b\cdot\vec{\nabla}_b U) \tag{2.195}$$

If we put $A = 1$ in Eq. (2.194), the continuity equation for the density $\rho = \Psi^*\Psi$ results:

$$\frac{d\langle 1\rangle}{dt} = \frac{d\langle\Psi|\Psi\rangle}{dt} = -\sum\oint\vec{j}_a\cdot d\vec{S}_a \tag{2.196}$$

with the usual definition (Ref. 33, p. 26) of the current:

$$\vec{j}_a = \frac{i\hbar}{2m_a}(\Psi\vec{\nabla}_a\Psi^* - \Psi^*\vec{\nabla}_a\Psi) \tag{2.197}$$

In operator form, Eq. (2.196) has exactly the same form as the continuity equation of a classical fluid,

$$\frac{\partial\rho}{\partial t} + \sum_a \vec{\nabla}_a\cdot\vec{j}_a = 0 \tag{2.198}$$

Equation (2.194) appears as a generalization of this last equation, since upon defining a "density" associated with the operator A,

$$\rho_A = \Psi^* A\Psi \tag{2.199}$$

and the corresponding current

$$\vec{j}_a^A = \frac{i\hbar}{2m_a}\{(A\Psi)\vec{\nabla}_a\Psi^* - \Psi^*\vec{\nabla}_a(A\Psi)\} \tag{2.200}$$

we can write Eq. (2.194) in operator form as

$$\frac{\partial\rho_A}{\partial t} + \sum\vec{\nabla}_a\cdot\vec{j}_a^A = \frac{1}{i\hbar}\Psi^*[A, H]\Psi \tag{2.201}$$

The right-hand side of this equation does not appear in classical mechanics and is purely quantum-mechanical in origin, like the noncommutability of the operators. We note that in stationary state $\partial \rho_A / \partial t = 0$; if, in addition, A is a constant of the motion ($[A, H] = 0$), the flux of current through the bounding surface vanishes. Eq. (2.201) was given by Srebrenik, Bader, and Tung Nguyen-Dang[45] (1978) in their discussion of the VT for molecular fragments (cf. Section III.D.3).

This and the preceding section have attempted to summarize and integrate several diverse approaches to determining the contribution of the boundary conditions to the virial theorem in quantum mechanics, but leave unresolved a number of issues: (1) the possibility of including in the potential-energy function the particle–wall interactions, as in the derivation (cf. Section II.A.2) of the $3PV$ term in the classical analog; (2) the effect of including an infinite wall discontinuity in the potential energy U of the Schrödinger equation, thus in effect incorporating the boundary condition in the differential equation; (3) the relation of the Hermiticity of the operators (e.g., ∇^2) to the completeness of the set of wavefunctions employed—for example, the sine wavefunctions of a particle in a box do not constitute a complete set, and in particular cannot describe any functions (e.g., a cosine function from $\nabla\psi$) that does not vanish on the boundary; (4) the general limitations and pitfalls of operator algebra conducted without consideration of the possible pathological behavior of the operands.

6. *Energy-Derivative Theorem*

A small change δH in the Hamiltonian H averaged over a stationary state ψ, assumed for convenience normalized to unity, yields by first-order perturbation theory the change in energy

$$\delta E = \langle \psi | \delta H | \psi \rangle \tag{2.202}$$

If δH results from a small variation of a parameter λ in the Hamiltonian $H(\lambda)$, the corresponding wavefunction is also a function of λ. But since the change in $\psi(\lambda)$ does not affect the energy in first order, we can write

$$\delta E(\lambda) = \langle \psi(\lambda) | \delta H(\lambda) | \psi(\lambda) \rangle$$

or

$$\frac{\partial E(\lambda)}{\partial \lambda} = \left\langle \psi(\lambda) \left| \frac{\partial H(\lambda)}{\partial \lambda} \right| \psi(\lambda) \right\rangle \tag{2.203}$$

This energy-derivative formula, which will be derived in detail below, has very wide applicability. It is often called the Hellmann–Feynman theorem

(HFT), although it seems to have been derived first by Pauli[46] in 1933 in a Handbuch article (Musher,[47] 1966).[‡] What Hellmann[48] (1937) and Feynman[49] (1939) originally and independently did was to apply this equation to molecules: by taking λ as a nuclear coordinate, they calculated, from simple electrostatic arguments, the force acting on a nucleus in a molecule. The resulting theorem is often called the electrostatic theorem.[35] This force approach to chemical bonding, as contrasted with the more current energy approach, has inspired since many quantum chemists, and a recent book has been devoted to it.[50]

The proof of Eq. (2.203) is obtained immediately by calculating directly the derivative of $E(\lambda) = \langle\psi(\lambda)|H(\lambda)|\psi(\lambda)\rangle$ with respect to λ; we shall see that this proof suffers the same limitation as for the HVT in the last section: for enclosed systems, the Hermiticity of H is not assured, and an extra surface integral may be required. We have

$$\frac{dE}{d\lambda} = \left\langle \frac{\partial\psi}{\partial\lambda}\middle| H \middle| \psi\right\rangle + \left\langle \psi \middle| \frac{\partial H}{\partial\lambda} \middle| \psi\right\rangle + \left\langle \psi \middle| H \middle| \frac{\partial\psi}{\partial\lambda}\right\rangle \tag{2.204}$$

and (2.203) arises when H is Hermitian with respect to ψ^* and $\partial\psi/\partial\lambda$, that is, when the last term can be written

$$\left\langle \psi \middle| H \middle| \frac{\partial\psi}{\partial\lambda}\right\rangle = \left\langle H\psi \middle| \frac{\partial\psi}{\partial\lambda}\right\rangle \tag{2.205}$$

Since $H\psi = E\psi$ and ψ is normalized, the first and last terms on the right-hand side of Eq. (2.204) together vanish. When Eq. (2.205) does not hold, we have by Green's theorem

$$\left\langle \psi \middle| H \middle| \frac{\partial\psi}{\partial\lambda}\right\rangle - \left\langle H\psi \middle| \frac{\partial\psi}{\partial\lambda}\right\rangle = \left\langle \psi \middle| T \middle| \frac{\partial\psi}{\partial\lambda}\right\rangle - \left\langle T\psi \middle| \frac{\partial\psi}{\partial\lambda}\right\rangle$$
$$= \sum \frac{\hbar^2}{2m_a} \int d\tau^{(a)} \oint d\vec{S}_a \cdot \left\{ \psi^* \vec{\nabla}_a \frac{\partial\psi}{\partial\lambda} - \frac{\partial\psi}{\partial\lambda} \vec{\nabla}_a \psi^* \right\} \tag{2.206}$$

[‡] Musher's article gives some interesting history of this theorem: Feynman did this work in an undergraduate thesis at MIT under Slater. Hellmann was a German physicist, dismissed by the Nazis in 1934 because his wife was Jewish. He then sought refuge in Russia, but was summarily executed as a German spy during the Stalinist purges of the late 1930s. He was, however, later "rehabilitated."

The HFT may thus be written

$$\frac{\partial E}{\partial \lambda}=\left\langle \frac{\partial H}{\partial \lambda}\right\rangle-\sum \frac{\hbar^2}{2m_a}\int d\tau^{(a)}\oint d\vec{S}_a\cdot\left\{\psi^*\vec{\nabla}_a\frac{\partial\psi}{\partial\lambda}-\frac{\partial\psi}{\partial\lambda}\vec{\nabla}_a\psi^*\right\} \tag{2.207}$$

or, in terms of the generalized current density defined by Eq. (2.200),

$$\frac{\partial E}{\partial \lambda}=\left\langle \frac{\partial H}{\partial \lambda}\right\rangle-i\hbar\sum\oint d\vec{S}_a\cdot\frac{\partial \vec{j}_a}{\partial\lambda} \tag{2.208}$$

If the term $\partial E/\partial\lambda$ can be considered as a generalized force (it is an ordinary force when λ is a coordinate, the negative of the pressure when λ is a volume, etc.), then $\lambda\,\partial E/\partial\lambda$ is a kind of generalized virial. It is thus not surprising to meet the HFT in conjunction with the VT or HVT, as we shall see in many applications (Sections III.B.2, III.D.4, III.E.2).

The next section features a remarkable example of the coupling of the HFT and VT. Before closing the present section, we note that there exist, of course, classical and relativistic analogs of the HFT (McKinley,[51] 1971) and that the HFT assumes different forms in different coordinate systems, the difference being a HVT (Epstein,[52] 1965). Indeed, let ψ, ψ' and H, H' be the wavefunctions and Hamiltonians in two coordinate systems related by the unitary transformation U. We have

$$\psi'=U\psi \tag{2.209}$$

$$H'=UHU^\dagger \tag{2.210}$$

and

$$UU^\dagger=1 \tag{2.211}$$

Taking the derivative of Eq. (2.210) with respect to a parameter λ gives

$$\frac{dH'}{d\lambda}=U\left\{\frac{dH}{d\lambda}+\left[U^\dagger\frac{dU}{d\lambda},H\right]\right\}U^\dagger \tag{2.212}$$

where we have used the λ derivative of Eq. (2.211). Thus, using Eq. (2.209), the difference between the HFT in the primed and unprimed coordinate systems can be written

$$\left\langle\psi'\left|\frac{dH'}{d\lambda}\right|\psi'\right\rangle-\left\langle\psi\left|\frac{dH}{d\lambda}\right|\psi\right\rangle=\left\langle\psi\left|\left[U^\dagger\frac{dU}{d\lambda},H\right]\right|\psi\right\rangle \tag{2.213}$$

which is the HVT for $U^\dagger dU/d\lambda$.

7. *Virial Theorem under External Constraints*

Argyres[53] (1967) noted the formal similarity of the VT in the three following situations:

1. Diatomic molecules with fixed nuclei (Slater,[43] 1933):

$$2\langle T\rangle + \langle U\rangle + R\frac{dE}{dR} = 0 \tag{2.214}$$

where R is the internuclear distance.

2. Interacting particles in a rigid box (de Boer,[54] 1949; Cottrell and Paterson,[44] 1951):

$$2\langle T\rangle + \sum\langle \vec{r}_a \cdot \vec{\nabla}_a U\rangle + l\frac{\partial E}{\partial l} = 0 \tag{2.215}$$

where l is a parameter defining the scale of the box.

3. Electron gas in a uniform background of positive neutralizing charges (March,[55] 1958):

$$2\langle T\rangle + \langle U\rangle + r_s\frac{dE}{dr_s} = 0 \tag{2.216}$$

where r_s is the radius of a sphere containing one electron.

We will find an extended VT including these 3 cases by combining two previously derived theorems with a surface term: the VT, Eq. (2.189), restricted to stationary states,

$$2\langle T\rangle - \sum\langle \vec{r}_b \cdot \nabla_b U\rangle = \sum \frac{\hbar^2}{2m_a}\int d\tau^{(a)} \oint d\vec{S}_a \cdot\left\{\left(\sum \vec{r}_b \cdot \vec{\nabla}_b \psi\right)\nabla_a \psi^* - \psi^* \vec{\nabla}_a\left(\sum \vec{r}_b \cdot \nabla_b \psi\right)\right\} \tag{2.217}$$

and the HFT, Eq. (2.207),

$$\frac{\partial E}{\partial \lambda} - \left\langle \frac{\partial H}{\partial \lambda}\right\rangle = \sum \frac{\hbar^2}{2m_a}\int d\tau^{(a)} \oint d\vec{S}_a \cdot \left\{ \frac{\partial \psi}{\partial \lambda}\vec{\nabla}_a \psi^* - \psi^* \vec{\nabla}_a \frac{\partial \psi}{\partial \lambda}\right\} \tag{2.218}$$

We imagine that we have N particles of masses m_a, in a volume V bounded by the surface S, in a stationary state

$$H\psi = E\psi \tag{2.219}$$

with $H = T + U_i + U_e$. The particles interact through the internal potential energy $U_i = U_i(\vec{r}_a)$ and are subjected to an external potential energy $U_e = U_e(\vec{r}_a, \lambda_k)$, functions of some parameters λ_k. For example, λ_1 may define the scale of the container (e.g., $\lambda_1^3 = V$).

We write the HFT, Eq. (2.218), for each parameter λ_k, using

$$\frac{\partial H}{\partial \lambda_k} = \frac{\partial U_e}{\partial \lambda_k} \tag{2.220}$$

then multiply each of these equations by the corresponding λ_k, sum these equations over k, and add the relation so obtained to the VT, Eq. (2.217); we end up with

$$2\langle T\rangle - \left\langle \left(\sum \vec{r}_b \cdot \nabla_b + \sum \lambda_k \frac{\partial}{\partial \lambda_k} \right) \right\rangle + \sum \lambda_k \frac{\partial E}{\partial \lambda_k}$$

$$= \sum \frac{\hbar^2}{2m_a} \int d\tau^{(a)} \oint d\vec{S}_a \cdot \{ \phi \vec{\nabla}_a \psi^* - \psi^* \vec{\nabla}_a \phi \} \tag{2.221}$$

in which ϕ is defined by

$$\phi = \left(\sum \vec{r}_b \cdot \vec{\nabla}_b + \sum \lambda_k \frac{\partial}{\partial \lambda_k} \right) \psi \tag{2.222}$$

To this point there has been no assumption concerning the boundary conditions on ψ, and we have seen (Sections II.B.5 and II.B.6) that surface terms such as the right-hand side of Eq. (2.221) do not automatically vanish; we have to examine the values of the functions in the integrand at the bounding surface. Nevertheless, it is desirable to use the extended VT, Eq. (2.221), because we shall see that its right-hand side vanishes in the three cases considered.

a. Rigid Molecule. In the Born–Oppenheimer approximation (treated in detail in Section III.D.1) U_i is the Coulomb repulsion of the electrons between themselves, notably a homogeneous function of degree -1 in the electronic coordinates; by Euler's theorem

$$\sum \vec{r}_b \cdot \vec{\nabla}_b U_i = -U_i \tag{2.223}$$

The external forces are due to the Coulomb potentials between electrons and nuclei,

$$U_e = -\sum_b \sum_n \frac{Z_n e^2}{|\vec{r}_b - \vec{R}_n|} \tag{2.224}$$

where $\vec{R}_n$ denotes the position of the nucleus of charge $Z_n e$. The potential U_e is homogeneous of degree -1 in both electronic and nuclear coordinates:

$$\left(\sum_b \vec{r}_b \cdot \vec{\nabla}_b + \sum_n \vec{R}_n \cdot \vec{\nabla}_n\right) U_e = -U_e \tag{2.225}$$

Since U_i is independent of the nuclear coordinates R_n, Eq. (2.225) holds also for U_i and thus, by addition, for U too; the generalized virial in Eq. (2.221) is thus just $-U$. For the bound states of a molecule, the bounding surface can be extended to infinity where ψ as well as its gradients vanish; the right-hand side of Eq. (2.221) is zero, and we are left with the molecular VT

$$2\langle T\rangle + \langle U\rangle + \sum_n \vec{R}_n \cdot \vec{\nabla}_n E = 0 \tag{2.226}$$

For diatomic molecules, this reduces to Slater's relation, Eq. (2.214), whose relevance to chemical bonding will be further discussed in Section III.D.3.

b. Particle in a Box. Let $l^3 = V$ be the volume of the box, and l the only parameter involved. If we require that ψ vanish on the surface of the box, there remains the term $\phi \vec{\nabla}_a \psi^*$ in the integrand of Eq. (2.221). We now prove that in this case it is ϕ (not $\vec{\nabla}_a \psi$ as in Section II.B.7.a) which vanishes when all coordinates $\{\vec{r}_a\}$ are on the wall. Defining new scaled coordinates

$$\vec{r}\,' = \frac{\vec{r}}{l} \quad (\text{all } a) \tag{2.227}$$

the function ϕ defined as

$$\phi(\vec{r}, l) = \sum \vec{r}_a \cdot \nabla_a \psi + l \frac{\partial \psi}{\partial l} \tag{2.228}$$

becomes a new function $f(\vec{r}, l)$ given by

$$f(\vec{r}\,', l) = \psi(l\vec{r}\,', l) = \psi(\vec{r}, l) \tag{2.229}$$

We differentiate this relation with respect to l:

$$\left(\frac{\partial f}{\partial l}\right)_{\vec{r}\,'} = (\nabla\psi)_l \left(\frac{\partial \vec{r}}{\partial l}\right)_{\vec{r}\,'} + \left(\frac{\partial \psi}{\partial l}\right)_{\vec{r}} = l^{-1} \sum \vec{r}_a \cdot \vec{\nabla}_a \psi + \frac{\partial \psi}{\partial l} = \frac{\phi}{l} \tag{2.230}$$

Now $(\partial f/\partial l)_{\vec{r}}$ is zero whenever $\vec{r}'$ is on the bounding surface, which completes the proof of the vanishing of ϕ on the boundary.

We may illustrate this process with a single particle in a one-dimensional box:

$$U(x)=\begin{cases}0, & 0<x<l\\ \infty & \text{outside}\end{cases} \tag{2.231}$$

The wave function is

$$\psi(x,l)=\left(\frac{2}{l}\right)^{1/2}\sin\frac{n\pi x}{l} \tag{2.232}$$

and

$$f(x',l)=\left(\frac{2}{l}\right)^{1/2}\sin n\pi x' \tag{2.233}$$

It is easy to check by direct differentiation that $(\partial f/\partial l)_{x'}$ is zero when $x'=0$ or l and that Eq. (2.230) holds in the form

$$\left(\frac{\partial f}{\partial l}\right)_{x'}=\frac{x}{l}\left(\frac{\partial\psi}{\partial x}\right)_l+\left(\frac{\partial\psi}{\partial l}\right)_x \tag{2.234}$$

The VT in a box [Eq. (2.215)] is thus derived.

In terms of the volume ($V=l^3$) and with the usual definition of the pressure‡

$$P=-\left(\frac{\partial E}{\partial V}\right)_{\text{oc.\#}} \tag{2.235}$$

it is readily seen that

$$3PV=-3\frac{\partial E}{\partial\ln V}=-\frac{\partial E}{\partial\ln l} \tag{2.236}$$

which makes contact with the classical VT, Eq. (2.49). (For a more general derivation of the quantum-mechanical VT with the $3PV$ term see Section III.B.2.)

‡The subscript is there to remind us that we perform a quantum-mechanical adiabatic (i.e., isentropic) transformation: while changing the volume we need to keep the occupation numbers fixed.

c. Homogeneous Electron Gas. In our third example, we consider the so-called "jellium model" of N electrons of charge $-e$ in a background of neutralizing positive charge, uniformly spread over the volume V. The average charge density of the electrons is $\rho = -eN/V$. In terms of the parameter r_s, we clearly have

$$V \sim r_s^3 \quad \text{and} \quad \rho \sim r_s^{-3} \tag{2.237}$$

As in the molecule case, the internal (interelectronic) potential satisfies

$$\sum \vec{r}_a \cdot \vec{\nabla}_a U_i = -U_i \tag{2.238}$$

but for U_e we have

$$U_e(\vec{r}_a, r_s) = U_{en} + U_{nn} = -e\sum \int d^3\vec{r}\,\frac{\rho}{|\vec{r}_a - \vec{r}|} + \frac{1}{2}\int d^3\vec{r}\int d^3\vec{r}\,'\,\frac{\rho^2}{|\vec{r}-\vec{r}\,'|} \tag{2.239}$$

where the second term is the self-energy of the background charge density. U_e is a homogeneous function of degree -1 in all its variables $(\vec{r}_a, r_s)$. Thus, as in Section II.B.7.a,

$$\left(\sum \vec{r}_a \cdot \vec{\nabla}_a + r_s \frac{\partial}{\partial r_s}\right) U_e = -U_e \tag{2.240}$$

as well as

$$-\left(\sum \vec{r}_a \cdot \vec{\nabla}_a + r_s \frac{\partial}{\partial r_s}\right) U = U \tag{2.241}$$

which greatly simplifies the virial of Eq. (2.221).

We now examine the right-hand side of Eq. (2.221), where

$$\phi(\vec{r}_a, r_s) = \left(\sum_a \vec{r}_a \cdot \vec{\nabla}_a + r_s \frac{\partial}{\partial r_s}\right)\psi \tag{2.242}$$

If we impose on the electrons no constraints (apart from their mutual repulsion and attraction to the positive background), that is, if we take an infinite normalization volume, then the surface integral vanishes, since ψ and its gradients go to zero at infinity. But one may object (Argyres,[53] 1967) to these free boundary conditions for an electron gas, for then the volume of the system is not well defined. It is more appropriate to take periodic boundary conditions. For convenience, we choose V to be a cube of side l and impose the periodicity condition

$$\psi(\vec{r}, l) = \psi(\vec{r}+\vec{l}_a, l) \tag{2.243}$$

We have introduced the notation

$$(\vec{r}+\vec{l}_a, l) = (\vec{r}_1, \ldots, \vec{r}_a + \vec{l}_a, \ldots, \vec{r}_N, l) \tag{2.244}$$

where $\vec{l}_a$ denotes the translation of all the components of $\vec{r}_a$ by the cube length, $\vec{l} = (l, l, l)$. To prove that ϕ, defined in Eq. (2.242), has the full periodicity of ψ—that is, that

$$\phi(\vec{r}, l) = \phi(\vec{r}+\vec{l}_a, l) \tag{2.245}$$

—we observe that partial differentiation of Eq. (2.243) gives

$$(\vec{\nabla}_b \psi)_{\vec{r}, l} = (\vec{\nabla}_b \psi)_{\vec{r}+\vec{l}_a, l} \tag{2.246}$$

for all b including $b = a$, and

$$l\left(\frac{\partial \psi}{\partial l}\right)_{\vec{r}, l} = \left(\vec{l}\cdot\vec{\nabla}_a \psi + l\frac{\partial \psi}{\partial l}\right)_{\vec{r}+\vec{l}_a, l} \tag{2.247}$$

In the last two equations the derivatives are evaluated at the values of the arguments shown in the subscripts. If we multiply Eq. (2.246) by $\vec{r}_b$, sum over all b, and add to it Eq. (2.247) we get

$$\phi(\vec{r}, l) = \phi(\vec{r}+\vec{l}_a, l) \tag{2.248}$$

since

$$l\frac{\partial \psi}{\partial l} = r_s \frac{\partial \psi}{\partial r_s} \tag{2.249}$$

because we can choose $l \propto r_s$. Thus ϕ and ψ have the same period and the right-hand side of Eq. (2.221) clearly vanishes; we have proved Eq. (2.216).

Another proof of Eq. (2.216), based on second quantization, was given by Kanazawa[56] (1967), and the extension to an inhomogeneous gas was recently studied by Ziesche[57] (1980). Vannimenus and Budd[58] (1977) used the

VT to calculate the surface energy of an electron gas, and Heinrichs[59] (1979) applied their result to the jellium model. Similar applications of the HFT were recently reviewed by van Himbergen and Silbey[60] (1979). Other applications of the VT to the solid state will be considered in Sections III.A.3 (cohesive energies of metals), III.B.4 (solids under pressure) and III.C (SCF and TF approximations).

We conclude Section II with a few relativistic generalizations of the VT.

C. The Relativistic Virial Theorem

As Minkowski[61] foresaw, "Henceforth space by itself, and time by itself, are doomed to fade away into mere shadows, and only a kind of union of the two will present an independent reality." All of the nonrelativistic classical and quantum-mechanical equations, invariant under the Galilean transformation, must be modified so as to be invariant under the Lorentz transformation. We shall limit the following discussion to special relativity (references to the VT in general relativity and applications to astrophysics are given in the Appendix), and we adopt the notation of Jackson[20] (1975), with a few modifications. Greek indices indicate the (Cartesian) components of four-vectors and run from 0 to 3, while Latin indices indicate the components of the space vectors, and run from 1 to 3. The contravariant coordinates of a four-dimensional (world) point are given by

$$x^{\mu} = (x^0, x^i) = (ct, \vec{r}) \tag{2.250}$$

The square of the differential line element (or interval) is

$$ds^2 = c^2\,dt^2 - dx_1^2 - dx_2^2 - dx_3^2 \equiv dx^{\mu}\,dx_{\mu} \tag{2.251}$$

in which the usual summation convention is employed. With the metric tensor of the form

$$g^{\mu\nu} = \begin{bmatrix} 1 & 0 & 0 & 0 \\ 0 & -1 & 0 & 0 \\ 0 & 0 & -1 & 0 \\ 0 & 0 & 0 & -1 \end{bmatrix} \tag{2.252}$$

raising or lowering a space index changes the sign of the component, while raising or lowering the time index leaves the component unchanged. Other quantities will be introduced as needed. We will prove the relativistic VT in the three following cases:

1. The classical particle.
2. An ensemble of classical particles in interaction with an electromagnetic field.
3. A Dirac particle in an electromagnetic field.

1. *The Classical Particle*

The relativistic analog of Newton's force law, Eq. (2.11), is simply

$$\frac{dp^\alpha}{d\tau} = G^\alpha \tag{2.253}$$

In this equation, τ is the proper time of the particle, that is, the time read by a clock moving with the particle; recalling that moving clocks go at a slower rate than those at rest:

$$d\tau = \frac{dt}{\gamma} = \frac{ds}{c} \tag{2.254}$$

where

$$\gamma = (1-\beta^2)^{-1/2} \tag{2.255}$$

and

$$\beta = \frac{v}{c} \tag{2.256}$$

The four-momentum p^α is given by

$$p^\alpha = m_0 u^\alpha = \left(\frac{K}{c}, \vec{\mathrm{p}}\right) \tag{2.257}$$

where m_0 is the rest mass, u^α the four-velocity

$$u^\alpha = \frac{dx^\alpha}{d\tau} = \gamma\frac{dx^\alpha}{dt} = \gamma(c, \vec{\mathrm{v}}) \qquad (u^\alpha u_\alpha = c^2) \tag{2.258}$$

and K the total relativistic energy of the free particle,

$$K = \gamma m_0 c^2, \quad \text{or} \quad \frac{K^2}{c^2} = p^2 + (m_0 c)^2 \tag{2.259}$$

The space part of the four-momentum is given by

$$\vec{\mathrm{p}} = \gamma m_0 \vec{\mathrm{v}} \tag{2.260}$$

and finally G^α is called the Minkowski force or four-force, being defined as

$$G^\alpha = \left(\gamma \vec{\mathrm{F}} \cdot \frac{\vec{\mathrm{v}}}{c}, \gamma \vec{\mathrm{F}}\right) \tag{2.261}$$

We separate Eq. (2.253) into its space and time components; the time part gives the time rate of change of the energy

$$\frac{dp^0}{d\tau} = G^0, \quad \text{that is,} \quad \frac{dK}{dt} = \vec{F} \cdot \vec{v} \tag{2.262}$$

and the space part is similar to Newton's law:

$$\frac{d\vec{p}}{dt} = \vec{F} \tag{2.263}$$

but now $\vec{p}$ is given by Eq. (2.260). To obtain the relativistic VT, we follow the classical derivation (cf. Section II.A.1) and multiply Eq. (2.263) by $x^i \equiv \vec{r}$:

$$\vec{r} \cdot \vec{F} = \vec{r} \cdot \frac{d\vec{p}}{dt} = \frac{d(\vec{r} \cdot \vec{p})}{dt} - \vec{v} \cdot \vec{p} \tag{2.264}$$

Taking the time average and assuming $\vec{r} \cdot \vec{p}$ bounded,‡ we obtain

$$\overline{\vec{v} \cdot \vec{p}} = \overline{-\vec{r} \cdot \vec{F}} \tag{2.265}$$

Since

$$\vec{v} \cdot \vec{p} = \gamma m_0 v^2 = m_0 c^2 (\gamma - \gamma^{-1}) = K + L_0 \tag{2.266}$$

where L_0 $(= -m_0 c^2 \gamma^{-1})$ is the Lagrangian of a free particle, Eq. (2.265) may be written as (Goldstein[62])

$$\bar{K} + \bar{L}_0 = \overline{-\vec{r} \cdot \vec{F}} \tag{2.267}$$

If we compare Eq. (2.267) with Eq. (2.6), twice the classical kinetic energy has been replaced by $K + L$. In the limit of low velocity, when $v \ll c$,

$$L_0 \cong -m_0 c^2 + \frac{m_0 v^2}{2} \tag{2.268}$$

$$K \cong m_0 c^2 + \frac{m_0 v^2}{2} \tag{2.269}$$

and Eq. (2.267) becomes Eq. (2.6):

$$\overline{mv^2} = 2\bar{T} = \overline{-\vec{r} \cdot \vec{F}} \tag{2.270}$$

‡Although for a material particle p, Eq. (2.260), appears to increase without limit as $v \rightarrow c$, an upper limit is imposed by the finiteness of the energy available.

2. *Ensemble of Classical Particles in an Electromagnetic Field*

We follow a derivation of Landau and Lifschitz[63] which treats the ensemble of particles as a continuous medium of mass density μ and charge density ρ. The discrete case is recovered by setting

$$\mu = \sum_a m_a \delta(\vec{r} - \vec{r}_a) \quad \text{and} \quad \rho = \sum_a e_a \delta(\vec{r} - \vec{r}_a) \tag{2.271}$$

but under the condition, as Shafranov[64] has noted, that the self-energy of the charged particles must be eliminated.

For the whole system of particles interacting with an electromagnetic field $(\vec{\mathscr{E}}, \vec{\mathscr{B}})$, the total energy–momentum tensor $\mathbf{S}$ is conserved, that is, its four-divergence is zero:

$$\partial_\alpha \mathbf{S}^{\alpha\beta} = 0 \tag{2.272}$$

Here the four-del, or quad, operator is denoted by

$$\partial_\alpha = \frac{\partial}{\partial x^\alpha} = \left(\frac{\partial}{\partial(ct)}, \vec{\nabla} \right) \tag{2.273}$$

The total energy–momentum tensor is (or can be made) symmetric and is represented by the 4-by-4 matrix

$$\mathbf{S} = \begin{bmatrix} w & c\vec{g} \\ c\vec{g} & \sigma \end{bmatrix} \tag{2.274}$$

Here S^{00} $(= w)$ is the total energy density, $(1/c)S^{0i}$ $(= \vec{g})$ is the total momentum density, and $S^{ij} = \sigma^{ij}$ is the three-dimensional total momentum flux dyadic or stress tensor.

$\mathbf{S}$ is the sum of the energy–momentum tensors for the field $\mathbf{S}_f$ and for the particles $\mathbf{S}_p$. The former is given by

$$S_f^{\alpha\beta} = \left(-F^{\alpha\gamma}F^{\beta}{}_{\gamma} + \tfrac{1}{4} g^{\alpha\beta} F_{\gamma\delta} F^{\gamma\delta} \right) / 4\pi \tag{2.275}$$

where the $F^{\alpha\beta}$ are the components of the electromagnetic field tensor

$$F_{\alpha\beta} = \partial_\alpha A_\beta - \partial_\beta A_\alpha \tag{2.276}$$

in terms of the potential four-vector

$$A^\alpha = (\phi, \vec{A}) \tag{2.277}$$

The energy–momentum tensor for the particles is simply

$$S_p^{\alpha\beta} = \mu(1-\beta^2)^{1/2} u^\alpha u^\beta \tag{2.278}$$

When we separate the conservation equation (2.275) into its time and space components, the time part expresses the conservation of energy, while the space part

$$\partial_\alpha S^{\alpha i} = 0 \tag{2.279}$$

expresses the conservation of momentum. As usual, the VT is derived by multiplying this last equation with the position vector x^i:

$$0 = x^i \partial_\alpha S^{\alpha i} = \partial_\alpha (x^i S^{\alpha i}) - (\partial_\alpha x^i) S^{\alpha i} = \partial_\alpha (x^i S^{\alpha i}) - S^{ii} \tag{2.280}$$

or

$$\frac{\partial(x^i S^{0i})}{\partial(ct)} + \partial_j (x^i S^{ji}) = S^{ii} \tag{2.281}$$

This is a preliminary differential version of the VT. Averaging on time, the first term on the left of Eq. (2.281) vanishes, assuming that $x^i S^{0i}$ remains finite. Since we have a continuous distribution, if we integrate also over all three-dimensional space, the second term on the left-hand side of Eq. (2.281) also vanishes, under the condition that $x^i S^{ji}$ is bounded; we are left with

$$\int d^3\vec{r}\, \bar{S}^{ii} = 0 \tag{2.282}$$

where the vinculum as usual connotes the time average. The energy–momentum tensor for the field, Eq. (2.275), is traceless, so that we have

$$S^{ii} = S^{\alpha\alpha} - S^{00} = S_p^{\alpha\alpha} - S^{00} \tag{2.283}$$

where S^{00} $(=w)$ is the total energy density and

$$S_p^{\alpha\alpha} = \mu c^2 (1-\beta^2)^{1/2} \tag{2.284}$$

In terms of the total energy

$$\bar{E} = \int d^3\vec{r}\, w \tag{2.285}$$

the VT for a continuous medium thus takes the form

$$\bar{E} = \int d^3\vec{r}\,\overline{\mu c^2(1-\beta^2)^{1/2}} \tag{2.286}$$

To go into the discrete case we set $\mu = \Sigma_a m_a \delta(\vec{r}-\vec{r}_a)$ to obtain

$$\bar{E} = \sum m_a c^2 \overline{\left(1-\frac{v_a^2}{c^2}\right)^{1/2}} \tag{2.287}$$

For this equation to be meaningful the left-hand side must be considered as the total energy minus the electromagnetic self-energy of the particles (Shafranov,[64] 1979). Indeed, if we do not subtract the self-energy, then since the energy density of the field is

$$U = \frac{\mathscr{E}^2 + \mathscr{B}^2}{8\pi} \tag{2.288}$$

and the energy per particle is $\gamma_a m_a c^2$, Eq. (2.287) becomes

$$\int d^3\vec{r}\,\bar{U} + \sum \overline{m_a c^2 \gamma_a} = \sum \overline{m_a c^2 \gamma_a^{-1}} \tag{2.289}$$

or

$$\int d^3\vec{r}\,\frac{\bar{\mathscr{E}}^2 + \bar{\mathscr{B}}^2}{8\pi} + \sum \overline{\gamma_a m_a v_a^2} = 0 \tag{2.290}$$

This is an impossible condition, since the left-hand side of (2.290) is the sum of two positive terms.

Instead of (2.290) we should write

$$\int d^3\vec{r}\,\frac{\overline{\mathscr{E}^2} + \overline{\mathscr{B}^2}}{8\pi} - \sum \int d^3\vec{r}\,\frac{\overline{\mathscr{E}_a^2} + \overline{\mathscr{B}_a^2}}{8\pi} + \sum \overline{\gamma_a m_a v_a^2} \tag{2.291}$$

It is the difference between the first two terms—equal to the total electromagnetic energy of the system of charges after the electromagnetic self-energy of the point charges has been subtracted—that goes over into the interaction potential U (a negative quantity) in the nonrelativistic limit. In this limit, we recover immediately the VT for the Coulomb potential, Eq. (2.42):

$$\bar{U} + 2\bar{T} = 0 \tag{2.292}$$

3. *Dirac Particle in an Electromagnetic Field*

Pauli[65] (1927) extended the Schrödinger equation to take the spin into account: the wavefunction became a spinor, a two-component wavefunction (for an electron). Dirac undertook a thoroughgoing revision to develop a new wave equation invariant under Lorentz transformation while still preserving its general form; the Dirac equation reads

$$H_D\Psi = i\hbar\frac{\partial\Psi}{\partial t} \tag{2.293}$$

where now Ψ is a four-component wavefunction or bispinor and H_0 is the Dirac Hamiltonian, namely

$$H_D = c\vec{\alpha}\cdot\vec{\mathrm{p}} + \beta mc^2$$

for a free particle,

$$H_D = c\vec{\alpha}\cdot\vec{\pi} + \beta mc^2 - e\phi$$

for an electron of change $-e$ in an electromagnetic field $(\phi, \vec{A})$. Here, $\vec{\pi}$ is the operator corresponding to the kinetic momentum‡

$$\vec{\pi} = \vec{\mathrm{p}} + \frac{e\vec{\mathrm{A}}}{c} \tag{2.294}$$

The $\vec{\alpha}$ and β represent four 4×4 matrices, Hermitian and anticommuting, obeying the relations

$$\begin{aligned} \alpha^i\alpha^k + \alpha^k\alpha^i &= 2\delta_{ik} \\ \alpha^i\beta + \beta\alpha^i = 0, &\qquad \beta^2 = 1 \end{aligned} \tag{2.295}$$

Symbolically,

$$\vec{\alpha} = \begin{pmatrix} 0 & \vec{\sigma} \\ \vec{\sigma} & 0 \end{pmatrix} \quad \text{and} \quad \beta = \begin{pmatrix} \mathbf{1} & 0 \\ 0 & -\mathbf{1} \end{pmatrix} \tag{2.296}$$

where $\vec{\sigma}$ stands for the three Pauli matrices

$$\sigma_1 = \begin{pmatrix} 0 & 1 \\ 1 & 0 \end{pmatrix}, \quad \sigma_2 = \begin{pmatrix} 0 & -i \\ i & 0 \end{pmatrix}, \quad \text{and} \quad \sigma_3 = \begin{pmatrix} 1 & 0 \\ 0 & -1 \end{pmatrix} \tag{2.297}$$

and **1** is the unit two-dimensional matrix.

‡Classically $\dot{\vec{\pi}} = -e(\vec{\mathscr{E}} + \vec{\mathrm{v}}\times\vec{\mathscr{B}}/c)$ with $\vec{\mathscr{E}} = -\vec{\nabla}\phi - \partial\vec{\mathrm{A}}/c\,\partial t$ and $\vec{\mathscr{B}} = \mathrm{curl}\,\vec{\mathrm{A}}$; $\vec{\mathrm{p}}$ $(= -i\hbar\vec{\nabla})$ is the momentum operator, conjugate to $\vec{\mathrm{r}}$.

The Dirac equation is thus constituted of four equations [the 4×4 unit matrix has been omitted from the right-hand side of Eq. (2.293)], and applies only to spin-$\frac{1}{2}$ particles (most of the known fermions). We consider an electron subject to an electromagnetic field and in a stationary state ψ of definite energy E:

$$H_D \psi = E\psi \tag{2.298}$$

Proceeding in parallel with the nonrelativistic classical- and quantum-mechanical derivations, the equation of motion for the operator $\vec{r} \cdot \vec{\pi}$ (cf. Section II.B.3) is

$$\frac{d(\vec{r} \cdot \vec{\pi})}{dt} = \dot{\vec{r}} \cdot \vec{\pi} + \vec{r} \cdot \dot{\vec{\pi}} \tag{2.299}$$

The velocity operator is just

$$\dot{\vec{r}} = \frac{[\vec{r}, H_D]}{i\hbar} = c\vec{\alpha} \tag{2.300}$$

since the $\vec{\alpha}$ are constant matrices. But since

$$\dot{\vec{\pi}} = \dot{\vec{p}} + \frac{e\dot{\vec{A}}}{c} \tag{2.301}$$

with

$$\dot{\vec{p}} = \frac{[\vec{p}, H_D]}{i\hbar} = e\vec{\nabla}\phi - e\vec{\nabla}(\vec{\alpha} \cdot \vec{A}) \tag{2.302}$$

and

$$\begin{aligned} \dot{\vec{A}} &= \frac{\partial \vec{A}}{\partial t} + \frac{[\vec{A}, H]}{i\hbar} = \frac{\partial \vec{A}}{\partial t} + \frac{[\vec{A}, c\vec{\alpha} \cdot \vec{p}]}{i\hbar} \\ &= \frac{\partial \vec{A}}{\partial t} + c\vec{\alpha} \cdot \vec{\nabla}\vec{A} \end{aligned} \tag{2.303}$$

we obtain

$$\dot{\vec{\pi}} = -e\left(-\frac{\partial \vec{A}}{c\, \partial t} - \vec{\nabla}\phi\right) - e\vec{\alpha} \times (\vec{\nabla} \times \vec{A}) = -e\vec{\mathscr{E}} - e\vec{\alpha} \times \vec{\mathscr{B}} = \vec{F} \tag{2.304}$$

Here we have defined the Lorentz force operator $\vec{F}$ and used the vector formula

$$\vec{\nabla}(\vec{\alpha}\cdot\vec{A}) = \vec{\alpha}\cdot\vec{\nabla}\vec{A} + \vec{\alpha}\times(\vec{\nabla}\times\vec{A}) \tag{2.305}$$

The equation of motion for $\vec{r}\cdot\vec{\pi}$ thus becomes

$$\frac{d(\vec{r}\cdot\vec{\pi})}{dt} = c\vec{\alpha}\cdot\vec{\pi} + \vec{r}\cdot\vec{F} \tag{2.306}$$

In a stationary state, since all average values are time-independent, the quantum-mechanical average of this equation yields

$$\langle c\vec{\alpha}\cdot\vec{\pi}\rangle = -\langle\vec{r}\cdot\vec{F}\rangle \tag{2.307}$$

This VT for the Dirac particle was fist derived by Gupta[66] in 1932 (see also Rose and Welton,[67] 1952; March,[68] 1953). An elegant proof of Eq. (2.307) by scaling (cf. Section II.B.2) has been given by McKinley[51] (1971) and Brack[69] (1983). The Appendix gives many references on the applications of the relativistic VT to particle physics.

In the non-quantum-mechanical limit, $c\vec{\alpha}$ reduces to a velocity and we recognize the classical relativistic VT, Eq. (2.265). In the nonrelativistic limit, two of the four components of the bispinor vanish and the two remaining go over into the two-component Pauli spinors (Schectman and Good,[2] 1957); from Eq. (2.307), we recover the quantum-mechanical VT, Eq. (2.142).

III. APPLICATIONS

A. The Virial Theorem as Qualitative Guide

Throughout history, the physical sciences have evolved sophisticated models to provide very detailed, quantitative descriptions of the properties and behavior of matter. Although the usefulness of quantitative models is not in question, in trying to take into account all relevant aspects of a physical situation, a model often loses both its simplicity and generality. As a counterpart, it is thus useful also to have simplified *qualitative* models that illuminate the essential points of a large range of phenomena. Such simplified models play the role of scientific caricatures that display only the most significant features—just as a master painter, with a few strokes of the brush, evokes a whole landscape or portrait. It is in this spirit that we offer this account of the VT as a qualitative guide, largely inspired by a recent article by Weisskopf.[70]

The electrical attraction between nuclei and electrons, together with the gravitational attraction between masses, is responsible for the most evident properties and aggregations of matter. Since both involve inverse-square forces, the VT simply reads‡ [cf. Eq. (2.37)]:

$$2\overline{T} = -\overline{U} \tag{3.1}$$

(for zero applied pressure on the envelope). This relation provides a key to the understanding of the mechanics of the solar system, of electrons in an atom, of atoms in molecules, of ions in ionic crystals, of electrons in metals, and so on, as we shall see.

1. Stability of Atoms

Consider first the hydrogen atom. Let the ground state be characterized by a single parameter, its radial dimension r. The potential energy is then of the order of magnitude

$$U \sim -\frac{e^2}{r} \tag{3.2}$$

The electron wavelength $\lambdabar$ extends roughly over the distance

$$\lambdabar \sim r \tag{3.3}$$

to which corresponds the de Broglie momentum,

$$p \sim \frac{\hbar}{r} \tag{3.4}$$

and thus the kinetic energy,

$$T \sim \frac{\hbar^2}{2mr^2} \tag{3.5}$$

where m is the mass of the electron.‡‡ The total energy is therefore

‡Here it is to be understood that, although the kinetic energy is not separable into parts respectively ascribable to the electrical (e) or gravitational (g) forces, the potential energy is composed of two such terms: $U = U_e + U_g$; in most cases of interest one or the other of these terms may be negligible.

‡‡Strictly speaking one should use the reduced mass of the electron–proton system, but for present purposes the difference is negligible.

approximately

$$E \sim -\frac{e^2}{r} + \frac{\hbar^2}{2mr^2} \tag{3.6}$$

which has its minimum value for the radius $r = a$, where

$$a = \frac{\hbar^2}{me^2} = 0.53\ \text{Å} \tag{3.7}$$

that is, just the Bohr radius.

Upon inserting this value back into the potential [Eq. (3.2)] and kinetic [Eq. (3.5)] energies, the VT [Eq. (3.1)] is automatically satisfied and the ground-state energy [Eq. (3.6)] corresponds exactly to the hydrogenic value:

$$E_0 = -\frac{e^2}{2a} = -R = -13.6\ \text{eV} \tag{3.8}$$

where R is the rydberg. Although hydrogen is the smallest of all atoms, its size and (ionization) energy represent characteristic orders of magnitude for all atoms.

For excited states, the potential-energy function is still Coulombic, but the kinetic energy must be modified because there are nodes in the wavefunction; for $n-1$ nodes the wavelength is $\lambda\!\!\!^{-} \sim r/n$ and the kinetic energy is

$$T \sim \frac{\hbar^2 n^2}{2mr^2} \tag{3.9}$$

Minimizing the total energy as before, we now obtain

$$r \sim n^2 a = \frac{n^2 \hbar^2}{me^2} \tag{3.10}$$

With this radius the VT [Eq. (3.1)] is again fulfilled and the energy levels of the H atom are recovered exactly:

$$E_0 = -\frac{e^2}{2n^2 a} = -\frac{R}{n^2} \tag{3.11}$$

In the helium atom there are two electrons and the kinetic energy is twice Eq. (3.5); the potential energy of the nuclear attraction is four times Eq. (3.2), because there are now two electrons subject to a doubly charged nucleus.

Account must be taken also of the electron–electron repulsion, for which the potential energy is

$$V_{ee} \sim \frac{ze^2}{r} \tag{3.12}$$

where r/z is the average distance between electrons. The total energy thus takes the form

$$E \sim -(4-z)e^2/r + \hbar^2/mr^2 \tag{3.13}$$

After minimization we get

$$E_0 \sim -2\left(2-\frac{z}{2}\right)^2 R \tag{3.14}$$

The experimental value being -5.81 rydberg,

$$\frac{z}{2} = 0.296 \tag{3.15}$$

The quantity $z/2$, by which the nuclear charge in Eq. (3.14) is effectively reduced, may thus be viewed alternatively as the consequence of partial screening of the nucleus by each electron; indeed, its experimental value is close to the $\frac{5}{16}$ screening obtained by the familiar quantum-mechanical charge-variation method.

Kregar and Weisskopf[71] present more detailed, but still simple, models for L-shell atoms and find surprisingly good agreement with the experimental values of the ionization energies and electron affinities.

In any case, insofar as the total energy of the atom can be written in the form

$$E \sim -\frac{A}{r} + \frac{B}{2r^2} \tag{3.16}$$

minimization yields

$$r \sim \frac{B}{A} \tag{3.17}$$

The VT [Eq. (3.1)] is satisfied, and the ground-state (minimum) energy is given by

$$E_0 = -\frac{A^2}{2B} = \frac{U_0}{2} = -T_0 \tag{3.18}$$

The VT explains the stability of atoms as the result of a balance between opposing effects. The Coulomb attraction that strives to reduce the radius has two effects:

On the one hand, it tends to reduce the energy, since the potential energy $-A/r$ becomes more negative with decreasing r.

On the other hand, it tends to increase the energy, since a shorter wavelength implies higher momentum and thus higher kinetic energy. Weisskopf[70] calls this latter effect the quantum resistance to compression or the Schrödinger pressure.

These two countervailing effects exist in any aggregation of charges, and will next be used to estimate the masses of stars.

2. *Stellar Masses*

The order of magnitude of the number N of nucleons in most ordinary stars is given by a surprisingly simple combination of fundamental constants:

$$N = sN_0, \qquad N_0 = \left(\frac{\hbar c}{GM^2}\right)^{3/2} = 2.2\times 10^{57} \tag{3.19}$$

where M is the mass of the proton, G the universal gravitational constant, and s a numerical factor lying between 0.1 and 100 (for the sun, a fairly small star, $s = 0.54$). Following Weisskopf,[70] we now undertake to derive this equation qualitatively.

As model for a star we consider a nonrotating sphere of radius R of uniform density composed of hydrogen only.‡ Further, we assume the temperature to be high enough so that all N hydrogen atoms are ionized, giving altogether $2N$ particles. In the formation of a star of mass NM through gravitational contraction to radius R, the gravitational potential energy per particle is decreased by

$$-\frac{U_g}{2N} \sim \frac{GM^2N}{2R} \tag{3.20}$$

This energy ~ 1 keV for the sun—which, of course, is how the temperature of the stellar core becomes high enough to support thermonuclear reactions; in comparison with this value, the Coulombic potential energy is negligible and will henceforth be disregarded.

‡In reality, stars consist almost entirely (96–99% in mass) of hydrogen and helium, of which 50–80% is hydrogen; and at the core the density may approach 100 times its average value.

In this aggregation of N electrons and N protons held together by gravity, the VT [Eq. (3.1)] thus requires that the kinetic energy be minus one-half the (always negative) gravitational potential energy. Insofar as the electron gas is nondegenerate, the kinetic energy is due to thermal motion: from the equipartition theorem at temperature Θ, we have $\overline{T}/2N = 3k\Theta/2$ per particle. The VT thus becomes, per particle,

$$-\frac{U_g}{2N} = \frac{GM^2N}{2R} = \frac{2T}{2N} = 3k\Theta \tag{3.21}$$

or

$$3Nk\Theta = \frac{GM^2N^2}{2R} = -E \tag{3.22}$$

where E is the total gravitational energy of the star.

Having become heated through gravitational contraction, the star radiates into space, its total energy becoming more and more negative. According to Eq. (3.22), this means an increasing temperature and a decreasing radius: the loss of energy makes the star hotter, so we have the seemingly paradoxical result that the specific heat of a star is negative. The explanation of this curious fact is that any change in total energy implies an equal but opposite change in kinetic energy. Although the radiation loss heats up the star, this is more than compensated by the decrease in potential energy attending the gravitational contraction.

Nevertheless, the temperature cannot increase indefinitely with continued contraction, for two reasons: either its increase is halted by reaching the ignition temperature Θ_i of the nuclear reactions, or, failing that, the continuing contraction produces such a high density that the quantum-mechanical kinetic energy of the increasingly degenerate electron gas displaces the thermal agitation as the dominant term in the kinetic energy.

In the first instance, after ignition occurs a steady state is ultimately reached, in which the generation of energy by thermonuclear reaction replaces the heat being lost by radiation, so that the gravitational energy no longer decreases. The star functions like a thermostat: if the rate of energy production slows, the star contracts and the temperature rises, thus accelerating the nuclear reaction; conversely, if the rate becomes too large, the star expands, cools, and reduces the reaction rate.

If the ignition temperature is not reached, the kinetic energy per (degenerate) electron is of order $\hbar^2/2mr^2$, where r is the radius of the volume per electron; this is twice the kinetic energy per particle, counting also the protons. Equating this quantity to its thermal counterpart defines a maximum temperature $\hat{\Theta}$ that cannot be exceeded by further contraction:

$$3k\hat{\Theta} \sim \frac{\hbar^2}{2mr^2} \tag{3.23}$$

Clearly, for thermonuclear reactions to occur requires $\hat{\Theta} > \Theta_i$; this provides a lower bound on N. In Eq. (3.21), we first replace the star radius by $R = N^{1/3}r$, so that

$$\frac{GM^2N}{2R} = \frac{GM^2N^{2/3}}{2r} = \frac{GM^2}{\hbar c}N^{2/3}\frac{\hbar c}{2r}$$
$$= \left(\frac{N}{N_0}\right)^{2/3}\frac{\hbar c}{2r} \tag{3.24}$$

where

$$N_0 \equiv \left(\frac{GM^2}{\hbar c}\right)^{3/2} \tag{3.25}$$

At $\hat{\Theta}$ we thus have

$$\left(\frac{N}{N_0}\right)^{2/3}\frac{\hbar c}{2r} \sim \frac{\hbar^2}{2mr^2}$$

or

$$3k\hat{\Theta} \sim \left(\frac{N}{N_0}\right)^{4/3}\frac{mc^2}{2} = \left(\frac{N}{N_0}\right)^{4/3} \times 255 \text{ KeV} \tag{3.26}$$

The Coulomb energy barrier $\varepsilon^{\ddagger}$ between two protons that must be overcome to start the nuclear reaction is about $mc^2/4 \cong 125$ KeV, but because there are always some very high-speed particles in the high-energy tail of the Maxwell velocity distribution, a significant reaction rate is achieved at very much lower temperatures, perhaps ~ 5 KeV. With this value of $k\Theta$, Eq. (3.26) yields, as the corresponding minimum value for N,

$$\frac{N}{N_0} \sim \left(\frac{15}{255}\right)^{3/4} = 0.12 \tag{3.27}$$

For large stars for which $N/N_0 \gg 1$, a more stringent upper limit is imposed upon the temperature by two effects. The first involves the "cross-over temperature" Θ_c at which the energy density in the radiation field, $u = a\Theta^4$, equals that of the particles, $3nk\Theta/2$. At a hydrogen plasma particle density $n \sim 10^{24}/\text{cm}^3$, $k\Theta_c \sim 2$ KeV. Because of its dependence upon the fourth power of the temperature, even a large increase in the radiation energy density u causes a relatively small change in temperature: a 16-fold increase in u corresponds to only a twofold increase in Θ.

The second effect comes about because the rate of the nuclear "bimolecular reaction" depends exponentially on the temperature,

$$\text{reaction rate} \sim n^2 \exp\left(-\frac{\varepsilon^\ddagger}{k\Theta}\right) \tag{3.28}$$

where the activation energy $\varepsilon^\ddagger$ is the Coulomb energy barrier mentioned above; this causes the very large and very hot stars to have very short lifetimes τ—which thus makes their population in the stellar community at any epoch rather small.‡ In comparison with our standard star (N_0, Θ_0, τ_0), the lifetime ratio is

$$\frac{\tau}{\tau_0} = \frac{N}{N_0} \exp\left[\frac{\varepsilon^\ddagger}{k\Theta_0}\left(\frac{\Theta_0}{\Theta} - 1\right)\right] \tag{3.29}$$

The temperature ratio is itself a function of the mass ratio μ ($\equiv N/N_0$), albeit—for reasons given above—a relatively insensitive one. In the temperature regime where radiation pressure is dominant in the stellar core, $\Theta \sim N^{1/6}$. Thus,

$$\frac{\tau}{\tau_0} = \mu \exp\left[\frac{\varepsilon^\ddagger}{k\Theta_0}(\mu^{-1/6} - 1)\right] \tag{3.30}$$

For illustration, Table II lists the lifetime ratios for stars of various mass ratios μ, taking the parameter $\varepsilon^\ddagger/k\Theta_0 \cong 20$. The lifetimes τ are normalized to an estimated lifetime of the sun ($\mu = 0.54$) of $\cong 12 \times 10^9$ years, so that $\tau_0 \cong 2 \times 10^9$ years.

Over the mass-ratio range $0.1 < N/N_0 < 100$ the stellar lifetime thus varies by six orders of magnitude. At the low end the low rate of energy production makes stars so faint as to be virtually invisible; and at the high end their lifetime is over in a cosmic twinkling.

‡Weisskopf derives an upper limit on N by including the radiation energy in the VT, which then takes the form $T = -\Lambda U$, with $\frac{1}{2} < \Lambda < 1$ [cf. Juttner[72] (1928), Danilow[73] (1930)].

TABLE II
Stellar Lifetime as a Function of Mass Ratio μ

$\mu \equiv N/N_0$	τ/τ_0	τ (years)
0.1	10^3	2×10^{12}
0.5	6	12×10^9
1	1	2×10^9
10	2×10^{-2}	4×10^7
100	2×10^{-3}	4×10^6

These crude calculations are, of course, to be taken seriously only as order-of-magnitude estimates that serve to illustrate the simplicity and perspicuousness with which the VT can illuminate the subject of astrophysics.

3. *Cohesive Energy of Monovalent Metals*

This section describes another qualitative application of the VT, namely, to obtain the cohesive energies of monovalent metals that—considering the simplicity of the model—are in surprisingly close agreement with experimental values. The treatment is similar to that of Frenkel[74] (1924, 1928), Fröman and Löwdin[75] (1962), and Rothwarf[76] (1969).

The cohesive energy of a metal is defined as the energy required to separate the bulk metal (m) into a gas (g) of its constituent atoms—that is, essentially the sublimation energy, ΔE^s:

$$\Delta E^s = E^g - E^m \tag{3.31}$$

Since the only forces operating in both the free atom and the metal are Coulombic, the VT of Eq. (3.1) can be used to equate the cohesive energy simply to the (negative) difference in kinetic energies:

$$\Delta E^s = T^m - T^g \tag{3.32}$$

In what follows the model is restricted to monovalent metals at 0 K, so that the focus is on the changes in kinetic energy of the electrons only. These changes, due essentially to the valence electrons, can be associated with two regions: outside the kernel, where the valence-electron wavefunction in the free atom has changed from an exponential decay to the lattice wavefunctions characteristic of the metallic conduction electrons; and inside the kernel, where the kernel electrons are affected only indirectly by the change in the wavefunction of the valence electron.

TABLE III
Comparison of Experimental Cohesive Energies in Monovalent Metals with the Predictions of Eq. (3.36)[a]

	Z	n	r (Å)	I (eV)	ΔE^s (eV) $0.6\mu_F$	Obs.	$0.6\mu_F - I/n^2$	δ (%)
(i)	(ii)	(iii)	(iv)	(v)	(vi)	(vii)	(viii)	(ix)
Li	3	2(s)	1.727	5.390	2.82	1.582	1.47	−6.7
Na	11	3	2.109	5.138	1.89	1.130	1.32	16.8
K	19	4	2.620	4.339	1.23	0.942	0.96	1.9
Rb	37	5	2.807	4.176	1.07	0.889	0.90	1.2
Cs	55	6	3.041	3.893	0.91	0.817	0.80	−2.0
Cu	29	4(p)	1.411	7.724	4.23	3.529	3.75	6.3
Ag	47	5	1.597	7.574	3.30	2.993	3.00	0.2
Au	79	6	1.593	9.22	3.32	3.569	3.06	−14.3

[a] Recalculated from A. Rothwarf[76]; with permission of North-Holland Publishing Co.

Outside the kernel the dominant contribution to the kinetic energy comes from the degeneracy of the "electron gas," whose levels are filled to the Fermi energy μ_F, given in terms of the maximum electron momentum P by

$$\mu_F = \frac{P^2}{2m} = \left(\frac{9\pi}{4}\right)^{2/3}\left(\frac{a}{r}\right)^2 \frac{\hbar^2}{2ma^2}$$

$$= 50.165\left(\frac{a}{r}\right)^2 \text{ eV} \tag{3.33}$$

where r is the radius of the equivalent-volume sphere per valence electron (i.e., the volume per atom of a monovalent metal), and a is the Bohr radius. The average kinetic energy per electron in the Fermi sea is $3\mu_F/5$. As seen by comparing columns (vi) and (vii) in Table III, this term alone accounts for the major part of the cohesive energy.

This zeroth approximation, however, can be improved by subtracting off in the region outside the kernel the kinetic energy in the tail of the free-atom wavefunction of the valence electron of principal quantum number n:

$$\psi \sim e^{-r/\lambda} \quad \text{with } \lambda = na \tag{3.34}$$

This kinetic energy is

$$\frac{p^2}{2m} \sim \frac{\hbar^2}{2ma^2n^2} \sim \frac{A}{n^2} \tag{3.35}$$

Also, account should be taken of the decrease in screening of the kernel electrons by the valence electrons in going from the free atom to the metal. Because the conduction electrons are distributed more uniformly throughout the atomic sphere, the kernel electrons in the metal experience a somewhat larger effective nuclear charge, which makes their potential energy still more negative than in the free atom and correspondingly increases their kinetic energy. This effect should be roughly proportional to the fraction of the nuclear charge Ze which the valence electron represents, say B/Z. But, at least for the low-Z elements where this effect is largest, $Z \sim n^2$, so that this term can be combined with that of Eq. (3.35) to give

$$\Delta E^s \sim 0.6\mu_F + \frac{Q}{n^2} \tag{3.36}$$

where $Q = B - A$. Rothwarf finds that the choice of the free-atom ionization potential I for the coefficient Q gives quite good agreement with the observed values of the sublimation energy, as shown in columns (vii) and (viii) of Table III. The deviations δ in column (ix) average to 0.4%, with a standard deviation of 9%.

Even the unimproved values ($\Delta E^s = -0.6\mu_F$) are quite good, considering that they contain no adjustable parameter. For polyvalent metals, however, these unimproved values can be in error by as much as a factor of 10; this is to be expected, since the increased interaction among the electrons affects the kinetic energy associated with the polyvalent free atom to a much greater extent than in the monovalent atom. According to Rothwarf,[76] an equation such as Eq. (3.36) with Q/n instead of Q/n^2, and Q being determined for each valence group, can yield agreement with the experimental cohesive energies to within $\cong 30\%$ for the metals in columns 2A–4A of the periodic table.

B. Equations of State

Recalling Eq. (2.49),

$$3PV = 2\overline{T} - \sum \left(\vec{r}_a \cdot \vec{\nabla}_a U \right) \tag{3.37}$$

evidently the VT provides a way to calculate the pressure of a system of particles when the average kinetic energy and internal virial are known. This section addresses the main applications of Eq. (3.37). The VT will first be shown to constitute a theoretical foundation for the virial equation of state for gases, which takes its name from the theorem, and then used to calculate the virial coefficients (Section III.B.1).

The VT, Eq. (3.37), is valid also in quantum mechanics with quantum-mechanical averages instead of time averages. In the early 1950s a question arose whether the pressure calculated from Eq. (3.37), called the kinetic pressure, should equal that given by quantum statistical mechanics, the thermodynamic pressure. By use of the HFT (Section II.B.6) and scaling (Section II.B.2), we show in Section III.B.2 that the kinetic and thermodynamic pressures are indeed equal.

Equation (3.37) of course applies to more than just the equation of state of gases. In Section III.B.3 we derive equations of state for plasmas (fully ionized gases) in three and two dimensions. The Debye–Hückel screened Coulomb potential is not homogeneous but still yields a simple and useful virial relation. The two-dimensional logarithmic potential is an interesting though nonphysical example. In both two and three dimensions, the VT provides alternative derivations for results that are usually obtained in a more complicated way by statistical mechanics.

Finally Eq. (3.37) applies equally well to condensed phases under pressure. Among its many applications to solids, we briefly review in Section III.B.4 a particularly simple model—having, however, a broad generality—that accounts for the equations of state of many metals and certain other solids. (See also Section III.C on approximate quantum-mechanical methods.)

1. The Virial Equation of State

a. History, Importance, and Limitations. The virial equation of state applies to gases at densities low compared with condensed states, and expresses the deviation of the compression factor $PV/Nk\Theta$ from ideality as a power series in the molecular number density ρ ($\equiv N/V$):

$$\frac{PV}{Nk\Theta} = \frac{\beta P}{\rho} = 1 + B_2\rho + B_3\rho^2 + B_4\rho^3 + \cdots \tag{3.38}$$

where $B_2, B_3, \ldots$ are the second, third,... virial coefficients. These coefficients depend on the temperature Θ ($\equiv 1/k\beta$) and are specific to the gas considered, but are independent of the pressure and the density.

Among the myriad of empirical and semiempirical equations of state for gases, the virial equation of state is unique in having a fundamental theoretical basis. There is a definite interpretation for each virial coefficient in terms of molecular properties: the second virial coefficient represents the deviation from ideality corresponding to interactions between pairs of molecules; the third virial coefficient, to three-molecule interactions; and so on. Moreover, there is a definitive prescription for evaluating successive coefficients experimentally from measurements at various densities,

$$B_2 = \lim_{\rho \to 0} \frac{\beta P/\rho - 1}{\rho}, \qquad B_3 = \lim_{\rho \to 0} \frac{\beta P/\rho - 1 - B_2\rho}{\rho^2}, \ldots \tag{3.39}$$

although practically such experimental evaluations are limited to the first few virial coefficients. The virial equation of state is one of the few direct bridges between experimental results and the theoretical intermolecular forces in gases. (For a detailed exposition see Mason and Spurling's monograph.[77])

Nevertheless, the virial series encounters some problems of divergence for gases at high densities and for plasmas and electrolyte solutions. Physically this divergence arises when the intermolecular interactions become long-ranged (for example, Coulomb forces) so that the assumption underlying the virial expansion (i.e., the successive reduction of a many-body problem to two-body, three-body,... problems) is not always valid. However, Mayer[78] (1950) showed how to combine positive and negative terms to attain convergence for an ionic solution. Brief mention will be made also of a virial-type expansion of solutions of nonelectrolytes developed by McMillan and Mayer[79] (1945; see also Hill,[80] 1959). The essential result of the McMillan–Mayer theory is that there is a one-to-one correspondence between the equations for imperfect gases and dilute solutions of nonelectrolytes. The pressure P of a gas is found to map into the osmotic pressure Π of the solution, which obeys a virial expansion of the form

$$\beta\Pi = c_2 + \sum_{j=2}^{N} B_j^*(\mu_1, \Theta) c_2^j \tag{3.40}$$

where c_2 is the concentration of the solute and μ_1 the chemical potential of the solvent.

Historically, infinite-series expansions in a form similar to Eq. (3.38) were proposed at least as early as 1885 by Thiensen.[81] However the major development came in 1901 from Kamerlingh Onnes,[82] who suggested the name virial coefficients—presumably having in mind their basis in Clausius' work, although there is no reference to the VT in the original paper. In sequence, we next show how the virial equation of state results from the VT by expanding the internal virial of Eq. (3.37) in terms of the density, and calculate explicitly the second virial coefficient. The other common approach to the derivation of the equation of state is based on one of the partition functions of statistical mechanics, the most natural to use being the grand partition function that automatically performs the decomposition of the N-body problem successively into $1, 2, 3, \ldots$-body problems.

b. Connection with the Virial Theorem. The VT, Eq. (3.37), is considerably simplified when the potential U can be written as a sum of pair potentials of the form

$$U = \sum_{a>b}\sum u_{ab}(r_{ab}) \tag{3.41}$$

As discussed in Section II.A.4, the sums in Eqs. (3.37) and (3.41) are to be understood at the level (see Table I) of the molecules (or atoms) of the gas, without consideration of their internal structures. The r_{ab} are the distances between the centers of mass of these molecules [cf. Eqs. (2.40) and (2.41)]. The assumption of Eq. (3.41) is valid for gases at sufficiently low densities that, in good approximation, we need consider only pairwise interactions that are unchanged by the presence of a third body. In this case, since all $N(N-1)/2$ pairs make equivalent contributions, we get for the intermolecular virial

$$\sum \vec{r}_a \cdot \vec{\nabla}_a U = \tfrac{1}{2}N(N-1)(\vec{r}_1\cdot\vec{\nabla}_1 + \vec{r}_2\cdot\vec{\nabla}_2)u_{12} = \tfrac{1}{2}N(N-1)r_{12}\frac{\partial u_{12}}{\partial r_{12}} \tag{3.42}$$

since

$$\vec{\nabla}_1 u_{12} = -\vec{\nabla}_2 u_{12} = \frac{\partial u_{12}}{\partial r_{12}}\frac{\vec{r}_{12}}{r_{12}} \tag{3.43}$$

Thus, at this point, the VT reads

$$3PV = 2\overline{T} - \tfrac{1}{2}N(N-1)\overline{r_{12}\frac{\partial u_{12}}{\partial r_{12}}} \tag{3.44}$$

To recover the equation of state, we now have to add the assumption that the system is at equilibrium. We can thus proceed further and evaluate the average kinetic energy $\overline{T}$ by replacing, as is legitimate at equilibrium, the time average with an ensemble average

$$\overline{T} = \frac{\int\int d\vec{p}^N\, d\vec{r}^N\, Te^{-\beta H}}{\int\int d\vec{p}^N\, d\vec{r}^N\, e^{-\beta H}} \tag{3.45}$$

where $d\vec{p}^N = d^3\vec{p}_1 \cdots d^3\vec{p}_N$, $d\vec{r}^N = d^3\vec{r}_1 \cdots d^3\vec{r}_N$, and $H\ (=T+U)$ is the Hamiltonian of the system. The integrations over the coordinates cancel, and the remaining integrals over momenta yield the well-known equipartition theorem

$$T = \frac{3Nk\Theta}{2} \tag{3.46}$$

that is, $k\Theta/2$ for each quadratic term in H. Note that this result is valid for both perfect and imperfect gases.

From Eqs. (3.44) and (3.46), we now have a pressure–volume–temperature relation

$$3PV = 3Nk\Theta - \tfrac{1}{2}N(N-1)\overline{r_{12}\frac{\partial u_{12}}{\partial r_{12}}} \tag{3.47}$$

As mentioned in Section II.A.2, in the absence of interactions between molecules we are left with the ideal-gas equation of state

$$PV = Nk\Theta \tag{3.48}$$

—a virial equation of state with the second and higher virial coefficients null.

In the presence of pair interactions the right-hand side of Eq. (3.47) can be expressed in terms of the radial distribution function $g(r)$.‡ Explicitly [cf. McQuarrie's pressure equation, Ref. 83, p. 262, Eq. (13.23)], we have

$$3PV = 3Nk\Theta - \tfrac{1}{2}\rho^2 V\int_0^\infty ru'(r)g(r)4\pi r^2\,dr \tag{3.49}$$

or

$$\frac{\beta P}{\rho} = 1 - \frac{\rho\beta}{6}\int_0^\infty g(r)u'(r)4\pi r^3\,dr \tag{3.50}$$

In order to make a connection with the virial expansion, Eq. (3.38), g is expanded in a power series of the density,

$$g = g_0 + g_1\rho + g_2\rho^2 + \cdots \tag{3.51}$$

and substituted into Eq. (3.50) to yield

$$\frac{\beta P}{\rho} = 1 - \frac{\rho\beta}{6}\sum_{j=0}^{\infty}\rho^j\int_0^\infty g_j u' 4\pi r^3\,dr \tag{3.52}$$

‡The radial distribution function is of central importance because it can be determined by X-ray diffraction and, in the pairwise-potential approximation, all thermodynamic functions can be expressed in terms of it.

By comparison with the virial equation of state, Eq. (3.38), we get an integral for each virial coefficient:

$$B_{j+2} = -\frac{\beta}{6}\int_0^\infty g_j u' 4\pi r^3\, dr \tag{3.53}$$

This same result, obtained here from the VT, is usually derived by using the quantum-mechanical grand-partition-function method [cf. Ono[84] (1951) and Kilpatrick[85] (1953)], then going to the classical limit [Kirkwood[86] (1933); cf. Mason and Spurling, Ref. 77, p. 21].

2. *Kinetic vs. Thermodynamic Pressure*

The VT, Eq. (3.37), is valid also in quantum mechanics (cf. Section II.B.7); thus the pressure calculated from it, sometimes called the kinetic pressure, should equal the pressure derived from statistical mechanics, the thermodynamic pressure. This may seem obvious, but at one time there was disagreement over their equivalence for quantum fluids, although there was some agreement that they coincide in the classical limit. Green[87] (1949) (see also Born and Green,[88] 1947) argued that the difference between the two pressures becomes appreciable at very low temperature, and used this result to explain the abnormal properties of He II, the only substance to remain liquid near 0 K due to quantum effects. It was soon shown (de Boer,[89] 1949; Yvon,[90] 1948; Riddell and Uhlenbeck,[91] 1950; Zwanzig,[92] 1950; Brown,[93] 1958) by modifying Green's treatment of the effect of the wall of the container, that the kinetic and thermodynamic pressures are indeed the same. To prove this equality, we start from the thermodynamic pressure defined in the canonical ensemble, and use the Hellmann–Feynman theorem (Section II.B.6) together with scaling (Section II.B.3) in this rather neat derivation of the VT.

For the canonical ensemble, the partition function is given by

$$Q = \sum \exp(-\beta E_i) \tag{3.54}$$

and is related to the Helmholtz free energy through

$$\beta A = -\ln Q \tag{3.55}$$

The pressure is defined as

$$P = -\left(\frac{\partial A}{\partial V}\right)_\beta = Q^{-1}\Sigma\left(-\frac{\partial E_i}{\partial V}\right)\exp(-\beta E_i) \tag{3.56}$$

In other words, the pressure is just the ensemble average of $-\partial E_i/\partial V$, in which the occupation numbers of all states are held constant. We need the dependence of the energy levels on the volume, which we make appear by the following device: we scale the coordinates with a characteristic length of the system, for example $V^{1/3}$; that is, we measure the distances in units of the cube root of the volume. The Hamiltonian is then an explicit function of V, and the energy derivative is given by the HFT, Eq. (2.203), with the volume as parameter:

$$\frac{\partial E_j}{\partial V} = \left\langle \frac{\partial H}{\partial V} \right\rangle_j = \left\langle \psi_j \left| \frac{\partial H}{\partial V} \right| \psi_j \right\rangle \tag{3.57}$$

where the energy E_j is a solution of

$$H\psi_j = E_j\psi_j \tag{3.58}$$

The new coordinates are defined by

$$r'_a = \frac{r_a}{V^{1/3}} \tag{3.59}$$

keeping the $\vec{r}'_a$ constant and varying the volume corresponds to dilating or contracting the system without any change in its relative shape or configuration. In the new coordinate system, we can write

$$H = -\frac{\hbar^2}{2V^{2/3}} \sum m_a^{-1} \nabla_a'^2 + U\{V^{1/3}\vec{r}'_a\} \tag{3.60}$$

where U refers only to the intermolecular forces; the wall forces are taken into account directly by requiring the wavefunction to vanish at the walls.

Differentiating H with respect to V and reverting to the original coordinates, we obtain

$$3V\frac{\partial H}{\partial V} = -2T + \sum \vec{r}_a \cdot \vec{\nabla}_a U \tag{3.61}$$

The HFT, Eq. (3.57), thus yields

$$3V\frac{\partial E_j}{\partial V} = \sum \left(-2\langle T \rangle_j + \langle \vec{r}_a \cdot \vec{\nabla}_a U \rangle_j \right) \tag{3.62}$$

If one defines a quantum-mechanical pressure operator P as $-\partial H/\partial V$, then its average value over state j gives $-\partial E_j/\partial V$, and Eq. (3.62) is the quantum-mechanical VT for an enclosed system, the classical analog of which is Eq. (3.37). Substituting back this result into the thermodynamic pressure [Eq. (3.56)] and rearranging, we obtain

$$3PV = Q^{-1}\sum \exp(-\beta E_j)\left[2\langle T\rangle_j - \sum\langle \vec{r}_a \cdot \vec{\nabla}_a U\rangle_j\right] \qquad (3.63)$$

or

$$3PV = 2\langle \bar{T}\rangle - \sum \overline{\vec{r}_a \cdot \vec{\nabla}_a U} \qquad (3.64)$$

where the vincula refer to ensemble averages, which are identical to time averages at equilibrium. Equation (3.64) is the extension to enclosed systems of the doubly averaged VT, Eq. (2.172). We have thus verified that the kinetic and thermodynamic pressures are equal (cf. Mason and Spurling, Ref. 77, p. 25).

3. *Three- and Two-Dimensional Plasmas*

a. Plasma Equation of State. Although the equation of state for a plasma is usually derived from statistical mechanics (see, for example, Lifshitz and Pitaevskii, Ref. 94, p. 239), here also the VT provides an alternative proof (Kelly,[95] 1963) that has the pedagogical advantage of not requiring complicated statistical-mechanical tools like the cluster integrals or the propagators.

In order to proceed, we invoke the Debye–Hückel[96] theory for the average potential $\Phi_s(r)$ surrounding a plasma ion of species s bearing charge $Z_s e$; one obtains a screened Coulomb potential [Ref. 94, Eq. (78.9)]

$$\langle \Phi_s(r)\rangle = \frac{Z_s e}{r} e^{-r/D} \qquad (3.65)$$

where D is the Debye (shielding) length given by

$$D^2 = \frac{k\Theta}{4\pi e^2 \sum n_s Z_s^2} \qquad (3.66)$$

Here n_s is the number density of ions of species s, and the summation is over all species s.

Recall the VT for a Coulombic potential:

$$3PV = 2\langle T\rangle + \langle U\rangle \qquad (3.67)$$

where $\langle T \rangle$ is given by the equipartition theorem, Eq. (3.46),

$$2\langle T \rangle = 3k\Theta V \sum n_s \tag{3.68}$$

and U is the total potential energy

$$U = \tfrac{1}{2} \sum Z_i e \Phi_i' \tag{3.69}$$

Here Φ_i' is the potential experienced by the ith ion due to the Coulomb field of all other plasma ions:

$$\Phi_i' = \sum_{j(\neq i)} \frac{Z_j e}{r_{ij}} \tag{3.70}$$

Replacing in Eq. (3.69) the sum over ions i by a sum over species s and averaging, U becomes

$$\langle U \rangle = \frac{1}{2} \left\langle \sum_i Z_i e \Phi_i' \right\rangle = \frac{1}{2} \sum_s n_s V Z_s e \langle \Phi_s' \rangle \tag{3.71}$$

The average potential $\langle \Phi_s' \rangle$ is

$$\langle \Phi_s' \rangle = \left[\langle \Phi_s(r) \rangle - \frac{Z_s e}{r} \right]_{r=0} \tag{3.72}$$

that is, the average potential acting on an ion of charge $Z_s e$ is the Debye–Hückel potential, Eq. (3.65), less the self-potential of the ion. One has [Kirkwood and Poirier[97] (1954)]

$$\langle \Phi_s' \rangle = -\frac{Z_s e}{D} \tag{3.73}$$

so that, using the definition Eq. (3.66) of D,

$$\langle U \rangle = -\frac{e^2 V}{2D} \sum n_s Z_s^2 \tag{3.74}$$

Putting Eqs. (3.68) and (3.74) into the VT, Eq. (3.67), gives the desired equation of state for a plasma:

$$P = k\Theta \sum n_s - \frac{e^3}{3} \left(\frac{\pi}{k\Theta} \right)^{1/2} \left(\sum n_s Z_s^2 \right)^{3/2} \tag{3.75}$$

Using Eq. (3.66) again, this may be written

$$P = k\Theta \sum n_s \left[1 - (18\mathcal{N})^{-1}\right] \tag{3.76}$$

where

$$\mathcal{N} = \frac{4\pi}{3} D^3 \sum n_s \tag{3.77}$$

is the total number of ions inside the "Debye sphere." This form shows that deviations from the ideal-gas law are small when $\mathcal{N}$ is large. For illustration, in a uni-univalent plasma, $D^2 \sim 1/n$ and $\mathcal{N} \sim D^3 n \sim n^{-1/2}$; thus large $\mathcal{N}$ implies low ion density. It is recalled that this same condition must be satisfied if the concept of a Debye shielding cloud is to be meaningful (Kelly,[95] 1963). As a note of historical interest, early treatments of the VT in the Debye–Hückel theory of electrolytes were given by von Gross and Halpern[98] (1925), Adams[99] (1926), van Rysselberghe[100] (1933), Fuoss[101] (1934), and Finkelstein[102] (1935). Even prior to the seminal paper of Debye and Hückel,[96] Milner[103] (1912) published an extended discourse on "the virial of a mixture of ions."

b. Two-Dimensional Plasma. A two-dimensional plasma consisting of ions of charges $Z_a e$ that attract and repel each other with a $1/r$ force law‡ possesses an exact equation of state (May,[104] 1967). Once again, although the statistical-mechanical treatment is quite complicated, the VT provides an elementary proof of this exact equation of state (Knorr,[105] 1970). Instead of

$$3PV = 3Nk\Theta - \sum \vec{r}_a \cdot \vec{\nabla}_a U \tag{3.78}$$

we must now write

$$2PV = 2Nk\Theta - \sum \vec{r}_a \cdot \vec{\nabla}_a U \tag{3.79}$$

since in two dimensions $\operatorname{div} \vec{r} = 2$ and each of the N particles has only two degrees of freedom.

‡Although this force law (which admits no screening) and the subsequent equation of state provide a simple mathematical illustration of the power of the VT, this model has apparently not yet found any significant physical application.

The potential is a sum over pairs of logarithmic pair potentials

$$U = \sum_{a>b}\sum U_{ab} \tag{3.80}$$

with

$$U_{ab} = -e^2 Z_a Z_b \ln r_{ab} \tag{3.81}$$

(The nonphysical dimensions of this potential energy would be improved by writing $v_{ab} = -(Z_a Z_b e^2/l)\ln r_{ab}$, where l has dimensions of a length.) The interaction virial is thus

$$\sum \vec{r}_a \cdot \vec{\nabla}_a U = \sum_{a>b}\sum \left(\vec{r}_a \cdot \vec{\nabla}_a U_{ab} + \vec{r}_b \cdot \vec{\nabla}_b U_{ab}\right) = -e^2 \sum_{a>b}\sum Z_a Z_b \tag{3.82}$$

since

$$\left(\vec{r}_a \cdot \vec{\nabla}_a + \vec{r}_b \cdot \vec{\nabla}_b\right) \ln r_{ab} = \frac{\vec{r}_a \cdot \vec{\nabla}_a r_{ab} + \vec{r}_b \cdot \vec{\nabla}_b r_{ab}}{r_{ab}} = 1 \tag{3.83}$$

where we have used

$$\vec{\nabla}_a r_{ab} = -\vec{\nabla}_b r_{ab} = \frac{\vec{r}_{ab}}{r_{ab}} \tag{3.84}$$

If we now consider the N particles to be $N/2$ electrons of charge $-e$ and $N/2$ ions of charge $+e$, so that there are 2 times $\frac{1}{2}[\frac{1}{2}N(\frac{1}{2}N-1)]$ combinations of equal charges and $N^2/4$ combinations of opposite charges, Eq. (3.82) becomes

$$\sum \vec{r}_a \cdot \vec{\nabla}_a U = \frac{Ne^2}{2} \tag{3.85}$$

Collecting our results in Eq. (3.79), we find

$$\frac{PV}{N} = k\Theta - \frac{e^2}{4} \tag{3.86}$$

which is May's result. This equation of state exhibits a condensation phenomenon for temperatures below

$$\Theta_c = e^2/4k \tag{3.87}$$

Unfortunately, a more detailed investigation reveals there is much more to the complete story [see Knorr[105] (1970) and references therein].

4. *Solids under Pressure*

When a substance is compressed, its quantum-mechanical energy levels are shifted to higher values and its kinetic energy increases; this behavior can be explained in terms of the VT for Coulomb forces, $3PV = 2T + U$. With $E = T + U$,

$$\begin{aligned} \overline{T} &= -E + 3PV \\ \overline{U} &= 2E - 3PV \end{aligned} \tag{3.88}$$

that is, both $\overline{T}$ and $\overline{U}$ are expressed in terms of experimentally measurable quantities. This result has been used to obtain the change in T and U with compression in solids (Hirschfelder, Curtiss, and Bird[106]) and for stress calculations in molecular dynamics (Tsai,[107] 1979); it is valid also in many approximate calculations: the self-consistent-field Hartree–Fock method including exchange and correlation effects (Ross,[108] 1969; Liberman,[109] 1971; see Section III.C.1), the muffin-tin approximation (Janak,[110] 1974), the Thomas–Fermi model (More,[111] 1979; see Section III.C.2); and the density-functional theory (Bartolotti and Parr,[112] 1980). Numerical computations have been performed mainly for transition metals (Pettifor,[113,114] 1971, 1977); Kakehashi,[115] 1980).

We shall describe here only one application of Eq. (3.88); it has a wide range of interest, since it yields a one-parameter equation of state valid for metals and some other solids (McMillan and Latter,[116] 1958; Libby and Libby,[117] 1972). We introduce the Helmholtz free energy A in Eq. (3.88), in place of E, to get

$$3PV = 2(A + \Theta S) - U = 2A - (U - 2\Theta S) \tag{3.89}$$

and take the isothermal volume derivative $(\partial/\partial V)_\Theta$, indicated by a prime; since

$$-A' = P \tag{3.90}$$

we obtain

$$3P'V + 3P = -2P - (U - 2\Theta S)' \tag{3.91}$$

or, rearranging,

$$(PV^{5/3})' = -\frac{V^{2/3}}{3}(U - 2\Theta S)' \tag{3.92}$$

Now, we make two assumptions (neither very restrictive, see below):

1. The potential energy U increases on compression inversely as the atomic distance; that is,

$$U = U_0 \left(\frac{V_0}{V} \right)^{1/3} \tag{3.93}$$

implying that

$$U' = \frac{U_0 V_0^{1/3}}{3V^{4/3}} \tag{3.94}$$

where the subscript zero refers to the uncompressed state.

2. The compression has small effect on the entropy S, so that we can neglect S'. (This is, of course, exact at absolute zero.)

When assumptions 1 and 2 are satisfied, their insertion into Eq. (3.92) gives, upon integration, our desired equation of state:

$$P = \frac{U_0}{3V_0} \left[\left(\frac{V_0}{V} \right)^{5/3} - \left(\frac{V_0}{V} \right)^{4/3} \right] \tag{3.95}$$

The isothermal bulk modulus, defined as

$$B_\Theta = -\left(\frac{\partial P}{\partial \ln V} \right)_\Theta \tag{3.96}$$

is calculated from Eq. (3.95) to be

$$3B_{\Theta_0} = \frac{U_0}{3V_0} \tag{3.97}$$

In terms of the bulk modulus at zero pressure B_{Θ_0}, Eq. (3.95) appears as a one-parameter equation of state,

$$P = 3B_{\Theta_0} \left[\left(\frac{V_0}{V} \right)^{5/3} - \left(\frac{V_0}{V} \right)^{4/3} \right] \tag{3.98}$$

and applies to numerous solids [all for which assumptions 1 and 2 are valid], although McMillan and Latter[116] (1958), in their original derivation of Eq. (3.98), limited their application to the alkali and alkaline-earth metals. Figure 1 [from Libby and Libby[117] (1972)], gives plots of P against the quantity

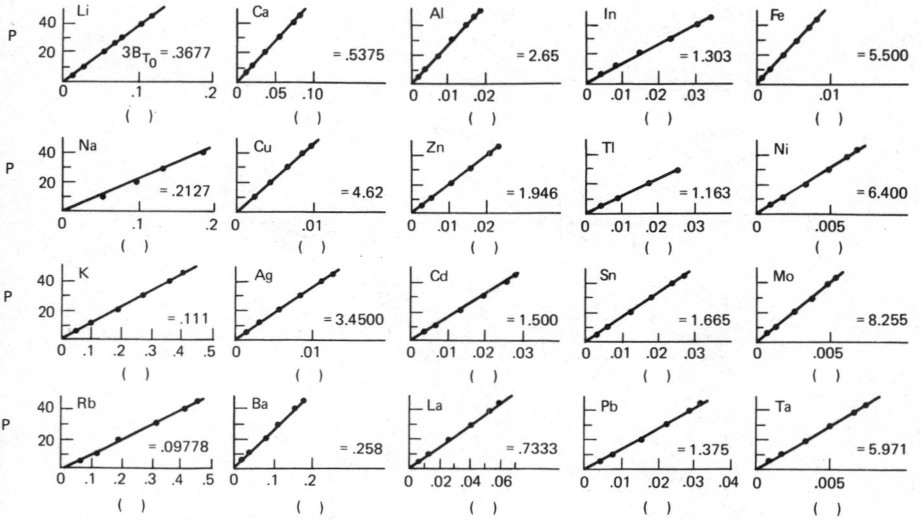

(a)

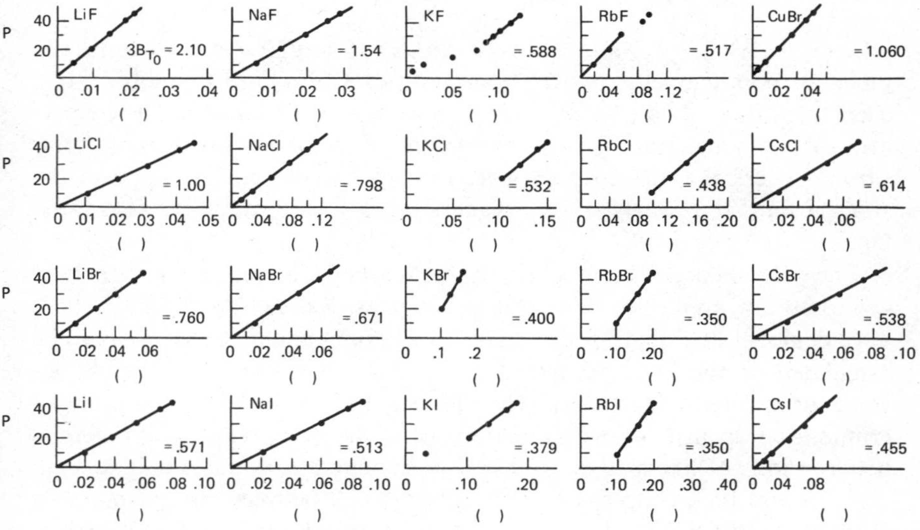

(b)

Fig. 1. One-parameter equation of state for solids, Eq. (3.52). [From Ref. 117: L. M. Libby and W. F. Libby, *Proc. Nat. Acad. Sci. U.S.A.* **69** (No. 11), 3305–3306 (1972), with permission.] Data for (*a*) metals, (*b*) alkali halides, (*c*) silver and ammonium halides and certain organic compounds.

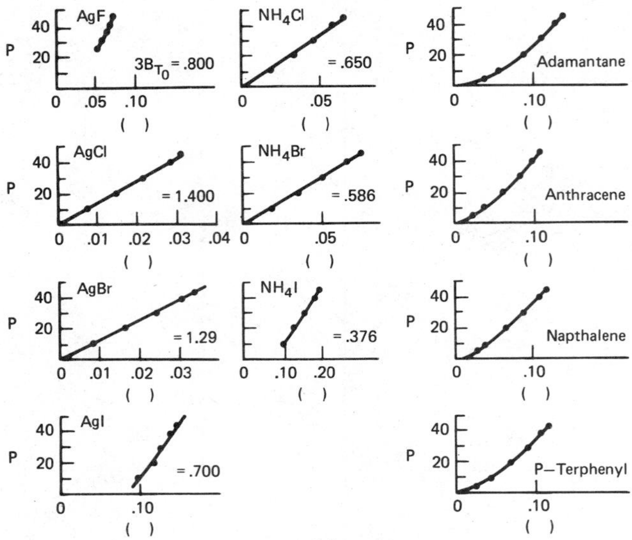

(c)

Fig. 1 (*Continued*).

[ ] $\equiv (V_0/V)^{5/3} - (V_0/V)^{4/3}$] for 20 metals (*a*) and 20 metallic halides (*b*) taken at room temperature (23°C) and pressures from 0 to 45 kilobars. It is clear from these data that Eq. (3.98) fits in most cases to within a few percent for these materials, the only exceptions being for cases with phase transitions, where the line does not pass through the origin. The slope of the straight line, $3B_{\Theta_0}$, is given (in megabars) in each case where the fit is satisfactory.

From the theory of ionic solids, the assumption 1 is reasonable in that the Coulomb law applies between charged spheres. Reiss et al.[118] (1961, 1962) have demonstrated that for fused salts the thermodynamic properties can be calculated on the assumption that the potential energy depends on the inverse mean interionic distance. The rationale for this lies in the natural discrimination against configurations involving the short-range pair potential between ions of like charge, so that such configurations make only a small contribution relative to the short-range potential between ions of opposite charge. For molecularly bonded solids this does not apply, and so we see in Fig. 1*c* that the straight-line relation of Eq. (3.98) fails in such compounds as anthracene, *p*-terphenyl, naphthalene, and adamantane.

The above discussion, due to Libby and Libby,[117] also provides an illustration of the usefulness of the model: taking the straight line for iron in Fig. 1*a*, from which $3B_{\Theta_0}$ is deduced to be 5.50 Mbar, they calculated the expected compression at the pressure of 3.8 Mbar, the pressure at the center of the earth. The result is 2.03 for V_0/V. This is to be compared with the "observed" value (Allen[119]), if the core be pure iron, of 2.20.

C. Hartree–Fock–Slater and Thomas–Fermi–Dirac Models

Only in the case of one-electron ("hydrogenic") ions can the Schrödinger equation be solved exactly. For two or more electrons one has to rely on approximation techniques to take into account the interelectronic repulsions in the Hamiltonian. The two basic approximation schemes are the variation and perturbation methods. Using these methods, the ground-state energy of the helium atom, for example, can be found with six-figure agreement between theory and experiment, a convincing proof of the validity of quantum mechanics: a variation calculation, corrected to take into account the nuclear motion, the relativistic effects, and the Lamb shift (effect of fluctuations in the vacuum field on the electron distribution), yields for the helium ground state 198310.665 ± 0.04 cm, in phenomenal agreement with the experimental value 198310.82 ± 0.15 cm (see Bethe and Jackiw[120]). For more complex atoms, this kind of calculation gets prohibitively difficult and other approaches have been sought.

The most accurate of these approaches, the method of the self-consistent field (SCF), will be outlined in Section III.C.1, including the Hartree (H) and Hartree-Fock (HF) methods, Slater's local-exchange approximation, and the X_α method.

The SCF calculations are themselves very cumbersome for many-electron atoms. For these, there is another approximate method whose value lies in its simplicity: the statistical model of Thomas and Fermi (TF) and its improvement by Dirac (D) to take into account exchange effects (Section III.C.2). The predictions of the TF and TFD models are admittedly less accurate than those of the SCF method, but the physical meaning does not get drowned in the computations.

In the SCF as well as in the TF and TFD approximations the VT holds, as we shall see; it is a strong attribute of these methods to respect the virial balance between kinetic and potential energies. Since the VT is an exact relation, it is a necessary—though not sufficient—condition for the accuracy of any model.

1. *Self-Consistent Field and Local-Exchange Approximations*

Although the SCF method applies equally to molecules, for clarity we begin with the simpler application to an atom. In first approximation (Hartree,[121] 1928), each electron—assumed distinguishable—is described by its own wavefunction, a spin–orbital, and is subject to the Coulomb field of the nucleus and the average field of all the other electrons. For electron i at position $\vec{r}_1$ the potential is given by

$$V_i(\vec{r}_1) = \frac{Z}{r_1} - \sum_{k \neq i} \int d\tau_2 \frac{|u_k(\vec{r}_2)|^2}{r_{12}} \tag{3.99}$$

where the u_k are the spin–orbitals of the electrons, so chosen as to form an orthonormal set:

$$\int d\tau_1\, u_i^*(1)\, u_j(1) = \delta_{ij} \tag{3.100}$$

Following Hartree, we use atomic units[‡] throughout. As in Section II, the symbol $\int d\tau$ means integration over all space and eventually summation over spin. For N electrons we thus have to solve N simultaneous nonlinear integrodifferential equations,

$$\left[-\tfrac{1}{2}\nabla^2 + V_i(\vec{r})\right] u_i(\vec{r}) = E_i u_i(\vec{r}) \tag{3.101}$$

The central-field approximation is used, in which $V_i(\vec{r})$ is replaced by its average over all directions:

$$V_i(\mathrm{r}) = \frac{1}{4\pi}\oint d\Omega\, V_i(\vec{r}) \tag{3.102}$$

By writing each spin–orbital as a product of a radial function and a spherical harmonic, the set of equations (3.101) can be solved numerically in a self-consistent manner. An initial potential is constructed by guessing a first set of orbitals (for example, hydrogenic wavefunctions); the differential equations are then solved, giving new orbitals and thus a new potential, and so on until the desired degree of self-consistency is reached, that is, until further iteration gives no substantial improvement. The Hartree method is equivalent to a variation calculation where the trial function is set equal to a simple product of orbitals and where the variation is performed by varying each orbital in an arbitrary way (but maintaining normalization).

The two main shortcomings of the Hartree method are that electron correlation and symmetry (antisymmetry of the wavefunction) effects are ignored. The former can be evaluated by perturbation theory. The latter is taken into account in what is called the Hartree–Fock method, by choosing

[‡] The unit of mass is the rest mass m of the electron; of charge, the magnitude of the electronic charge e; of length, the Bohr radius of the hydrogen atom $a = \hbar/me^2$. Consequently, the unit of energy is $e^2/a = 1\ au = 2$ rydbergs.

as variational function a Slater determinant (Slater,[122] 1930; Fock,[123] 1930):

$$\psi = (N!)^{-1/2}\begin{vmatrix} u_1(1) & u_1(2) & \cdots & u_1(N) \\ u_2(1) & u_2(2) & \cdots & u_2(N) \\ \vdots & \vdots & & \vdots \\ u_n(1) & u_N(2) & \cdots & u_N(N) \end{vmatrix}$$

or

$$\begin{aligned}\psi &= (N!)^{-1/2}\sum_p (-1)^p \prod_{i=1}^{N} u_{\mathrm{Pi}}(i) \\ &= (N!)^{-1/2}\sum_p (-1)^p \prod_{i=1}^{N} u_i(\mathrm{Pi}) \end{aligned} \tag{3.103}$$

where the sum extends over all permutations Pi of $(1,2,\ldots, N,)$, and $(-1)^p = +1$ or -1 according as the permutation is even or odd. This determinantal wavefunction takes care of the antisymmetry and thus of the Pauli exclusion principle: if two spin–orbitals are the same, two rows of the determinant are equal and $\psi = 0$; if two electrons are in the same position, two columns are equal and again $\psi = 0$; and on interchange of two rows or columns the determinant changes sign.

The problem is to minimize the variational integral,

$$\delta\langle\psi|H|\psi\rangle = 0 \tag{3.104}$$

subject to arbitrary variations of the one-electron wavefunctions u_k; H is the Hamiltonian in atomic units:

$$H = \sum\left[-\left(\tfrac{1}{2}\nabla_i^2 + \frac{Z}{r_i}\right)\right] + \sum_{i<j}\frac{1}{r_{ij}} \tag{3.105}$$

These variations yield the so-called Hartree–Fock equations:

$$\begin{aligned} -\tfrac{1}{2}\nabla_1^2 u_i(\vec{r}_1) - \frac{Z u_i(\vec{r}_1)}{r_1} + \left[\sum_j \int d\tau_2 \frac{|u_j(\vec{r}_2)|^2}{r_{12}}\right] u_i(\vec{r}_1) \\ - \sum_j \delta(m_{s_i}, m_{s_j})\left[\int d\tau_2 \frac{u_j^*(\vec{r}_2)\, u_i(\vec{r}_2)}{r_{12}}\right] u_j(\vec{r}_1) = \epsilon_i u_i(\vec{r}_1) \end{aligned} \tag{3.106}$$

The last term on the left-hand side, where the summation is over only electrons of the same spin, is referred to as the exchange term. Note that the $i = j$ terms in the two summations over j cancel each other: the introduction of the exchange term takes care of the self-energy of the electron. After making this cancellation, it can be seen by comparison with the Hartree equations, (3.99) and (3.101), that the Hartree–Fock (HF) equations differ by the extra term

$$\sum_{j \neq i} \delta(m_{s_j}, m_{s_i}) \left[\int d\tau_2 \frac{1}{r_{12}} u_j^*(\vec{r}_1) u_i(\vec{r}_2) \right] u_j(\vec{r}_1) \tag{3.107}$$

which arises from the use of a determinantal function instead of a single product of orbitals.

The exchange term involves a nonlocal potential, that is, it can be written as

$$-\int U(\vec{r}_1, \vec{r}_2) u_i(\vec{r}_1)\, d\tau_2 \tag{3.108}$$

where

$$U_i(\vec{r}_1, \vec{r}_2) = \frac{1}{r_{12}} \sum_{j=1}^{N} \delta(m_{s_i}, m_{s_j}) u_j^*(\vec{r}_2) u_j(\vec{r}_1) \tag{3.109}$$

is a nonlocal potential that satisfies the Hermiticity condition

$$U(\vec{r}_1, \vec{r}_2) = U^*(\vec{r}_2, \vec{r}_1) \tag{3.110}$$

Defining the number density of electrons by

$$n(\vec{r}_2) = \sum_j |u_j(\vec{r}_2)|^2 \tag{3.111}$$

and the effective potential V_x by

$$V_x(\vec{r}_1) u_i(\vec{r}_1) = -\int U(\vec{r}_1, \vec{r}_2) u_i(\vec{r}_2)\, d\tau_2 \tag{3.112}$$

the HF equation becomes

$$\left[-\tfrac{1}{2}\nabla_1^2 - \frac{Z}{r_1} + \int d\tau_2 \frac{1}{r_{12}} m(\vec{r}_2) + V_x(\vec{r}_1) \right] u_i(\vec{r}_1) = \epsilon_i u_i(\vec{r}_1) \tag{3.113}$$

It was shown by Koopmans[124] (1933) that ϵ_i represents the energy of removal of electron i. The HF method, with all the refinements of configuration interactions, is the most accurate method to obtain the energy levels and wavefunctions of atoms and ions. Nevertheless, the presence of the nonlocal potential makes the computations very laborious; it is thus often advantageous to employ less accurate methods to reduce the extensive numerical calculations.

Slater[125] (1951) developed an approximation to replace the nonlocal potential in the HF equation by the local potential

$$V_{xs} = -\alpha\left[\frac{3n(r)}{\pi}\right]^{1/3}, \quad \text{where } \alpha = \tfrac{3}{2} \tag{3.114}$$

This potential is obtained in the same way as the exchange term in the TFD theory (cf. Section III.B.2), the electrons being identified with a zero-temperature Fermi gas with plane waves as orbitals. The Hartree–Fock–Slater (HFS) equations obtained using the local-exchange approximation are as simple to solve as the Hartree equations (3.101). Hartree[126] (1957) pointed out that the HFS results would agree better with the more precise HF ones if the coefficient $\alpha = \frac{3}{2}$ were smaller. Gaspar[128] (1954) and Kohn and Sham[127] (1965) showed that by calculating the exchange energy first using the free-electron approximation and then applying the variational principle, the coefficient α becomes unity. One thus gets results in better agreement with the true HF values (Cowan et al.,[129] 1966; Tong and Sham,[130] 1966). Numerical tables of HFS ($\alpha = \frac{3}{2}$) have been published by Herman and Skillman[131] (1963), and many calculations using the Kohn–Sham value ($\alpha = 1$) and other modifications are given by Cowan et al.[129,132] (1966, 1967), Griffin et al.[133] (1971), and Liberman[134] (1968, 1970). The total binding energy of atoms can be obtained within about 0.1 rydberg of the true HF values if one operates with care (cutting the tails of the electron density, etc.).

Although the determinantal wavefunctions constructed from the solutions of the HF equation satisfy the VT (Hill,[135] 1937), those constructed from the SCF scheme in the presence of the Slater local potential do not. However, the Kohn–Sham–Gaspar ($\alpha = 1$) version does satisfy the VT (Ross,[136] 1969; Sham,[137] 1970; Holland,[138] 1980) and is thus on a somewhat sounder footing.

Nevertheless, Berrondo and Goscinski[139] (1969) varied α as a parameter in Slater's scheme and fixed its value so as to fulfill the VT; this is what is usually called the X_α method (X for exchange). Slater and Wood[140] (1971) applied the X_α methods to the computation of the total binding energies and ionization potentials for the first transition series, with results that agree well with HF computations. Slater[141] (1972) checked directly the validity of the

Hellmann–Feynman and virial theorems in the X_α method; VT and HFT are presented as "two of the most powerful theorems applicable to molecules and solids." Gopinathan and Whitehead[142] (1976) have recently proposed a correction scheme for improving energy estimates made by the SCF–X_α method; they make use of the Hellmann–Feynman theorem to derive conditions under which their technique may be expected to apply. The method itself is based on the VT.

The VT can also be verified to hold in the SCF–Kohn–Sham scheme for solids under pressure (Liberman,[143] 1971; Ross,[136] 1980). The computations are much simplified in Slater's muffin-tin approximation (Janak,[110] 1974). Finally, the VT is also valid and useful within the SCF pseudopotential method, where the effect of the atomic potentials on the electrons is assimilated to that of a weak pseudopotential (Ihm, Zunger, and Cohen,[144] 1979; Szasz,[145] 1980).

2. *The Statistical Model*

Numerical calculations with the SCF method discussed in Section III.C.1 are very cumbersome, especially for atoms with many electrons. For these, there exists a simpler method, a semiclassical statistical model originally and independently developed by Thomas[146] (1927) and Fermi[147] (1928), based on the Fermi–Dirac statistics. Admittedly less accurate than the Hartree–Fock SCF, the Thomas–Fermi (TF) model still gives a fair approximation, particularly when the exchange correction (Dirac,[148] 1930) is included. Besides, the model has been provided a new significance by the recent proof (Lieb and Simon,[149] 1973, 1976, 1977) that in the limit of infinite atomic number it corresponds to the exact solution for the nonrelativistic atom. It thus stands in relation to the many-electron system as the hydrogen atom stands to the few-electron system.

In this subsection, we first describe a few generalities of the TF and TFD models (Section III.C.2.a); then calculate the relations between the energy components of the TF atom (b); and finally check that the VT holds in both TF and TFD theories (c).

a. Generalities. The statistical model of the atom treats the electrons semiclassically as an ideal Fermi–Dirac gas of charge density $\rho = -en$: the number density $n = N/\Omega$ (Ω = volume of the N-electron atom) is given by

$$n = 2\left(\frac{4\pi}{3}\right)\left(\frac{\hat{p}}{2\pi\hbar}\right)^3 = \frac{\hat{p}^3}{3\pi^2\hbar^3} \tag{3.115}$$

where $\hat{p}$ is the maximum momentum at the top of the Fermi sea, the factor 2 the electron spin degeneracy, and $(2\pi\hbar)^3$ the volume per (translational) quantum state in phase space.

The central-field approximation assumes that each electron moves in a spherically symmetric potential including the field of the nucleus and, in an average way, the interaction with all the other electrons (the exchange effect will be considered later on):

$$V(r) = V_n(r) + V_e(r) = \frac{Ze}{r} + V_e(r) \tag{3.116}$$

and

$$V_e(r) = -4\pi e \int d^3r' \frac{n(\vec{r}')r'^2}{|\vec{r}-\vec{r}'|} \tag{3.117}$$

This semiclassical model requires that V satisfy the Poisson equation of electrostatics,

$$\nabla^2 V = 4\pi e n \tag{3.118}$$

with the following boundary conditions appropriate to an ion of atomic number Z containing N electrons and thus bearing charge $z = Z - N$:

$$V(r) \to \begin{cases} Ze/r & \text{as } r \to 0 \quad (3.119) \\ ze/r & \text{as } r \to \infty \quad (3.120) \end{cases}$$

The energy of an electron is

$$E = \frac{p^2}{2m} - eV \tag{3.121}$$

and at the top of the Fermi sea we have

$$\frac{\hat{p}^2}{2m} - eV = \mu = -e\hat{V} \tag{3.122}$$

where μ is the electron chemical potential. This implies that the validity of the model is restricted to $V \geq \hat{V}$, which determines the radius $\hat{r}$ of the atom:

$$V(\hat{r}) = \hat{V} \tag{3.123}$$

In order to have continuous potential and field strength, the boundary condition at large distance can be written, instead of Eq. (3.120),

$$V(\hat{r}) = \frac{ze}{\hat{r}}, \qquad \left(\frac{dV}{dr}\right)_{\hat{r}} = -\frac{ze}{\hat{r}^2} \tag{3.120$'$}$$

From Eqs. (3.122) and (3.115) we get

$$\hat{p} = [2m(eV+\mu)]^{1/2} = [2me(V-\hat{V})]^{1/2} \tag{3.124}$$

$$n = \frac{(2me)^{3/2}(V-\hat{V})^{3/2}}{3\pi^2\hbar^3} \tag{3.125}$$

Using the spherical symmetry and the constancy of μ, the Poisson equation (3.118) becomes

$$e\nabla^2 V = \nabla^2(eV+\mu) = \frac{d^2[r(eV+\mu)]}{r\,dr^2} = \frac{4e^2}{3\pi\hbar^3}[2m(eV+\mu)]^{3/2} \tag{3.126}$$

Introducing in this last equation the dimensionless function ϕ, the ratio of the maximum kinetic energy to the Coulomb attraction with the nucleus:

$$\phi = \frac{eV+\mu}{Ze^2/r} = \frac{re(V-\hat{V})}{Ze^2} \tag{3.127}$$

and the scaled distance

$$x = \frac{r}{\alpha} \tag{3.128}$$

where

$$\alpha = 2^{-2}\left(\frac{9\pi^2}{2}\right)^{1/3} Z^{-1/3}a = 0.8853Z^{-1/3}a \tag{3.129}$$

a being the Bohr radius ($a = \hbar^2/me^2$), we obtain the universal TF equation, for all atoms, within this approximation:

$$\phi'' = \frac{\phi^{3/2}}{x^{1/2}} \tag{3.130}$$

In terms of the new variables, the boundary conditions [Eqs. (3.119) and

(3.120′)] become

$$\phi(0)=1 \tag{3.131}$$

$$\phi(\hat{x})=0 \quad \text{and} \quad \hat{x}\phi'(\hat{x})=-\frac{z}{Z} \tag{3.132}$$

where $\hat{x}=\hat{r}/\alpha$.

We can check directly that these equations give the correct number of electrons inside the sphere of radius r. Using Eq. (3.127) in Eq. (3.125), we obtain

$$n=\frac{32Z^2}{9\pi^3}\left(\frac{\phi(x)}{x}\right)^{3/2} \tag{3.133}$$

Invoking the differential Eq. (3.130), and with the help of an integration by parts, we get

$$\begin{aligned} 4\pi\int_0^r n(r)r^2\,dr &= Z\int_0^{\hat{x}} dx\,x^{1/2}\phi^{3/2} \\ &= Z\int_0^{\hat{x}} dx\,x\phi'' \\ &= Z\left[\hat{x}\phi'(\hat{x})+\phi(0)\right] \\ &= Z-z=N \end{aligned} \tag{3.134}$$

where the last line results from the boundary conditions [Eqs. (3.131) and (3.132)].

The TF equation (3.130) subject to the boundary conditions [Eqs. (3.131), (3.132)] has been solved approximately, both quasianalytically and numerically (see the references at the end of this subsection). Its various solutions are characterized by their initial slopes $\phi'(0)=-B$. This slope is negative for all solutions of interest, as shown in Fig. 2.

The solution that vanishes asymptotically when x tends to ∞ corresponds to the neutral free atom; its slope at the origin has the value

$$\phi_0'(0)=-1.588=-B_0 \tag{3.135}$$

In this case, the radius is infinite and the chemical potential is zero. The asymptotic behavior of the standard solution $\phi_0(x)$ goes like $144/x^3$; this, incidentally, is an exact solution of the TF equation, but cannot hold over the whole radius range on the physical ground that it diverges at the origin.

Again with reference to Fig. 2, lying below the neutral-atom curve are solutions that vanish for some finite value $\hat{x}$ of x. These represent positive

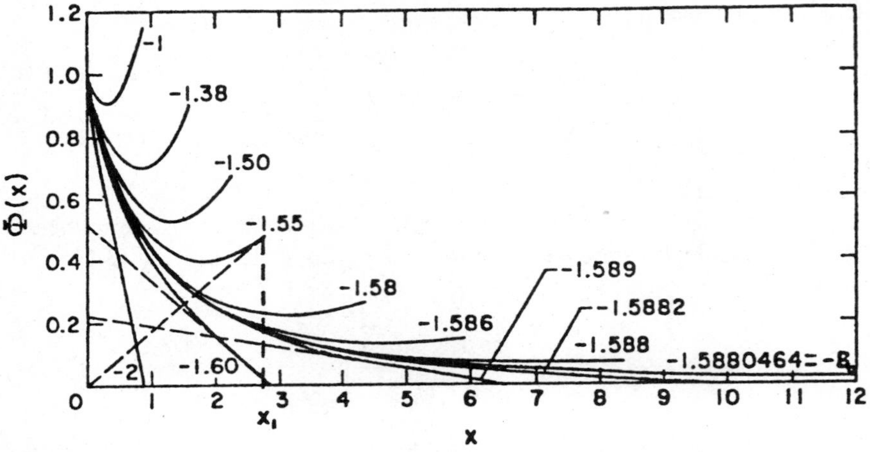

Fig. 2. Solutions of the Thomas–Fermi equation (Eq. 3.130) labeled with their initial slopes. [From Ref. 150: E. U. Condon and H. Odabasi, *Atomic Structure*, Cambridge, U.P., 1980, p. 453 (with permission).]

ions with a definite radius $\hat{r}$ and negative chemical potential, given from Eqs. (3.122) and (3.120) by

$$\mu = -\frac{ze^2}{\hat{r}} \tag{3.136}$$

that applies also to the neutral atom. Negative ions are not treated in the TF model, nor in the TFD model below.

There are also solutions nowhere zero and diverging at infinity (slopes larger than $-B_0$). For these solutions, at the point x_1 (see Fig. 2) where

$$x_1\phi'(x_1) = \phi(x_1) \tag{3.137}$$

it can be seen from Eq. (3.134) that the total charge inside the sphere of radius x_1 is zero. Graphically Eq. (3.137) shows that the tangent to the curve at this point passes through the origin. The curve cut at x_1 defines $\phi(x_1)$ and thus the density n for a neutral atom at whose boundary the density is not zero. For these conditions the model evidently represents a condensed state under various degrees of compression (Slater and Krutter,[151] 1935; Feynman et al.,[152] 1949). The TF model involves no cohesive forces, but this defect is (at least partially) removed by including the effect of exchange.

Dirac[148] (1930) introduced the exchange effect in the TF model. Its inclusion amounts to adding to the electronic energy, Eq. (3.122), a term linear in the maximum momentum $\hat{p}$ [cf. Eq. (3.124)]:

$$\frac{\hat{p}^2}{2m} - \frac{e^2}{\pi\hbar}\hat{p} - eV = \mu \tag{3.138}$$

where $\hat{p}$ is still related to the density n given by Eq. (3.115). Instead of the universal TF equation we thus have to solve the TFD equation

$$\phi''(x) = x\left\{\left(\frac{\phi(x)}{x}\right)^{1/2} + \epsilon\right\}^3 \tag{3.139}$$

where x is the scaled radius as defined earlier, but now

$$\phi = r\frac{\mu + eV + \tau^2}{Ze^2} \tag{3.140}$$

with $\tau^2 = e^2/2\pi^2 a = R/\pi^2$ and $\epsilon^2 = \tau^2\alpha/Ze^2 = 0.2118Z^{-2/3}$.

Besides the Dirac correction there are many other modifications and improvements of the TF model. We refer the reader to the recent reviews of the subject (Bethe and Jackiw, Ref. 120, Chapter 3; Condon and Odabasi, Ref. 150, Section 18) and to Gombás's[153] classical monograph on the subject. We merely mention here some of the major modifications before coming back, in the next subsections, to the VT.

The TF density is defective at both small and large radii: it goes to zero as $1/r^6$ as $r \to \infty$, instead of exponentially; and it varies rapidly (as $r^{-3/2}$) near the nucleus, whereas it should be constant. The correct behavior at both extremities was worked out by Weizsäcker[154] (1935) by modifying the electron wavefunction: instead of the plane wave $\exp(i\vec{k}\cdot\vec{r})$, he used the form $(1 + \vec{\alpha}\cdot\vec{r})\exp(i\vec{k}\cdot\vec{r})$. (See also Plaskett,[155] 1953; Ref. 150, p. 461.)

Even with these corrections, however, the TF density is not adequate, since it does not reflect the shell structure ($K, L, M, \ldots$) of the atom (whereas the HF density reproduces it quite accurately). Gombás et al.[156] (1960) managed to incorporate this shell structure in the statistical model by treating the electrons of different shells separately from the beginning.

Teller[157] (1962; see also Sheldon,[158] 1955) showed that there is no binding in a TF molecule.

Extensions to nonzero temperature have been made by Brachman[159] (1951) for the TF model and by Cowan and Ashkin[160] (1957) for the TFD model.

Among the many numerical calculations in the literature, we cite particularly those of Latter[161] (1955, 1956) for compressed TF atoms. Latter[162]

TABLE IV
Comparison of Hartree–Fock and Thomas–Fermi–Dirac Energy Levels for Silver[a]

State	Energy [rydbergs]	
	HF	TFD
1s	1828	1805
2s	270	263
2p	251	245
3d	29.8	29.2
4s	8.46	7.95

[a]Reprinted by permission from Bethe and Jackiw *Intermediate Quantum Mechanics*, 2ed., copyright 1968. The Benjamin/Cummings Publishing Co., Inc. Advanced Book Program. (Ref. 120, p. 97); after Latter.[162]

(1955) also solved the radial one-electron Schrödinger equation for all normally occupied orbitals n, l for many atoms, using the potential given by the TFD theory. These solutions compare quite well with the HF results (see Table IV), and for atoms for which no HF solutions are available Latter's are the best existing results.

b. Energy Components and Atomic Number Z. It is easy to show (for example, see Parr et al.[163]) that for the TF atom, we have the following relation between energy components:

$$\frac{5T}{3} + U_{en} + 2U_{ee} = \mu N \tag{3.141}$$

Thus, for a neutral free TF atom ($\mu = 0$),

$$\frac{5T}{3} + U_{en} + 2U_{ee} = 0 \tag{3.142}$$

We shall check in Section III.C.2.c that the VT holds in the TF approximation, so that we have the independent constraint

$$2T = -U_{en} - U_{ee} = -2E \tag{3.143}$$

Since U_{en}, for example, can be calculated directly from the volume integral,

$$U_{en} = -\int_0^\infty d\tau \frac{Ze^2 n}{r} = Z^2 e^2 \phi'(0) \tag{3.144}$$

each component can be found exactly (e.g., Flügge[164]), and a first approximation to the total binding energy (the energy to strip the atom of all its electrons) results:

$$E = -0.7687Z^{7/3} \text{ a.u.} = -20.93Z^{7/3} \text{ eV} \tag{3.145}$$

The $Z^{7/3}$ dependence of the total TF binding energy could have been predicted from a simple qualitative argument (Landau and Lifschitz[165]): from the VT, E is one-half the electrostatic energy, and since the characteristic dimension of the TF atom goes like $Z^{-1/3}$, the electrostatic energy of Z electrons at a mean distance $Z^{-1/3}$ from a nucleus of charge Z must go as $ZZ/Z^{-1/3} = Z^{7/3}$.

The values given by Eq. (3.145) are some 10% higher than the experimental binding energies for heavier atoms, some 30% higher for lighter atoms. Surprisingly, the binding energies for light atoms are found to increase very smoothly, almost exactly as $Z^{7/3}$, but with a different coefficient (Young[166], 1929; Allard,[167] 1948):

$$E = 15.73Z^{7/3} \text{ eV} \tag{3.146}$$

For heavier atoms, Foldy[168] (1951) recommends

$$E = 13.60Z^{12/5} \text{ eV} \tag{3.147}$$

Equation (3.145) is in fact the first term of a series in powers of $Z^{1/3}$; the two next leading corrections, one proportional to Z^2 due to the failure of the TF density near the nucleus, the other proportional to $Z^{5/3}$ due to the exchange effect, have been estimated by Scott[169] (1952). Recently Schwinger[170] (1980, 1981) gave a clear-cut derivation of these corrections, taking the VT as his point of departure. The series obtained,

$$E = -0.76874Z^{7/3} + 0.5Z^2 - 0.2699Z^{5/3} \text{ a.u.} \tag{3.148}$$

reproduces perfectly the values of nonrelativistic binding energies for neutral atoms as computed by the HF method (Plindov and Dmitrieva,[171] 1978). For light atoms ($5 \leq Z \leq 55$), the error does not exceed 1%, and for heavy atoms, from $Z = 56$ on, the error is 0.1%.

The TF expansion [Eq. (3.148)] can be compared with the exact nonrelativistic expansion

$$E = E_0Z^2 + E_1Z + E_2 + E_3Z^{-1} + \cdots \tag{3.149}$$

resulting from considering the interelectronic repulsion as a perturbation and using $1/Z$ as a natural expansion parameter (Hylleraas,[172] 1929; Löwdin,[28] 1959). Note that once the series for the energy is known, the series for the energy components can be found using the HFT (cf. Section II.B.6) with Z as parameter and the VT. The HFT gives

$$U_{en} = Z\frac{dE}{dZ} \tag{3.150}$$

and with the VT we obtain

$$T = -E \tag{3.151}$$

$$U_{ee} = 2E - Z\frac{dE}{dZ} \tag{3.152}$$

These relations have recently been used to explain certain regularities observed in the HF expectation values of terms such as $\langle 1/r_i \rangle$ and $\langle 1/r_{ij} \rangle$ along isoelectronic sequences (Boyd,[173] 1978; Cohen,[174] 1979; Murdoch and Magnoli,[175] 1982). We shall see other applications of the HFT and VT (more exactly HVT) in connection with perturbation theory in Section III.E. We now turn to the proof of the VT in the TF theory.

c. The Virial Theorem in Statistical Models. It is by no means obvious that the TF model should obey the VT in view of the simplifying assumptions involved. Fock[176] (1932) made the first investigation on this question using the same method that he used to prove the general quantum-mechanical VT (cf. Section II.B.2): scaling the coordinates and applying the variational principle. Instead of the wavefunction, we here scale the density, defining

$$n_\lambda(r) = \lambda^3 n(\lambda r) \tag{3.153}$$

the arbitrary parameter λ being cubed to assure the normalization condition

$$\int d\tau\, n_\lambda(r) = N \tag{3.154}$$

Putting this scaled density into each energy component we obtain

$$T(\lambda) = \lambda^2 T, \qquad U_{en}(\lambda) = \lambda U_{en}, \qquad U_{ee}(\lambda) = \lambda U_{ee} \tag{3.155}$$

Upon requiring that the total energy be a minimum at $\lambda = 1$,

$$0 = \left[\frac{\partial E(\lambda)}{\partial \lambda}\right]_{\lambda=1} = 2T + U_{ee} + U_{en} \tag{3.156}$$

it is clear that one obtains immediately the VT corresponding to Coulomb forces.

Jensen[177] (1933) showed that there were difficulties involved in extending Fock's treatment to many centers. But Duffin[178] (1935) gave a satisfactory derivation of the VT for TF systems with many electrons and nuclei. Generalization of these results when exchange is taken into account (TFD theory) have been worked out by Jensen[179] (1934) and March[180] (1952).

As was first pointed out by Slater and Krutter[151] (1935), the TF method is of considerable value for discussion of the behavior of matter under pressure. Hence it is of interest to investigate the form of the VT when the atom considered is not free but has a pressure at the boundary. For the single-nucleus case, this has been carried out by Slater and Krutter[151] (1935) and Jensen[181] (1939).

We provide here a general treatment from which all existing results can be obtained as particular cases (March,[182] 1952), first for TF and then for TFD theory. In conclusion, we discuss the limitation and utility of the TFD model (McMillan,[183,184] 1955, 1958).

The VT in the TF Theory. For an ideal Fermi–Dirac gas, the average kinetic energy of an electron is given by (cf. Section III.A.3)

$$\begin{aligned} t &= \frac{3}{5}\frac{\hat{p}^2}{2m} \\ &= c_k n^{2/3} \end{aligned} \tag{3.157}$$

where we have used Eq. (3.115) and where the kinetic coefficient is

$$c_k = \frac{3(3\pi^2\hbar^3)^{2/3}}{10m} \tag{3.158}$$

The total kinetic energy thus has the form $\propto \int d\tau\, n^{5/3}$. The kinetic energy can be related to the pressure at the boundary of the atom through Gauss's theorem: for any function $f(\vec{r})$,

$$\oint d\vec{S}\cdot\vec{r} f = \int d\tau\, \vec{r}\cdot\vec{\nabla} f + 3\int d\tau\, f \tag{3.159}$$

Applied to the kinetic energy density, for which $f = c_k n^{5/3}$, one obtains

$$\begin{aligned} 3T &= c_k \oint d\vec{S}\cdot\vec{r} n^{5/3} - c_k \int d\tau\, \vec{r}\cdot\vec{\nabla} n^{5/3} \\ &= 3c_k n_b^{5/3}\Omega - \frac{3}{2}\int d\tau\, n\vec{r}\cdot\vec{\nabla}(eV) \end{aligned} \tag{3.160}$$

where Ω is the volume of the system and n_b the value of the density at the boundary. The first term on the right of Eq. (3.160) is easily shown to be proportional to the pressure

$$P = -\left(\frac{\partial \epsilon}{\partial v}\right)_s \tag{3.161}$$

where v $(=1/n)$ is the specific volume and ϵ the average energy of an electron, namely

$$\epsilon = c_k n^{2/3} - eV \tag{3.162}$$

The pressure at the boundary is thus

$$P = n^2\left(\frac{\partial \epsilon}{\partial n}\right)_s = \frac{2c_k n_b^{5/3}}{3} \tag{3.163}$$

and the virial relation begins to emerge:

$$2T = 3P\Omega - e\int d\tau\, n\vec{r}\cdot\vec{\nabla}V \tag{3.164}$$

To evaluate the last integral, we recall from Eq. (3.116) that V is the total potential, $V = V_e + V_n$, but now V_n is produced by several nuclei, supposed fixed, of respective charges $Z_k e$ and positions $\vec{R}_k$: $V_n = \Sigma_k Z_k e/|\vec{r} - \vec{R}_k|$. Since V_n is homogeneous of degree -1 in $\vec{r}$ and all the $\vec{R}_k$'s, by Euler's theorem (cf. Section II.B.7.a)

$$\vec{r}\cdot\vec{\nabla}V_n + \sum_k \vec{R}_k\cdot\vec{\nabla}_k V_n = -V_n \tag{3.165}$$

The nuclear contribution to the integral in Eq. (3.160) is thus

$$-e\int d\tau\, n\vec{r}\cdot\vec{\nabla}V_n = e\int d\tau\, nV_n + e\sum_k \int d\tau\, n\vec{R}_k\cdot\vec{\nabla}_k V_n = -U_{en} - \overline{\sum \vec{R}_k\cdot\vec{F}_k} \tag{3.166}$$

where we recognize that the interaction potential energy between the nuclei and the electron charge density $-en$ is $U_{en} = -e\int d\tau\, nV_n$ and that the total force on nucleus k is $\vec{F}_k = \vec{\nabla}_k(-eV_n)$. It remains to evaluate

$$-e\int d\tau\, n\vec{r}\cdot\vec{\nabla}V_e \tag{3.167}$$

To this end we shall first derive what is sometimes called Duffin's lemma. The potential acting on an electron due to the other electrons is

$$V_e(\vec{r}) = \int d\tau' \frac{(-e)n(\vec{r}')}{|\vec{r}-\vec{r}'|} \tag{3.168}$$

and the electronic interaction potential energy is given by

$$U_{ee} = -\frac{e}{2}\int d\tau\, n(\vec{r})V_e(\vec{r}) = \frac{e^2}{2}\iint d\tau\, d\tau' \frac{n(\vec{r})n(\vec{r}')}{|\vec{r}-\vec{r}'|} \tag{3.169}$$

We define, with Duffin[178] (1935), the scaled potential energy

$$\begin{aligned} U_{ee}(\lambda) &= -\frac{e}{2}\int d\tau\, n(\vec{r})V_e(\lambda\vec{r}) = \frac{e^2}{2}\iint d\tau\, d\tau' \frac{n(\vec{r})n(\vec{r}')}{|\lambda\vec{r}-\vec{r}'|} \\ &= \frac{e^2}{2\lambda}\iint d\tau\, d\tau' \frac{n(\vec{r})n(\vec{r}')}{|\vec{r}-\vec{r}'/\lambda|} \\ &= -\frac{e}{2\lambda}\int d\tau'\, n(\vec{r}')V_e\left(\frac{\vec{r}'}{\lambda}\right) \end{aligned} \tag{3.170}$$

where to avoid problems we restrict the arbitrary parameter λ to be positive definite. We thus have

$$V_e(\lambda\vec{r}) = \frac{V_e(\vec{r}/\lambda)}{\lambda} \tag{3.171}$$

The derivative of this relation with respect to λ yields

$$\vec{\nabla}_{\lambda\vec{r}}V_e(\lambda\vec{r})\cdot\vec{r} = -\lambda^{-2}V_e\left(\frac{\vec{r}}{\lambda}\right) + \lambda^{-1}\vec{\nabla}_{\vec{r}/\lambda}V_e\left(\frac{\vec{r}}{\lambda}\right)\cdot(-\lambda^{-2}\vec{r}) \tag{3.172}$$

and by choosing $\lambda = 1$,

$$2\vec{r}\cdot\vec{\nabla}V_e(\vec{r}) = -V_e \tag{3.173}$$

The integral (3.167) is thus simply $-U_{ee}$.

Collecting our results in Eq. (3.160), the general form of the VT for TF systems conforms to the usual one,

$$2T + U_{ee} + U_{en} = 3P\Omega - \sum_k \overline{\vec{R}_k\cdot\vec{F}_k} \tag{3.174}$$

where we recognize the first term on the right as the surface term, while the second term is interpreted as the virial of the forces that hold the nuclei fixed (cf. the Born–Oppenheimer approximation in Section III.D.1).

The VT in TFD Theory. When the effect of exchange is introduced in the TF model (Dirac,[148] 1930), the energy is decreased through addition of the (negative) exchange energy:

$$E = T + U + A = c_k \int d\tau\, n^{5/3} - \int d\tau\, neV - c_e \int d\tau\, n^{4/3} \qquad (3.175)$$

where the coefficient in the kinetic energy is still given by Eq. (3.158), while the coefficient in the exchange energy A is

$$c_e = \frac{3e^2}{4}\left(\frac{3}{\pi}\right)^{1/3} \qquad (3.176)$$

Instead of Eq. (3.133), the density is given by

$$n = \left(\frac{3}{5c_k}\right)^{3/2} \chi^{3/2} \qquad (3.177)$$

where $\tau^2 = me^4/2\pi^2\hbar^2$ and we have introduced the quantity

$$\chi = \left[(\mu + eV + \tau^2)^{1/2} + \tau\right]^2 \qquad (3.178)$$

for future convenience. We start, as in the TF treatment above, from the kinetic energy

$$T = c_k \int d\tau\, n^{5/3} = \left(\frac{3}{5c_k}\right)^{3/2} \frac{3}{5} \int d\tau\, \chi^{5/2} \qquad (3.179)$$

or, by use of the identity in Eq. (3.159),

$$T = \frac{1}{5}\left(\frac{5}{3c_k}\right)^{3/2} \left\{ \oint d\vec{S}\cdot\vec{r}\chi^{5/2} - \int d\tau\, \vec{r}\cdot\vec{\nabla}\chi^{5/2} \right\}$$

$$= \frac{c_k}{3} \oint d\vec{S}\cdot\vec{r}n^{5/3} - \frac{1}{2}\int d\tau\, n\vec{r}\cdot\vec{\nabla}\chi \qquad (3.180)$$

since

$$\vec{\nabla}\chi^{5/2} = \tfrac{5}{2}\chi^{3/2}\vec{\nabla}\chi = \frac{5}{2}\left(\frac{5c_k}{3}\right)^{3/2} n\vec{\nabla}\chi \qquad (3.181)$$

We also have

$$\vec{\nabla}\chi = \vec{\nabla}(eV) + 2\tau\vec{\nabla}\chi^{1/2} \tag{3.182}$$

so that Eq. (3.180) becomes

$$2T = \frac{2c_k}{3}\oint d\vec{S}\cdot\vec{r}n^{5/3} - \int d\tau\, n\vec{r}\cdot\vec{\nabla}(eV) - 2\tau\int d\tau\, n\vec{r}\cdot\vec{\nabla}\chi^{1/2} \tag{3.183}$$

As previously we have

$$-\int d\tau\, n\vec{r}\cdot\vec{\nabla}(eV) = -U_{ee} - U_{en} - \overline{\sum\vec{R}_k\cdot\vec{F}_k} = -U - \overline{\sum\vec{R}_k\cdot\vec{F}_k} \tag{3.184}$$

By use of the identity $\chi^{3/2}\vec{\nabla}\chi^{1/2} = \vec{\nabla}\chi^2/4$ and of Eq. (3.177), the other volume integral in Eq. (3.183) becomes

$$2\tau\int d\tau\, n\vec{r}\cdot\nabla\chi^{1/2} = \frac{c_e}{3}\left\{\oint d\vec{S}\cdot\vec{r}n^{4/3} - 3\int d\tau\, n^{4/3}\right\} \tag{3.185}$$

Collecting the results given by Eqs. (3.185) and (3.184) in Eq. (3.183), the VT reads

$$\begin{aligned} 2T + U + A &= \frac{2c_k}{3}\oint d\vec{S}\cdot\vec{r}n^{5/3} - \frac{c_e}{3}\oint d\vec{S}\cdot\vec{r}n^{4/3} - \overline{\sum\vec{R}_k\cdot\vec{F}_k} \\ &= 3P\Omega - \overline{\sum\vec{R}_k\cdot\vec{F}_k} \end{aligned} \tag{3.186}$$

since with the average total energy ϵ per electron,

$$\epsilon = c_k n^{2/3} - c_e n^{1/3} - eV \tag{3.187}$$

the pressure, Eq. (3.161), is now [instead of Eq. (3.163)]

$$P = \frac{2c_k}{3}n_b^{5/3} - \frac{c_e}{3}n_b^{4/3} \tag{3.188}$$

Equation (3.186) is the VT for TFD systems and differs from the VT for TF system only by the presence of the exchange energy A. The coefficient 1 in front of A could have been easily derived by use of the scaled density, Eq. (3.153), since we have $A(\lambda) = \lambda A$.

It may be noted that the Dirac exchange correction is calculated using plane waves, and thus assumes that the electron density is both uniform and isotropic. It is therefore inconsistent to treat the Dirac model as though it applied to a free isolated atom, which has no surrounding electron density with which the interior electrons can undergo exchange. Instead, we must regard even the zero-pressure TFD atom as embedded in a condensed (e.g., metallic crystal) phase. Even then, the Dirac prescription seriously underestimates the exchange correction.

Far from being uniform and isotropic, the electron density in an atom varies markedly over (radial) distances comparable to the electron wavelength—within which exchange can occur. Since the electron density is least at the atomic boundary, electrons in this region—which determine the effective pressure—are in exchange mainly with inner regions of higher electron density. Dirac's plane-wave exchange calculations thus generally underestimate the exchange correction. This results in TFD atomic volumes at zero pressure that are considerably larger than the observed normal volumes for metals, with the notable exception of the alkali and alkaline-earth metals, for which the relative smallness of the atomic kernels coupled with the large volume of the s orbitals of the valence electrons makes the plane-wave approximation quite reasonable. For other metals, a semiempirical approximation that multiplies the Dirac exchange correction by a parameter λ allows the selection of λ so as to fit the experimental zero-pressure atomic volumes. The resulting values of the electron chemical potential μ ($\equiv -1.29\lambda^2 eV$) then agree semiquantitatively with the experimental photoelectric work function (McMillan,[183] 1955).

A further application to the equation of state (McMillan,[184] 1958) is based on the same assumption that preserves the form of the Dirac exchange correction but alters the coefficient so as to fit the observed zero-pressure volumes. Then, assuming that the maximum electron momentum is not appreciably altered from its value in the TF model, the pressure is given by

$$p = p_{\mathrm{TF}}\left[1-\left(\frac{p^{\circ}_{\mathrm{TF}}}{p_{\mathrm{TF}}}\right)^{1/5}\right] \tag{3.189}$$

where p_{TF} is the pressure predicted by the TF model for a given atomic volume, and the sign $^{\circ}$ refers to the normal atomic volume.

From Eq. (3.189) one easily obtains for the zero-temperature, zero pressure compressibility in terms of the usual TF scaled variables,

$$\beta^{\circ} \equiv -\left(\frac{d\ln V}{dp}\right)^{\circ} = -5\,\frac{\left[d\ln(ZV)/d\left(p_{\mathrm{TF}}/Z^{10/3}\right)\right]^{\circ}}{Z^{10/3}} \tag{3.190}$$

The quantity in brackets is a universal function of ZV° which was evaluated

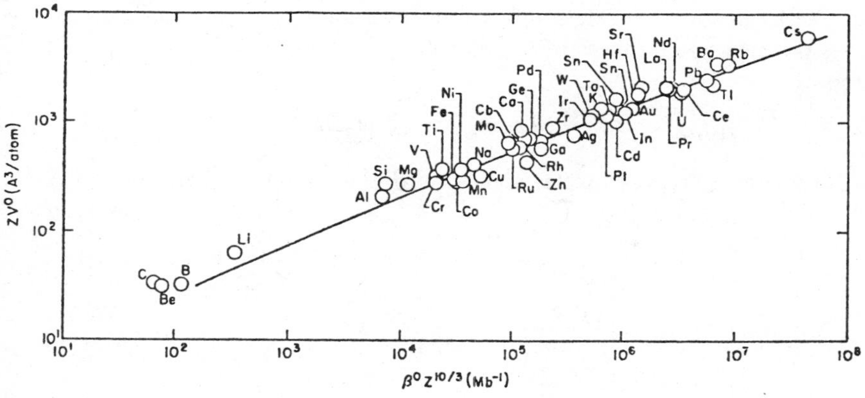

Fig. 3. Approximate compressibilities of the elements on the statistical model. [From Ref. 184: W. G. McMillan, *Phys. Rev.* **111**, 479 (1958), with permission of the American Physical Society.]

from Latter's[161] TF solutions and is shown as the curve in Figure 3. Also shown are the experimental values of $\beta^\circ Z^{10/3}$. It is evident that the TF curve represents a reasonable average of the experimental values, the mean fractional derivation being within 50%.

D. The Chemical Bond

This subsection on applications of the VT to molecules begins with a description of the Born–Oppenheimer approximation (Section III.D.1). We then examine the form of the VT in this approximation; the result has already been given (Section II.B.7.a), but we present here an alternative derivation, based on scaling (Section III.D.2). There follows a discussion of chemical-bond formation in the light of the VT and of the virial partitioning of molecules into atomic fragments (Section III.D.3). Finally, in Section III.D.4, a new application of the VT is given: using the VT for the nuclear-motion Schrödinger equation and the Hellmann–Feynman theorem with the reduced mass as parameter, we derive an analytic formula for the isotope effect of diatomic molecules for an arbitrary interatomic potential function.

1. The Born–Oppenheimer Approximation

Because the mass of an atomic nucleus is very large compared with that of an electron, the characteristic speeds of electrons are much greater than those of the nuclei—which can thus in first approximation be considered at rest. This makes possible the separation of the Schrödinger equation into electronic and nuclear parts (Born and Oppenheimer,[185] 1927).

For a neutral molecule composed of several nuclei of various charges $Z_\alpha e$ and positions $\vec{R}_\alpha$, the wavefunction $\psi\{\vec{r}_i; \vec{R}_\alpha\}$ must satisfy the eigenvalue equation

$$H\psi\{\vec{r}_i; \vec{R}_\alpha\} = E\psi\{\vec{r}_i; \vec{R}_\alpha\} \tag{3.191}$$

wherein the $\vec{r}_i$ represent electronic coordinates. The total molecular Hamiltonian H is given by

$$\begin{aligned} H &= T_n + T_e + U_{nn} + U_{ee} + U_{ne} \\ &= -\sum \frac{\hbar^2}{2m_\alpha}\nabla_\alpha^2 - \frac{\hbar^2}{2m}\sum \nabla_i^2 \\ &\quad + \sum_{\alpha>\beta}\sum \frac{Z_\alpha Z_\beta e^2}{R_{\alpha\beta}} + \sum_{i>j}\sum \frac{e^2}{r_{ij}} - \sum_i \sum_\alpha \frac{Z_\alpha e^2}{r_{i\alpha}} \end{aligned} \tag{3.192}$$

Although this Hamiltonian neglects spin–orbit, spin–spin, and other relativistic corrections and assumes that no external field is present, it is still too complicated to solve exactly. However, the relative slowness of the massive nuclei can be exploited by regarding the nuclei as fixed while the electrons carry out their rapid motion; classically speaking, during the time of a cycle of electronic motion the change in nuclear configuration is negligible.

Using $(m/m_\alpha)^{1/4}$ as an expansion parameter, Born and Oppenheimer[185] showed that the wavefunction is separable into an electronic factor $\psi_e\{\vec{r}_i; \vec{R}_\alpha\}$ and a nuclear factor $\psi_n\{\vec{R}_\alpha\}$,

$$\psi\{\vec{r}_i; \vec{R}_\alpha\} = \psi_e\{\vec{r}_i; \vec{R}_\alpha\}\psi_n\{\vec{R}_\alpha\} \equiv \psi_e\psi_n \tag{3.193}$$

so that the solution of Eq. (3.191) can be divided into two stages. In the first stage, one solves the Schrödinger equation for the electronic motion

$$H_e\psi_e = E_e\psi_e \tag{3.194}$$

where $H_e = T_e + U_{ne} + U_{ee}$ and E_e is a function of the coordinates of the nuclei, $E_e\{\vec{R}_\alpha\}$. By treating the nuclear positions $\vec{R}_\alpha$ as parameters, the only variables in Eq. (3.194) are the electronic positions $\vec{r}_i$. For fixed nuclei the nuclear interaction U_{nn} is constant and can be added to both sides of Eq. (3.194) with the sole effect of shifting the electronic energy:

$$(H_e + U_{nn})\psi_e = V\psi_e \tag{3.195}$$

where $V = V(\vec{R}_\alpha)$ is the electronic energy including the internuclear potential energy:

$$V = E_e + U_{nn} \tag{3.196}$$

Assuming that the electronic Schrödinger equation, Eq. (3.194) or (3.195), is solved (which is quite an assumption[186]), we consider the nuclear motion in the second stage of the Born–Oppenheimer approximation. According to our picture, when the nuclear configuration changes slightly, say from $\vec{R}_\alpha$ to $\vec{R}'_\alpha$, the electrons immediately adjust to this change, the electronic wavefunction changing from $\psi_e\{\vec{r}_i; \vec{R}_\alpha\}$ to $\psi_e\{\vec{r}_i; \vec{R}'_\alpha\}$ and the electronic energy from $V\{\vec{R}_\alpha\}$ to $V\{\vec{R}'_\alpha\}$. This electronic energy varies smoothly as a function of the parameters defining the nuclear configuration, $\{\vec{R}_\alpha\}$, and $V\{\vec{R}_\alpha\}$ becomes, in effect, the potential energy for the nuclear motion. Hence the Schrödinger equation for the nuclear motion reads

$$(T_n + V)\psi_n = E\psi_n \tag{3.197}$$

where the $\vec{R}_\alpha$ are now variables and E is the total molecular energy.

Although the Born–Oppenheimer separation is an excellent approximation, some feel that the rationale behind it is not clear; for example, Condon[187] (1947) says, "I have never felt that I properly understood it," and Essén[188] (1977) argues that "it is the form of the interaction between the particles that is responsible for the separation; the smallness of m/m_α is irrelevant." Using the VT as a starting point for his discussion, Essén purports to show that coordinates of collective and individual motion are natural coordinates for the approximate separation, rather than nuclear and electronic coordinates; his treatment is claimed to be not specially designed for molecules but for any system of particles with Coulomb interactions.

The link between the VT and the Born–Oppenheimer approximation has been studied for a few orders of perturbation by Oliver and Weislinger[189] (1977).

In Section III.D.4, we shall give the VT corresponding to the nuclear-motion Schrödinger equation, which appears to be a new result. This nuclear-motion VT is then used, in conjunction with the Hellman–Feynman theorem, to derive an integral formula for the derivative of the energy with respect to the reduced mass as parameter in a diatomic molecule.

For now, we return to the electronic Schrödinger equation (3.195) to get the electronic VT for molecules by use of scaling, still in the Born–Oppenheimer (or adiabatic) approximation.

2. *Molecular Virial Theorem for Fixed Nuclei*

The general proof of the VT by scaling (Section II.B.2) can be extended to the case of fixed nuclei, but now it is preferable, in our definition of the scaled wavefunction ψ_λ, to scale both electronic and nuclear coordinates. We thus define the normalized scaled wavefunction

$$\psi_\lambda = \lambda^{3N/2}\psi\{\lambda\vec{r}_i; \lambda\vec{R}_\alpha\} \tag{3.198}$$

where λ is an arbitrary parameter. The average values, denoted below with a bar, are integrals over electronic coordinates only, so that for T_e and U_{ee} the nuclear configuration does not play any part. The total scaled energy is then

$$E_e\{\lambda; \lambda\vec{R}_\alpha\} = \lambda^2\overline{T}_e\{\lambda\vec{R}_\alpha\} + \lambda\left[\overline{U}_{ee}\{\lambda\vec{R}_\alpha\} + \overline{U}_{en}\{\lambda\vec{R}_\alpha\}\right] \quad (3.199)$$

The derivative of this energy with respect to λ is set equal to zero for the exact ($\lambda = 1$) wavefunction, leading to

$$2\overline{T}_e + \overline{U}_{ee} + \overline{U}_{en} = -\sum \vec{R}_\alpha \cdot \vec{\nabla}_\alpha E_e \quad (3.200)$$

or, including U_{nn},

$$2\overline{T}_e + \overline{U} = -\sum \vec{R}_\alpha \cdot \vec{\nabla}_\alpha V \quad (3.201)$$

This is the same molecular VT that was derived in Section II.B.7.a, Eq. (2.226), by another method. Compared with the atomic VT, there is an extra term on the right-hand side of (3.200) or (3.201). The physical interpretation of this extra virial is the essence of the Born–Oppenheimer approximation: it represents the virial of the forces necessary to hold the nuclear positions fixed.

The electronic energy V is in general a function of the internuclear distances $R_{\alpha\beta}$, and we can write instead of Eq. (3.201)

$$2\overline{T}_e + \overline{U} + \sum_{\alpha > \beta}\sum R_{\alpha\beta} \frac{\partial V}{\partial R_{\alpha\beta}} = 0 \quad (3.202)$$

This relation can be further simplified if the nuclear configuration can be specified by a set of internuclear distances R_μ and angular coordinates θ_ν (Nelander,[190] 1969); the molecular VT can thus be written

$$2\overline{T}_e + \overline{U} + \sum R_\mu \frac{\partial V}{\partial R_\mu} = 0 \quad (3.203)$$

The proof is immediate by scaling, the important point being that the angular coordinates are unchanged by scaling and therefore make no contribution to the sum, which is now over internuclear distances only. For diatomic molecules, there is only one internuclear distance and consequently only one term in the sum. This makes the VT for diatomic molecules particularly simple; it will be used as a starting point for our discussion of chemical bonding in the next section.

To conclude this subsection, we recall that any wavefunction that is properly scaled can be made to fulfill the VT (see Section II.B.2 and Refs. 28, 39, 191). Nevertheless, as recently emphasized by Magnoli and Murdoch[192] (1982), when the scaling procedure is applied to SCF wavefunctions (Section II.C.1), the VT can be satisfied but self-consistency is lost. Although scaling generally has a small effect on the total energy, the effects on the energy components $(T, U_{ee}, U_{en}, U_{nn})$ can be two to three orders of magnitude larger and in the range of several eV. Consequently, for applications where the energy components are useful, it is highly desirable to obtain wavefunctions that both satisfy the VT and are self-consistent. Magnoli and Murdoch report a simple extrapolation technique that requires only one integral evaluation and two SCF cycles to achieve convergence. They apply their technique to atoms and small molecules.

3. *Chemical Bonding and Virial Fragments*

For a diatomic molecule, the electronic energy V is a function of the internuclear distance R only, and, as first derived by Slater[43] (1933), the VT of Eq. (3.203) reads

$$2\bar{T}_e + \bar{U} + R\frac{dV}{dR} = 0 \tag{3.204}$$

The dependence of the average kinetic and potential energies on R is known if $V(R)$ is given:

$$\bar{T}_e = -V - R\frac{dV}{dR} = -\frac{d(RV)}{dR} \tag{3.205}$$

$$\bar{U} = 2V + R\frac{dV}{dR} \tag{3.206}$$

We now use this result to discuss separately the changes in $\bar{T}_e$ and $\bar{U}$ that occur during the formation of a covalent chemical bond in a diatomic molecule. The VT is the only way to obtain this information.

At infinite separation, V_∞ being the total energy of the two separated atoms, the atomic VT is given by

$$\begin{aligned}(\bar{T}_e)_\infty &= -V_\infty \\ \bar{U}_\infty &= 2V_\infty\end{aligned} \tag{3.207}$$

At the equilibrium distance R_e, dV/dR is required to vanish for V to be a minimum, so that from Eqs. (3.205) and (3.206),

$$\begin{aligned} (\bar{T}_e)_{eq} &= -V_{eq} \\ \bar{U}_{eq} &= 2V_{eq} \end{aligned} \tag{3.208}$$

Subtracting Eq. (3.207) from Eq. (3.208),

$$\begin{aligned} (\bar{T}_e)_{eq} - (\bar{T}_e)_\infty &= -(V_{eq} - V_\infty) > 0 \\ \bar{U}_{eq} - \bar{U}_\infty &= 2(V_{eq} - V_\infty) < 0 \end{aligned} \tag{3.209}$$

where the inequalities are required for stability of the molecule against dissociation. We have already reached an important and nonobvious conclusion: the formation of a stable chemical bond is accompanied by an increase of kinetic energy; but this is more than compensated by the decrease in potential energy, which is twice as large.

Further, for large R the two neutral atoms interact through the London dispersion potential $V(R) = V_\infty - A/R^6$ (A constant), so that from Eqs. (3.205) and (3.206) we can write

$$\begin{aligned} \bar{T}_e &= (\bar{T}_e)_\infty - \frac{5A}{R^6} \\ \bar{U} &= \bar{U}_\infty + \frac{4A}{R^6} \end{aligned} \tag{3.210}$$

As R decreases, $\bar{T}_e$ at first decreases and $\bar{U}$ increases; combining this result with what was said above, we deduce that, between R_e and infinity, $\bar{U}$ must go through a maximum and $\bar{T}_e$ through a minimum. Figure 4 illustrates the typical variation of $\bar{T}_e$, $\bar{U}$, and V as functions of R; this figure is for a generic diatomic molecule, but resembles the known curves for H_2 and H_2^+ (Kolos and Wolniewicz,[193] 1964).

After the initial increase in $\bar{U}$ and decrease in $\bar{T}_e$, as R continues to decrease, the kinetic energy keeps increasing due to the squeezing of the electrons into a smaller region. Because the electrons feel more and more attraction from the nuclei, the potential energy continues to decrease until it reaches a minimum at an internuclear distance somewhat less than R_e, inside which it then goes to infinity as R approaches zero, due to the internuclear repulsion term.‡

‡For a more complete discussion of chemical bonding in terms of the VT, see Levine[35] and Cohen-Tannoudji, Diu, and Laloë;[194] the chemical bond is also often discussed in terms of the electrostatic Hellmann–Feynman theorem introduced in Section II.B.6. A recent book edited by Deb[195] reviews the numerous articles on the subject, and Levine[35] gives a good introduction to the electrostatic Hellmann–Feynman theorem.

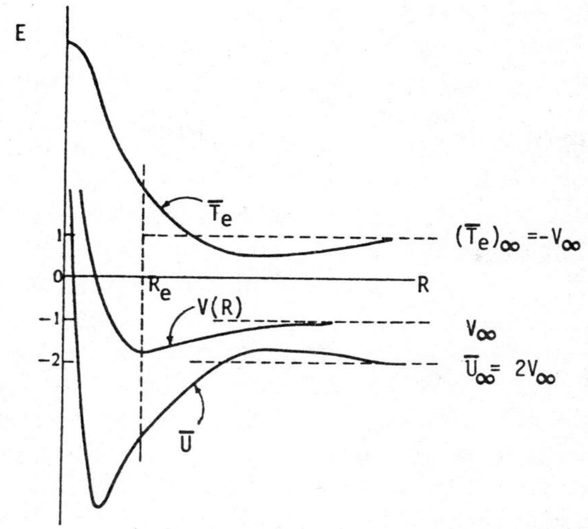

Fig. 4. Variations of the average kinetic ($\overline{T}_e$), potential ($\overline{U}$), and total ($\overline{V}$) energies of a typical diatomic molecule; the unit of energy is taken as V_∞, the energy of the separated atoms. [From Ref. 35: Ira N. Levine, *Quantum Chemistry*, Second Edition, copyright © 1974 by Allyn and Bacon, Inc., p. 370 (Reprinted with permission).]

The VT permits the calculation also of the vibrational force constants of chemical bonds. Taking the derivative of Eqs. (3.205) and (3.206) with respect to R, we obtain at equilibrium $[(dV/dR)_{R_e} = 0]$

$$\left(\frac{d\overline{U}}{dR}\right)_{R_e} = -\left(\frac{d\overline{T}_e}{dR}\right)_{R_e} = R_e k_e \tag{3.211}$$

The curvature, $k_e = (d^2V/dR^2)_{R_e}$, is always positive, and we check that at equilibrium $\overline{T}_e$ decreases and $\overline{U}$ increases. If the curve for the potential energy V (Fig. 4) of a molecule is known, k_e can be found from it by measuring the tangent at R_e of $\overline{U}$ or of $\overline{T}_e$, since from Eq. (3.211) the tangents at this point are equal and opposite. The anharmonic force constant

$$l_e = \left(\frac{d^3V}{dR^3}\right)_{R_e} \tag{3.212}$$

results from differentiating Eq. (3.205) twice. (We note that the calculations are simpler with the kinetic energy because it is a one-electron operator.) At equilibrium,

$$R_e l_e + 3k_e = -\left(\frac{d^2\overline{T}_e}{dR^2}\right)_{R_e} \tag{3.213}$$

Higher derivatives of V at R_e can be evaluated similarly. The subject of the derivation of force constants by use of the VT has recently been reviewed by Goodisman.[196] We shall say more about the $V(R)$ curve for a diatomic molecule and its relation to spectroscopic constants in Section III.D.4.

A central question in theoretical chemistry is to relate the properties of a molecule to those of its constituent atoms. Thus in valence-bond theory, the molecule wavefunction is approximated by a product (or sum of products) of atomic orbitals. Moffitt[197] (1951, 1953) pioneered this "atoms-in-molecules approach," as he called it, in the calculation of molecular binding energies.

Bader and coworkers[198] approached this question from the other side, so to speak, and tried to demonstrate the physical basis for the definition of an atom in a molecule starting from experimental molecular properties, such as the charge density. They spatially divided the charge distribution and energy of a molecule, drawing the partitioning surfaces so that the gradient of the electron charge density, $\nabla\rho$, has zero flux across them. The kinetic and potential energies of the molecular fragments so defined satisfy the VT; each fragment is "quantum-mechanically self-contained," as reflected in the constancy of both the electronic population and virial for the same fragment in different molecular systems. The fragment VT sets the condition to be obeyed for the transferability of fragment properties between different molecules, a cornerstone of chemistry. It is found that the fundamental basis for additivity is the result of the transferability of atomlike fragments rather than of "bond" contributions. Fig. 5 illustrates the virial partitioning for three different molecules.

A related problem arising in the treatment of substituent effects on rates and equilibria of chemical reactions is to evaluate the changes in energy and geometric structure that occur when a fragment is removed from a molecule and replaced by a different one. Murdoch[9] (1982) has recently approached this question from an *ab initio* viewpoint using SCF Hartree–Fock theory. He presents an original theory of substitution and introduces the notion of the "hemistructural" relationship, which is based on the HFT (cf. Section II.B.6). Murdoch concludes that group additivity is observed because compensating changes in electronic structure take place in the two molecular fragments. Such a result is a fundamental departure from existing ideas on substituent effects and suggests that models of molecular structure based on interactions between transferable groups may require significant revision.

4. *The Isotope Effect in Diatomic Molecules*

a. Virial Theorem for Nuclear Motion. We have already considered the first stage of the Born–Oppenheimer approximation, the electronic motion, from which the VT for the electrons of diatomic molecules was found to be Eq.

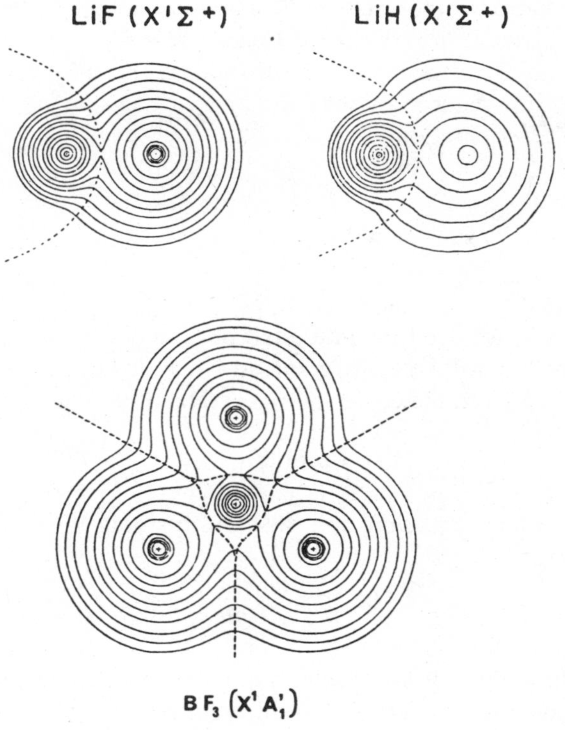

Fig. 5. Contour plots of the total electronic charge distribution for LiF, LiH, and BF_3, showing the virial partitioning surfaces (dashed lines) in atomic-like fragments. [From Ref. 195: R. F. W. Bader, in *The Force Concept in Chemistry*, B. M. Deb (Ed.), copyright © 1981 by Van Nostrand Reinhold Company, p. 128, reprinted by permission of Van Nostrand Reinhold Company; and Ref. 199: S. Srebrenik, R. F. W. Bader and T. Tung Nguyen Dang, *J. Chem. Phys.* **68**, 3667 (1978), with permission of the authors and the American Institute of Physics.]

(3.204). We now focus on the nuclear motion, that is, the second stage of the Born–Oppenheimer approximation. The (ground-state) electronic energy V (including the internuclear potential) is taken as the effective potential in the Schrödinger equation for the motion of the nuclei [cf. Eq. (3.197)]:

$$[T_n + V(R)]\psi_n = E\psi_n \tag{3.214}$$

Here T_n is the sum over nuclei of the kinetic energy operators, ψ_n the nuclear wavefunction, and E the total energy of the molecule including translational and internal (vibrational and rotational) energies.

For two nuclei α and β, Eq. (3.214) constitutes a two-particle problem that can be separated into two one-particle problems by use of internal coordinates: T_n becomes the sum of the translational kinetic energy of the center of mass associated with the total mass ($M = m_\alpha + m_\beta$) and the internal kinetic energy T_μ due to rotation and vibration (for fixed center of mass), associated with the reduced mass of the molecule [$\mu = m_\alpha m_\beta/(m_\alpha + m_\beta)$]. We thus have the separation

$$\psi_{n,\text{tot}} = \psi_{n,\text{trans}}\psi_{n,\text{int}} \tag{3.215}$$

$$E_{\text{tot}} = E_{\text{trans}} + E_{\text{int}} \tag{3.216}$$

The translational motion simply adds the quantity E_{trans} to the molecular energy; but since we are here interested only in the internal motion, the translational energy will be omitted and all remaining quantities refer to the motion of the nuclei in the center-of-mass system, for which[200]

$$\left[T_\mu + V(R)\right]\psi_n = E\psi_n \tag{3.217}$$

This is a one-variable equation, and the corresponding VT is simply

$$2\langle T_\mu\rangle_n = \left\langle R\frac{dV}{dR}\right\rangle_n \tag{3.218}$$

In this result the subscript n indicates that the average values are taken with the nuclear wavefunction. This VT is an exact relation within the limits of the accuracy of the Born–Oppenheimer approximation. $V(R)$ is not in general a homogeneous function of R, and we cannot invoke Euler's theorem to simplify the right-hand side of Eq. (3.218). This right-hand side will be evaluated approximately in Section III.D.4.d below, but first the Hellmann–Feynman theorem is used to transform the left-hand side.

b. Hellmann–Feynman theorem with Parametrized Reduced Mass. The other basic idea in this application is to exploit the Hellmann–Feynman theorem, Eq. (2.203), using the reduced mass as the variational parameter; this yields immediately

$$\frac{\partial E}{\partial \mu} = \left\langle \frac{\partial T_\mu}{\partial \mu}\right\rangle_n \tag{3.219}$$

since the electronic energy V is independent of μ. Moreover, with $T_\mu = -(\hbar^2/2\mu)\nabla^2$,

$$\frac{\partial T}{\partial \mu} = -\frac{T_\mu}{\mu} \tag{3.220}$$

Inserting this result in Eq. (3.219) and then in the VT [Eq. (3.218)], we obtain the relation of interest for the isotope effect in diatomic molecules:

$$\frac{\partial E}{\partial \mu} = -\frac{\langle R(dV/dR)\rangle_n}{2\mu} \tag{3.221}$$

Two isotopic molecules have different reduced masses and different energies (i.e., different vibrational and rotational energy levels, reflected in their band spectra). For a given nuclear wavefunction this formula gives the variation of E with the reduced mass; for small changes of E and μ, (3.221) can be written

$$\frac{\Delta E}{\Delta \mu} = -\frac{\langle R(dV/dR)\rangle_n}{2\mu} \tag{3.222}$$

The problem is now to evaluate the right-hand side of Eq. (3.221) or (3.222), which presumes a knowledge of the nuclear wavefunction and of the potential V.

c. *The Potential Energy of Diatomic Molecules.* The general behavior of the $V(R)$ curve can be seen in Figure 4 of the previous subsection. Many empirical potential functions have been proposed, most of them containing three parameters to be evaluated from spectroscopic data using the relations

$$\begin{aligned} V(\infty) - V(R_e) &= D_e \\ V'(R_e) &= 0 \\ V''(R_e) &= k_e \end{aligned} \tag{3.223}$$

where D_e is the dissociation energy and k_e the quadratic (or harmonic) force constant. The most famous of the empirical equations for V is the Morse function

$$V(R) = D_e\left\{1 - \exp\left[-\left(\frac{k_e}{2D_e}\right)^{1/2}(R - R_e)\right]\right\}^2 \tag{3.224}$$

There also exists a numerical method developed by Rydberg, Klein, and Rees (the RKR method) to evaluate $V(R)$ from experimental data, its limitation being that it requires molecular states for which sufficient vibrational levels have been spectroscopically determined. Steel et al.[201] (1962) have tested the predictions of several empirical potential-energy functions against the RKR curves for 19 diatomic electronic states. The Morse function turns out to be less accurate than the Varshin function

$$V(r) = D_e\left\{1 - \frac{R_e}{R}\exp\left[-\beta\left(R^2 - R_e^2\right)\right]\right\}^2 \tag{3.225}$$

More attractive in the present application, because of its simplicity, is the Fues potential (Fues,[202] 1926), used by Parr and Borkman[203] (1968) as a point of departure for modeling potential-energy curves for many diatomic molecules. The Fues potential has the form

$$V(R) = w_0 + \frac{w_1}{R} + \frac{w_2}{R^2} \tag{3.226}$$

with $k_e = -w_1/R_e^3 = 2w_2/R_e^4$. This potential can be derived by perturbation theory (Parr and White,[204] 1968) with perturbation parameter $\lambda = 1 - R_e/R$, the series expansion being truncated after the λ^2 term. A compelling argument in favor of this potential is that, by use of the electronic VT, a physical interpretation can be assigned to the parameters w_1 and w_2. We get easily

$$\begin{aligned} \langle U \rangle &= 2w_0 + \frac{w_1}{R} \\ \langle T_e \rangle &= -w_0 + \frac{w_2}{R^2} \end{aligned} \tag{3.227}$$

thus the Coulomb-like term w_1/R corresponds to the potential energy, while w_2/R^2 belongs only to the kinetic-energy part. Moreover, the relations between spectroscopic force constants (quadratic k_e, cubic l_e, quartic $m_e, \ldots$) predicted from the Fues potential are in good agreement with the experimental values. For example (Parr and White,[204] 1968), the Fues potential implies

$$\left(\frac{k_e m_e}{l_e^2}\right)^{1/2} = 1 \tag{3.228}$$

For a large number of species this quantity averages about 0.86. We shall use later in our discussion [Section III.D.4.d below] the Fues-potential relation between the cubic [$l_e = V'''(R_e)$] and quadratic [$k_e = V''(R_e)$] force constants, namely

$$R_e l_e = -6k_e \tag{3.229}$$

in order to make numerical estimates.

Back to our present problem: numerical integration of $\langle R(dV/dR)\rangle_n$ is possible in specific cases where the $V(R)$ curve for the state of interest is known; but to check the validity of our original formula, Eq. (3.221), and to carry out the calculation analytically, we use the wavefunctions and potential of the harmonic oscillator; that is, we expand $V(R)$ in a Taylor series about R_e truncated after the second-derivative term.

d. Harmonic-Oscillator Approximation. Setting $R - R_e = q$, with $V(R) = V(R_e) + \frac{1}{2}k_e q^2$, we obtain immediately

$$R\frac{dV}{dR} = R_e k_e q + k_e q^2 \tag{3.230}$$

The solutions for the harmonic oscillator[205] are known exactly: for the vibrational quantum number v ($v = 0, 1, 2, \ldots$), they are

$$E_v = (v + \tfrac{1}{2})\hbar\omega$$
$$|v\rangle \equiv \left(\frac{\alpha}{\pi}\right)^{1/4}(2^v v!)^{-1/2} e^{-\alpha q^2} H_v(\alpha^{1/2} q) \tag{3.231}$$

where $\alpha = \mu\omega/\hbar$; ω is the classical vibration frequency

$$\omega = \left(\frac{k_e}{\mu}\right)^{1/2} \tag{3.232}$$

The H_v are Hermite polynomials that obey the recursion relation

$$xH_v(x) = vH_{v-1}(x) + \tfrac{1}{2}H_{v+1}(x) \tag{3.233}$$

the first few being

$$H_0 = 1, \qquad H_1 = 2x, \qquad H_2 = 4x^2 - 2 \tag{3.234}$$

The wavefunctions $|v\rangle$ form a complete orthonormal set and, like the Hermite polynomials, have the parity of v. Using $|v\rangle$ as the nuclear wavefunction together with (3.230), the virial averaged over the state $|v\rangle$ becomes

$$\left\langle R\frac{dV}{dR}\right\rangle_n = R_e k_e \langle v|q|v\rangle + k_e \langle v|q^2|v\rangle \tag{3.235}$$

The first term on the right vanishes, since it is the integral of an odd function on the symmetrical interval $(-\infty, +\infty)$. We now calculate the second integral, $\langle v|q^2|v\rangle$, by use of the recursion relation of Eq. (3.233) with $x = \alpha^{1/2}q$:

$$q|v\rangle = \left(\frac{v}{2\alpha}\right)^{1/2} |v-1\rangle + \left[\frac{v+1}{2\alpha}\right]^{1/2} |v+1\rangle \tag{3.236}$$

and the adjoint relation (all quantities are real),

$$\langle v|q = \left(\frac{v}{2\alpha}\right)^{1/2} \langle v-1| + \left[\frac{v+1}{2\alpha}\right]^{1/2} \langle v+1| \tag{3.237}$$

The product of these two last relations, taking into account the orthonormality of the wavefunctions, gives

$$\langle v|q^2|v\rangle = (v+\tfrac{1}{2})\alpha^{-1} = (v+\tfrac{1}{2})\frac{\hbar\omega}{k_e} \tag{3.238}$$

In the harmonic-oscillator approximation, we thus obtain

$$\left\langle R\frac{dV}{dR}\right\rangle_n = (v+\tfrac{1}{2})\hbar\omega = E_v \tag{3.239}$$

e. Discussion. Inserting this last result into Eq. (3.221), we get

$$\frac{\partial E_v}{\partial \mu} = -\frac{E_v}{2\mu} \tag{3.240}$$

or, in terms of the frequency,

$$\frac{\partial \ln \omega}{\partial \ln \mu} = -\frac{1}{2} \tag{3.241}$$

This result is also easily obtained by differentiating logarithmically the expression [Eq. (3.232)] for the frequency. We have thus verified the consistency of Eq. (3.221).

One might hope to get improvement over Eq. (3.241) without more calculation, by including another term in the development of $V(R)$,

$$V(R) = V(R_e) + \tfrac{1}{2}k_e q^2 + \tfrac{1}{6}l_e q^3 \tag{3.242}$$

and keeping the same harmonic-oscillator wavefunctions. Since, for parity reasons, $\langle v|q^3|v\rangle = 0$ as well as $\langle v|q|v\rangle = 0$, we have no other integral to calculate:

$$R\frac{dV}{dR} = R_e k_e q + \left[k_e + \frac{R_e l_e}{2}\right]q^2 + \frac{l_e q^3}{2} \tag{3.243}$$

Equation (3.221) becomes

$$\frac{\partial \ln \omega}{\partial \ln \mu} = -\frac{1}{2}\left(1 + \frac{R_e l_e}{2k_e}\right) \tag{3.244}$$

The intrinsically negative quantity $R_e l_e / 2k_e$ appears as a correction term to Eq. (3.241). But comparisons with the experimental constants show that this term is much too large to be regarded as a small correction; for example, using Eq. (3.229)—that is, the prediction of the Fues potential, which is approximately correct for many molecules—we have $R_e l_e / 2k_e = -3$. The sign of the derivative of ω with respect to μ should thus become positive; this is physically incorrect, since heavier isotopes vibrate at lower frequencies $(\partial \omega / \partial \mu < 0)$. Usually the zeroth-order wavefunctions give the correct first-order correction to the energy (or frequency), but not in this case, since the anharmonic perturbation is not small compared to the harmonic term, as indicated by Eq. (3.229).

It remains that Eq. (3.221) [or Eq. (3.222)] appears to be correct, and numerical calculations of the integral in specific cases should give good predictions of the isotopic energy shift given the change in reduced mass, or of an isotope mass given the energy shift (e.g., from band spectra).

E. The Hypervirial Theorem

1. Review of Recent Developments

If ψ is a stationary-state eigenfunction of a Hamiltonian H, and A an arbitrary time-dependent operator involving coordinates q_i and momenta p_i $(i = 1, 2, \ldots, n)$, the Hypervirial theorem for A [HVT(A)] states that (Hirschfelder,[17] 1960)

$$\langle \psi | [A, H] | \psi \rangle = 0 \tag{3.245}$$

As shown in Section II.B.3 [see Eq. (2.179)], the HVT is an immediate consequence of the Heisenberg equation of motion, since any stationary-state expectation value is independent of time; it has a classical analog in terms of Poisson brackets [see Eq. (2.95) of Section II.A.7].

The HVT provides a wide variety of generalizations of the VT [VT = HVT($\Sigma q_i p_i$); cf. Sections II.A.7 and II.B.3]; for enclosed systems, the HVT requires an extra term (in Section II.B.5 it was found that this extra term can be written as a generalized current density; see also Fernândez and Castro,[206] 1981); the HVT has also been extended to open systems (Schweitz,[207] 1980).

Like the VT, various HVTs have been used as constraints to improve variational wavefunctions (Epstein and Hirschfelder,[38] 1962; Hirschfelder and Coulson,[37,208] 1962; Vetchinkin,[209] 1964; Coulson,[210] 1965; Robinson,[211] 1965; Epstein,[212] 1969), often in conjunction with scaling (Banerjee,[213] 1978; Castro,[214] 1978; Castro and Fernández,[215] 1980); with their help, one can calculate certain average values without integral computations (Epstein et al.,[216] 1962, 1967, 1975), and the Breit-type recoil correction for pionic atoms (Friar,[217] 1980).

The off-diagonal generalizations of Eq. (3.245), namely

$$\langle \psi_1 | [A, H] | \psi_2 \rangle = (E_2 - E_1) \langle \psi_1 | A | \psi_2 \rangle \tag{3.246}$$

are also useful as constraints in calculating approximate wavefunctions and transition-moment integrals (Chong and Benston,[218] 1967, 1968; Yue and Chong,[219] 1969; Bradley and Hughes,[220] 1967, 1969; Coulson and Nash,[221] 1972, 1974; Biemont and Godefroid,[222] 1978; Banerjee,[223] 1977; Sakamoto and Terasaka,[224] 1979).

In the next subsection we examine what appear to be some of the most promising applications of the HVT. The use of the HVT$(q^\alpha p)$, α integral, together with the Hellmann–Feynman theorem (HFT) allows, in some cases, development of the perturbation expansion of the energy and of average values of coordinates, raised to any power, in terms of the unperturbed energy only. This eliminates the tedious calculations of sums of products of matrix elements in the traditional perturbation theory. The method has been applied successfully to the anharmonic oscillator, which, together with several other one-dimensional problems, will be reviewed in detail.

Finally, in the last subsection we derive a very general sum rule that has applications in radiation theory and can be considered as a HVT.

2. *Perturbation Theory without Wavefunctions*

Ordinary (Rayleigh–Schrödinger) perturbation theory starts from the eigenfunctions $|l\rangle$ and eigenvalues E_l for the unperturbed Hamiltonian H_0. The Schrödinger equation, $H|\psi\rangle = E|\psi\rangle$, with the perturbed Hamiltonian $H = H_0 + \lambda V$, is solved by expanding the energy E and wavefunction $|\psi\rangle$ in power series in the perturbation parameter λ. For example, the first- and second-order terms in the energy series of a perturbed level, say l, are given by

$$E_l^{(1)} = V_{ll} = \langle l | V | l \rangle \tag{3.247}$$

$$E_l^{(2)} = \sum_{m \neq l} \frac{V_{ml} V_{ml}^*}{E_l - E_m} \tag{3.248}$$

These—and especially higher-order corrections—require making complicated calculations of matrix elements and summing over products of matrix elements (the sum in $E^{(2)}$ is over the complete set of $|l\rangle$'s). In practice, this perturbation is rarely carried beyond second order.

In this subsection, a set of HVTs is combined with the HFT (cf. Section II.B.6) to obtain, for the anharmonic oscillator and certain other problems, all the corrections to the energy in terms of the unperturbed energy without using the wavefunction. In addition, perturbation expansions for expectation values of an arbitrary power of the position coordinate are found to any order, again expressed solely in terms of the unperturbed energy.

a. Formulation of the Problem. For simplicity of the discussion we limit ourselves to a one-dimensional problem with Hamiltonian

$$H = \frac{p^2}{2m} + U(q) \tag{3.249}$$

The results given in this subsection can easily be extended to many degrees of freedom.

Setting $A = q^\alpha q^\beta$ (α and β positive integers) and employing the Hamiltonian of Eq. (3.249) in the HVT of Eq. (3.245), we obtain the HVT$(q^\alpha q^\beta)$ as

$$\langle q^\alpha [p^\beta, U]\rangle + \frac{\alpha i\hbar}{m}\langle q^{\alpha-1}p^{\beta+1}\rangle + \frac{\hbar^2\alpha(\alpha-1)}{2m}\langle q^{\alpha-2}p^\beta\rangle = 0 \tag{3.250}$$

The most useful of these relations correspond to $\beta = 1$ (the VT is recovered with $\alpha = \beta = 1$), for which

$$-i\hbar\left\langle q^\alpha \frac{\partial U}{\partial q}\right\rangle + \frac{i\hbar\alpha}{m}\langle q^{\alpha-1}p^2\rangle + \frac{\hbar^2\alpha(\alpha-1)}{2m}\langle q^{\alpha-2}p\rangle = 0 \tag{3.251}$$

A more desirable form of Eq. (3.251) results from eliminating the momenta by use of the Schrödinger equation (which introduces the energy of interest) and of the HVT(q^α):

$$2\langle q^{\alpha-1}p\rangle = i\hbar(\alpha-1)\langle q^{\alpha-2}\rangle \tag{3.252}$$

We note in passing that this last equality transforms any average value of the form $q^N p$ into q^{N-1}. The HVT in Eq. (3.251) thus becomes

$$2\alpha\langle q^{\alpha-1}\rangle E = \left\langle q^\alpha \frac{\partial U}{\partial q}\right\rangle + 2\alpha\langle q^{\alpha-1}U\rangle - \frac{\alpha(\alpha-1)(\alpha-2)\hbar^2}{4m}\langle q^{\alpha-3}\rangle \tag{3.253}$$

Equation (3.253) is an important HVT; it can be used, for example, to work out a recurrence relation between expectation values of powers of the radial distance for the nonrelativistic and relativistic hydrogen atom (Epstein et al.,[216] 1962, 1967, 1975). It can also be easily shown to be equivalent to the Schrödinger equation (Hirschfelder,[17] 1960) in the sense that it suffices to have the HVT[$f(q)p$] valid for an arbitrary function f of the coordinate to derive the Schrödinger equation (with the help of a few integrations by parts); a similar result holds for the HVT([[$f(q)$, H], H]) (Coulson,[210] 1965). Both results are limited to real wavefunctions (no magnetic field).

Back to our perturbation problem, the potential is written as a sum of an unperturbed part and a small perturbation:

$$U(q)=U_0(q)+\lambda V \tag{3.254}$$

Thus the HFT [Eq. (2.203)] with the perturbation strength λ as a parameter becomes

$$\frac{\partial E}{\partial \lambda}=\langle V\rangle \tag{3.255}$$

The HVT [Eq. (3.253)] and HFT [Eq. (3.255)] are applied next to the anharmonic-oscillator problem (Swenson and Danforth,[225] 1972).

b. The Anharmonic Oscillator. We consider a general anharmonic oscillator with Hamiltonian

$$H=\frac{p^2}{2m}+\frac{kq^2}{2}+\lambda q^{\gamma} \tag{3.256}$$

or, in reduced units ($m=1, k=1, \hbar=1$),

$$H=\frac{p^2+q^2}{2}+\lambda q^{\gamma} \tag{3.257}$$

The unit of energy is thus also unity [$\hbar\omega=(k/m)^{1/2}=1$], and the zero-order energy is simply

$$E_v=v+\tfrac{1}{2}, \qquad v=0,1,2,\ldots \tag{3.258}$$

With the Hamiltonian of Eq. (3.257), the HVT [Eq. (3.253)] and the HFT [Eq. (3.255)] read

$$\begin{aligned}2\alpha E\langle q^{\alpha-1}\rangle &= (\alpha+1)\langle q^{\alpha+1}\rangle+\lambda(2\alpha+\gamma)\langle q^{\alpha+\gamma-1}\rangle\\ &\quad -\frac{\alpha(\alpha-1)(\alpha-2)}{4}\langle q^{\alpha-3}\rangle\end{aligned} \tag{3.259}$$

and

$$\langle q^{\gamma} \rangle = \frac{\partial E}{\partial \lambda} \tag{3.260}$$

We now assume the existence of expansions in powers of λ for E and q^m:

$$E = E^{(0)} + \lambda E^{(1)} + \lambda^2 E^{(2)} + \cdots + \lambda^i Q^i + \cdots \tag{3.261}$$

$$q^m = Q_m^{(0)} + \lambda Q_m^{(1)} + \lambda^2 Q_m^{(2)} + \cdots + \lambda^j Q_m^{(j)} + \cdots \tag{3.262}$$

$E^{(0)}$ is given by Eq. (3.258), and, supposing that both perturbed and unperturbed wavefunctions are normalized to unity, we have

$$Q_0^{(i)} = \delta_{i0} \tag{3.263}$$

since $\langle q^0 \rangle = 1 = \Sigma \lambda^i Q_0^{(i)}$ must be true for all λ including $\lambda = 0$.

Inserting the series given by Eqs. (3.261) and (3.262) into the HVT [Eq. (3.259)] and equating equal powers of λ, we obtain

$$Q_{\alpha+1}^{(l)} = \frac{2\alpha}{\alpha+1} \left\{ \sum_{i=0}^{l} E^{(i)} Q_{\alpha-1}^{(l-i)} \right\} + \frac{\alpha(\alpha-1)(\alpha-2)}{4(\alpha+1)} Q_{\alpha-3}^{(l)} - \frac{2\alpha+\gamma}{\alpha+1} Q_{\alpha+\gamma-1}^{(l-1)} \tag{3.264}$$

for all positive integers l and α, with the understanding that for all $l = 0$ the last term does not contribute.

Substituting now the series given by Eqs. (3.261) and (3.262) into the HFT [Eq. (3.260)], we get

$$\sum_{i=0}^{\infty} \lambda^j Q_{\gamma}^{(j)} = E^{(1)} + 2\lambda E^{(2)} + \cdots + i\lambda^{i-1} E^{(i)} + \cdots \tag{3.265}$$

leading to

$$Q_{\gamma}^{(j)} = (j+1) E^{(j+1)}, \qquad j = 0, 1, 2, \ldots \tag{3.266}$$

This last equation is very important in this application; it states that the energy correction at a given order is proportional to the correction, at the preceding order, of $\langle q^{\gamma} \rangle$, the expectation value of the perturbing potential.

From the HVT of Eq. (3.264), the HFT of Eq. (3.266), and the normalization condition of Eq. (3.263), we now show that all the $Q_j^{(i)}$ and $E^{(i)}$ can be determined in terms of the unperturbed energy $E^{(0)}$ given by Eq. (3.258) (we have suppressed, for clarity, the label v of the specific state). Let us examine first the zeroth-order equations [$l=0$ in Eq. (3.264)]; the normalization condition yields

$$Q_0^{(0)}=1 \tag{3.267}$$

and we have successively for $\alpha=1,2,3,\ldots,\alpha,\ldots$

$$\begin{aligned}
&Q_1^{(0)}=0\\
&Q_2^{(0)}=E^{(0)}Q_0^{(0)}=E^{(0)}\\
&Q_3^{(0)}=\tfrac{4}{3}E^{(0)}Q_1^{(0)}=0\\
&Q_4^{(0)}=\tfrac{3}{2}E^{(0)}Q_2^{(0)}+\tfrac{3}{8}Q_0^{(0)}=\tfrac{3}{8}E^{(0)2}+\tfrac{3}{8}\\
&\vdots\\
&Q_{\alpha+1}^{(0)}=\frac{2\alpha}{\alpha+1}E^{(0)}Q_{\alpha-1}^{(0)}+\frac{\alpha(\alpha-1)(\alpha-2)}{4(\alpha+1)}Q_{\alpha-3}^{(0)}
\end{aligned} \tag{3.268}$$

Obviously we obtain all the $Q^{(0)}$ in terms of $E^{(0)}$ (note that all $Q_{2j+1}^{(0)}=0$, as expected from the parity of the zeroth-order wavefunctions), and if we keep going until $\alpha=\gamma-1$, we finally obtain $E^{(1)}$ using Eq. (3.266):

$$E^{(1)}=Q_\gamma^{(0)}=\frac{2(\gamma-1)E^{(0)}Q_{\gamma-2}^{(0)}}{\gamma}+\tfrac{1}{4}(\gamma-1)(\gamma-2)(\gamma-3)Q_{\gamma-4}^{(0)} \tag{3.269}$$

Next examining the first order, the normalization condition in Eq. (3.263) gives $Q_0^{(1)}=0$ and, with $l=1$, Eq. (3.264) takes the form

$$\begin{aligned}
Q_{\alpha+1}^{(1)}=&\frac{2\alpha}{\alpha+1}\left(E^{(0)}Q_{\alpha-1}^{(1)}+E^{(1)}Q_{\alpha-1}^{(0)}\right)\\
&+\frac{\alpha(\alpha-1)(\alpha-2)}{4(\alpha+1)}Q_{\alpha-3}^{(1)}-\frac{2\alpha+\gamma}{\alpha+1}Q_{\alpha+\gamma-1}^{(0)}
\end{aligned} \tag{3.270}$$

All the $Q_i^{(1)}$ are determined in terms of previously known quantities and ultimately of $E^{(0)}$; when we reach $Q_j^{(1)}$, we know $E^{(2)}$ by the HFT.

It is thus clear that for arbitrary γ, the power of the anharmonic term, all the $Q_j^{(l)}$ and $E^{(l)}$ can be exactly determined as functions of $E^{(0)}$ only. For example, taking a quartic perturbation ($\gamma=4$) we find

$$\begin{aligned}
&E_v^{(1)}=\tfrac{3}{2}\left(v^2+v+\tfrac{1}{2}\right)\\
&E_v^{(2)}=\tfrac{1}{8}(34v^3+51v^2+59v+21)
\end{aligned} \tag{3.271}$$

These agree with Flügge's Eqs. 35.8 and 35.14 (p. 81-2 in Ref. 226). Caswell[227] (1979), using a computer, solved this quartic anharmonic oscillator to the sixth order.

c. State of Progress. Swenson and Danforth[225] (1972) where the first‡ to use the method described above to obtain, for the anharmonic oscillator, the perturbation expansions of the energy and of expectation values of arbitrary power of the coordinate (incidentally these expansions are divergent), in terms of the unperturbed energy and without calculation of the wavefunction (see also Dmitrieva and Plindov,[230] 1980; Richardson and Blankenbecler,[231] 1979; Blankenbecler, DeGrand, and Sugar,[232] 1980). These results were applied to the vibrations and the rotations of diatomic molecules (Tipping,[233] 1973) and to lattice field vibrations (Richardson and Blankenbecler,[234] 1979). The method was extended to velocity-dependent anharmonicity by Maduemezia[235] (1973, 1974).

Killingbeck[236] (1978) found that the same approach can be adapted to the problem of a hydrogen atom with a radial perturbation, λr^{γ}, including states of angular momentum l, where the centrifugal term $l(l+1)/2r^2$ is present. Grant and Lai[237] (1979) used the HVT–HFT method to calculate to the 20th order the energy of atoms with a screened Coulomb potential of the form

$$V_{\rm sc}(r) = -\frac{Z}{r}\sum_{k=0}^{\infty}\lambda^k V_k r^k \tag{3.272}$$

(Z the nuclear charge; V_k constants alternating in sign and decreasing with k; λ the screening parameter $\lambda = \lambda_0 Z^{1/3}$).

Renormalized perturbation series for the hydrogen atom can also be obtained for the HVT–HFT method (Killingbeck,[238] 1981).

3. *Sum Rule*

In this subsection, we calculate a general sum rule that has many applications in radiation theory (Jackiw,[239] 1967). We seek a closed form for the sum

$$S_n^p = \sum_k \left[\frac{E_k - E_n}{\hbar}\right]^p A_{kn}A_{kn}^{\dagger} = \sum_k \omega_{kn}^p |A_{kn}|^2 \tag{3.273}$$

where A can be any operator and the summation extends over the complete

‡Previous related work was done by Bartell[228] (1963) and Bonham and Su[229] (1966).

set of eigenstates $|k\rangle$ of some Hamiltonian. Hausdorff's formula[240] reads

$$e^{itH/\hbar}Ae^{-itH/\hbar} = \sum_{p=0}^{\infty} \frac{(it/\hbar)^p A_p}{p!} \tag{3.274}$$

wherein the A_p are commutators defined by the recursion

$$\begin{aligned} A_0 &= A \\ A_p &= [H, A_{p-1}] \end{aligned} \tag{3.275}$$

It results immediately from Eq. (3.274) that

$$\langle n|A^{\dagger}e^{itH/\hbar}Ae^{-itH/\hbar}|n\rangle = \sum_{p=0}^{\infty} \frac{\left(\frac{it}{\hbar}\right)^p \langle n|A^{\dagger}A_p|n\rangle}{p!} \tag{3.276}$$

On the other hand, inserting the completeness relation into the left-hand side of Eq. (3.276), we obtain

$$\begin{aligned} \langle n|A^{\dagger}e^{itH/\hbar}Ae^{-itH/\hbar}|n\rangle &= \sum_k \langle n|A^{\dagger}e^{itH/\hbar}|k\rangle\langle k|Ae^{-itH/\hbar}|n\rangle \\ &= \sum_k e^{itE_k/\hbar}A^{\dagger}_{nk}e^{-itE_n/\hbar}A_{kn} \\ &= \sum_k e^{it\omega_{kn}}A^{\dagger}_{nk}A_{kn} \\ &= \sum_k \sum_p \frac{(it)^p \omega_{kn}^p |A_{kn}|^2}{p!} \\ &= \sum_p \frac{(it)^p S_n^p}{p!} \end{aligned} \tag{3.277}$$

By comparison with Eq. (3.276), we have thus achieved a closed expression for the sum S_n^p:

$$S_n^p \equiv \sum_{\alpha} \omega_{\alpha n}^p |A_{\alpha n}|^2 = \frac{\langle n|A^{\dagger}A_p|n\rangle}{\hbar^p} \tag{3.278}$$

To obtain A_p itself one needs to commute A with H p times; however, Eq. (3.278) can be simplified by use of the equality

$$\langle n|A^{\dagger}A_p|n\rangle = \langle n|A^{\dagger}_m A_{p-m}|n\rangle, \qquad p \geq m \tag{3.279}$$

To prove Eq. (3.297) we note that by definition of A_p

$$\begin{aligned}
\langle n|A^\dagger A_p|n\rangle &= \langle n|\big(A^\dagger HA_{p-1} - A^\dagger A_{p-1}H\big)|n\rangle \\
&= \langle n|A^\dagger HA_{p-1}|n\rangle - E_n\langle n|A^\dagger A_{p-1}|n\rangle \\
&= \langle n|A^\dagger HA_{p-1} - HA^\dagger A_{p-1}|n\rangle \\
&= \langle n|[A^\dagger, H]A_{p-1}|n\rangle \\
&= \langle n|[H, A]^\dagger A_{p-1}|n\rangle \\
&= \langle n|A_1^\dagger A_{p-1}|n\rangle
\end{aligned} \tag{3.280}$$

Repeating this procedure m times evidently yields Eq. (3.279). We thus have many formally different expressions for S_n^p:

$$\begin{aligned}
S_n^p &= \hbar^{-p}\langle n|A^\dagger A_p|n\rangle = \hbar^{-p}\langle n|A_1^\dagger A_{p-1}|n\rangle = \cdots \\
&= \hbar^{-p}\langle n|A_{p-1}^\dagger A_1|n\rangle = \hbar^{-p}\langle n|A_p^\dagger A|n\rangle
\end{aligned} \tag{3.281}$$

The most useful expression for practical purposes is the one for which A needs to be commuted with H the least number of times; following Jackiw,[239] we choose

$$S_n^{2p} = \frac{\langle n|A_p^\dagger A_p|n\rangle}{\hbar^{2p}} \tag{3.282}$$

and

$$\begin{aligned}
S_n^{2p+1} &= \langle n|A_p^\dagger A_{p+1}|n\rangle = \langle n|A_{p+1}^\dagger A_p|n\rangle \\
&= \frac{\langle n|\big(A_p^\dagger A_{p+1} + A_{p+1}^\dagger A_p\big)|n\rangle}{2}
\end{aligned} \tag{3.283}$$

If A is Hermitian ($A = A^\dagger$), we have

$$\begin{aligned}
A_1^\dagger &= (HA - AH)^\dagger = AH - HA = -A_1 \\
A_2^\dagger &= (HA_1 - A_1H)^\dagger = A_2
\end{aligned} \tag{3.284}$$

and more generally

$$A_p^\dagger = (-1)^p A_p \tag{3.285}$$

Equations (3.282) and (3.283) thus reduce to

$$S_n^{2p} = \frac{(-1)^p \langle n | (A_p)^2 | n \rangle}{\hbar^{2p}} \tag{3.286}$$

$$S^{2p+1} = \frac{\langle n | \{(-1)^p A_p A_{p+1} + (-1)^{p+1} A_{p+1} A_p\} | n \rangle}{2\hbar^{2p+1}}$$

$$= (-1)^p \frac{\langle n | [A_p, A_{p+1}] | n \rangle}{2\hbar^{2p+1}} \tag{3.287}$$

These sum rules can be considered as HVTs (Swenson,[241] 1970); like the HVT, they may be used as constraints to improve approximate wavefunctions. Their main applications lie in radiation theory, by setting $A = \vec{r}$ or $\exp(i\vec{k}\cdot\vec{r})$ (a phase factor). In the first case, one obtains in particular the Thomas–Reiche–Kuhn sum rule for oscillator strengths, which played an important role in the early development of quantum mechanics. In the second case, one of the relations yields Bethe's generalization of the Thomas–Reiche–Kuhn sum rule, which can be used to calculate the energy loss of charged particles passing through matter (see Jackiw,[239] 1967; Bethe and Jackiw, Ref. 120, Chapter 11). Robinson[242] (1977) generalized these sum rules to velocity-dependent potentials, and Epstein[243] (1973) discussed their validity for variational wavefunctions. Burnel and Caprasse[244] (1980) recently derived similar sum rules to be applied to the quarkonium system.

F. Moments of the Fokker–Planck Equation

As explained in developing the HVT (Section II.A.7), one can generate any number of exact relations by taking the moment of any fundamental equation with an arbitrary dynamical variable.

In this subsection, we illustrate the method of moments by a systematic application to the Fokker–Planck equation; that is, we multiply the Fokker–Planck equation by certain simple dynamical variables and integrate over the whole phase space. We shall thus obtain time-dependent generalizations of the VT and also of the fluctuation–dissipation theorem (Section III.F.4).

The Fokker–Planck equation is a kind of generalized diffusion equation that describes Markov processes and, in particular, the Brownian motion of particles immersed in a fluid. Before introducing the Fokker–Planck equation by sketching its proof (Section III.F.3), we mention a few generalities concerning Brownian motion (Section III.F.1) and its treatment based on the Langevin equation, a less general treatment than the one based on the

Fokker–Planck equation. After presenting the Langevin equation, we shall digress slightly in Section III.F.2 to show how Einstein's relation between the diffusion coefficient and mobility can be derived in a way analogous to the classical proof of the VT (Section II.A.1), the genesis of the method of moments.

1. *Brownian Motion*

Brownian motion was discovered in 1828 by the botanist Robert Brown, who observed that pollen grains dropped in water appear animated by a perpetual irregular movement. It is now well known that this movement results from impacts of the surrounding liquid molecules on a small but macroscopic body suspended in solution (e.g., the pollen grain). Each such impact changes the path of the Brownian particle, but since, for typical liquids, there are about 10^{21} collisions per second, there is no hope of describing the detailed motion. What we see under the microscope is a random statistical behavior. In fact, the Brownian motion itself constitutes a prototype problem, and the stochastic equations that apply to it are valid for other random problems. In particular, the analogous electrical problem provides an analysis of the Johnson noise which limits the signal-to-noise ratio in all electronic instruments.

Historically, Einstein[245] (1905, 1906) and then Langevin[246] (1908) developed the theory of Brownian motion, and Perrin[249] verified experimentally their theoretical predictions (see Section III.F.2). Nevertheless, a deeper understanding of Brownian motion is given by the theory of Markov processes that leads to the Fokker–Planck equation (Section III.F.3).

2. *Langevin Equation*

The Langevin equation partitions the force acting on a Brownian particle into two parts, so that Newton's second law can be written (McQuarrie[83], p. 452)

$$\dot{\vec{v}} = -\beta\vec{v} + \vec{A}(t) + \vec{g} \tag{3.288}$$

where $\vec{v}$ is the velocity of the Brownian particle and $\vec{g}$ is the acceleration due to external forces. The two first terms originate from different influences of the environment:

The quantity $-\beta\vec{v}$ is a dynamical friction or drag term proportional to the velocity; from Stokes's law, for a spherical particle of radius a and mass m in a fluid of viscosity η, the frictional coefficient β is given by

$$\beta = \frac{6\pi\eta a}{m} \tag{3.289}$$

The quantity $\vec{A}(t)$ is a fluctuating or random acceleration (Lifshitz and Pitaevskii, Ref. 247, p. 362) characteristic of the Brownian motion. $\vec{A}(t)$ is assumed to be:

1. Independent of the velocity $\vec{v}$.
2. Rapidly varying compared to $\vec{v}$.

More precisely, we distinguish two very different time scales: first, the period that characterizes the very rapid fluctuations of the small random accelerations $\vec{A}(t)$, namely $\sim 10^{-21}$ second in a typical liquid; and second, the (much longer) period between large changes in the velocity $\vec{v}$ of the Brownian particle. Between these two extremes there exist time intervals Δt, large with respect to the first but small with respect to the second, such that $\vec{v}(t+\Delta t)$ differs very little from $\vec{v}(t)$ although during Δt many (uncorrelated) fluctuations of $\vec{A}(t)$ occur. These assumptions, although plausible, are best justified *a posteriori* by the agreement of their consequences with experiment.

Being a stochastic relation, the Langevin equation cannot be solved as a usual differential equation; rather it requires finding the conditional probability density $\rho(\vec{r},\vec{v},t;\vec{r}_0,\vec{v}_0)$ that gives the probability of the particle having position $\vec{r}$ with velocity $\vec{v}$ at time t, given that it is at $\vec{r}_0,\vec{v}_0$ at time $t=0$. This solution has been given in a classic paper by Chandrasekhar[248] (1943).

We just report here an earlier paper of Langevin[246] (1908), who derived Einstein's relation between the diffusion coefficient and mobility by a proof similar to Clausius' original derivation of the VT (1870; cf. Section II.A.1), that is, by taking the moment of Newton's second law, in this case the Langevin equation, with the position vector of the particle: we obtain from Eq. (3.288) (without external force):

$$\vec{r}\cdot\ddot{\vec{r}} = -\beta\vec{r}\cdot\dot{\vec{r}} + \vec{r}\cdot\vec{A}(t) \tag{3.290}$$

or

$$\frac{1}{2}\frac{d^2(\vec{r}^2)}{dt^2} - \dot{\vec{r}}^2 = -\tfrac{1}{2}\beta\frac{d(\vec{r}^2)}{dt} + \vec{r}\cdot\vec{A}(t) \tag{3.291}$$

Taking the ensemble average (denoted by brackets) of this relation and arguing that

$$\langle \vec{r}\cdot\vec{A}(t)\rangle = 0 \tag{3.292}$$

because of the "irrégularité des actions complémentaires de $\vec{A}$" (in modern language, $\vec{r}$ and $\vec{A}$ are uncorrelated), we have, defining $\alpha \equiv \langle d(\vec{r}^2)/dt\rangle$,

$$\frac{1}{2}\frac{d\alpha}{dt} - \frac{3k\Theta}{m} = -\tfrac{1}{2}\beta\alpha \tag{3.293}$$

where we have used the equipartition theorem (Section II.A.3)

$$\langle \tfrac{1}{2} m \dot{\vec{r}}^2 \rangle = \tfrac{3}{2} k\Theta \tag{3.294}$$

The solution of Eq. (3.293) is

$$\alpha = c e^{-\beta t} + \frac{6k\Theta}{m\beta} \qquad (c \text{ constant}) \tag{3.295}$$

The first term represents a transient term that goes to zero rapidly ($\beta \sim 10^7$ $\sec^{-1}$), so that in steady state, we can write

$$\frac{\langle d\vec{r}^2 \rangle}{dt} = \frac{6k\Theta}{m\beta} \tag{3.296}$$

or

$$\langle \vec{r}^2 \rangle = \frac{6k\Theta}{m\beta} t \tag{3.297}$$

This relation was derived by Einstein in 1905 and its accuracy, was checked experimentally to within 0.5% by Perrin[249] in 1910. Perrin measured $\langle x^2 \rangle$ by following under a microscope the motion of single particles of mastic, recording their positions at equally spaced time intervals; knowing the mass of the Brownian particle and the viscosity and density of the surrounding fluid [β is given by Eq. (3.289)], he calculated a precise value for the Boltzmann constant.

The diffusion equation for a species of concentration c is given by

$$\frac{\partial c}{\partial t} = D \nabla^2 c \tag{3.298}$$

where D is the diffusion coefficient, and leads to the relation

$$\langle \vec{r}^2 \rangle = 6Dt \tag{3.299}$$

By comparison with Eq. (3.297) we obtain for the diffusion coefficient

$$D = \frac{k\Theta}{m\beta} \tag{3.300}$$

With the definition of the mobility μ as

$$\mu = \frac{1}{m\beta} \tag{3.301}$$

[from Eq. (3.289), the mobility is the drift velocity per unit force], we finally get what is often called the Stokes–Einstein relation

$$D = k\Theta\mu \tag{3.302}$$

The results in Eqs. (3.297) and (3.302) are the simpler forms of an important theorem of fluctuation theory, the fluctuation–dissipation theorem; we shall see in Section III.F.4 below that the time-dependent generalization of the fluctuation–dissipation theorem can be deduced from the Fokker–Planck equation, by taking the moment of this equation with the square of the momentum. We now outline the derivation of the Fokker–Planck equation.

3. *Sketch of the Derivation of the Fokker–Planck Equation*

The basis of the proof of the Fokker–Planck equation lies in the fundamental equation of the theory of Markov processes, the Chapman–Kolmogorov equation (McQuarrie, Ref. 83, p. 456; Chandrasekhar,[248] 1943). This equation embodies the assumption that the behavior of a system at a certain time depends only on its instantaneous state at that time and not on its past history (in probabilities, a Markov chain is a series of trials where the outcome of a trial depends uniquely on the present state of this trial). The knowledge of the conditional probability density function $\rho(t)$ is enough to determine $\rho(t+\Delta t)$.

The Chapman–Kolmogorov equation reads

$$\rho(\vec{r},\vec{v},t+\Delta t) = \iint \rho(\vec{r}-\Delta\vec{r},\vec{v}-\Delta\vec{v},t)\,\psi(\vec{r}-\Delta\vec{r},\vec{v}-\Delta\vec{v};\Delta\vec{r},\Delta\vec{v})\,d^3\Delta\vec{r}\,d^3\Delta\vec{v} \tag{3.303}$$

where $\psi(\vec{r},\vec{v};\Delta\vec{r},\Delta\vec{v})$ is the transition probability, that is, the probability that $\vec{r}$ and $\vec{v}$ undergo in time Δt small changes from $\vec{r}-\Delta\vec{r},\vec{v}-\Delta\vec{v}$ to $\vec{r},\vec{v}$. We have

$$\Delta\vec{r} = \vec{v}\Delta t \tag{3.304}$$

and, using the Langevin equation (3.288),

$$\dot{\vec{v}} = -\beta\vec{v} + \vec{A}(t) + \vec{g} \tag{3.305}$$

$$\dot{\vec{v}}\Delta t = \Delta\vec{v} = -(\beta\vec{v}-\vec{g})\,\Delta t + \vec{A}\,\Delta t \tag{3.306}$$

so that we can write

$$\psi(\vec{r},\vec{v};\Delta\vec{r},\Delta\vec{v}) = \psi(\vec{r},\vec{v};\Delta\vec{v})\,\delta(\Delta\vec{r}-\vec{v}\Delta t) \tag{3.307}$$

which allows us to integrate immediately in Eq. (3.303) over $\Delta\vec{r}$.

The various functions in the Chapman–Kolmogorov equation are then expanded in Taylor series in $\Delta\vec{v}$ and $\Delta\vec{r}$, keeping only terms to second-order, and the solutions of the Langevin equation are finally used to calculate the various moments; the Fokker–Planck equation thus reads (McQuarrie, Ref. 83, Eq. 20-33; Chandrasekhar, 1943, Ref. 248, Eq. 249)

$$\frac{\partial \rho}{\partial t} + \vec{v}\cdot\vec{\nabla}\rho + \vec{g}\cdot\vec{\nabla}_{\vec{v}}\rho - \beta\vec{\nabla}_{\vec{v}}\cdot(\rho\vec{v}) - \frac{\beta k\Theta}{m}\vec{\nabla}^2_{\vec{v}}\rho = 0 \tag{3.308}$$

The friction coefficient β, Eq. (3.289), is related to the diffusion constant D by Eq. (3.302), but it will be more convenient for us to define a new diffusion constant d as

$$d \equiv \beta k\Theta m \tag{3.309}$$

In addition, to simplify future calculation we shall restrict the discussion to a phase space of only two dimensions (q, p) (the generalization to several dimensions is straightforward). The Fokker–Planck equation thus governs the probability density $\rho(q, p; t)$, and reads

$$\frac{\partial \rho}{\partial t} + \frac{p}{m}\frac{\partial \rho}{\partial q} + \frac{\partial[\rho(-\beta p + F)]}{\partial p} - \frac{\partial^2 \rho}{\partial p^2} d = 0 \tag{3.310}$$

where now

$$F = mg \tag{3.311}$$

4. *Moments of the Fokker–Planck Equation*

Following de la Peña[250] (1980), we multiply Eq. (3.311) by an integrable but otherwise arbitrary function $f(q, p)$, then integrate over the whole phase space to get (after some simple integration by parts)

$$\frac{d\langle f\rangle}{dt} = \left\langle \frac{p}{m}\frac{\partial f}{\partial q}\right\rangle + \left\langle(-\beta p + F)\frac{\partial f}{\partial p}\right\rangle + \left\langle \frac{\partial^2 f}{\partial p^2}\right\rangle d \tag{3.312}$$

with the usual definition

$$\langle f\rangle = \int\!\!\int dp\, dq f\rho \tag{3.313}$$

Eq. (3.312) is the equation of motion that gives the time evolution of the average value of any dynamical variable $f(q, p)$; all of our results will be deduced from it.

We substitute successively $f = q,\ p,\ q^2,\ pq$, and p^2 in Eq. (3.312):

$$f = q: \qquad \frac{d\langle q\rangle}{dt} = \frac{\langle p\rangle}{m} \tag{3.314}$$

$$f = p: \qquad \frac{d\langle p\rangle}{dt} = \langle -\beta p + F\rangle \tag{3.315}$$

$$f = q^2: \qquad \frac{d\langle q^2\rangle}{dt} = \frac{2\langle qp\rangle}{m} \tag{3.316}$$

$$f = qp: \qquad \frac{d\langle qp\rangle}{dt} = \frac{\langle p^2\rangle}{m} - \beta\langle qp\rangle + \langle qF\rangle \tag{3.317}$$

$$f = p^2: \qquad \frac{d\langle p^2\rangle}{dt} = 2d - 2\beta\langle p^2\rangle + 2\langle pF\rangle \tag{3.318}$$

Equations (3.314) and (3.315) are just the average of the Langevin equation (3.288), rewritten in the form

$$m\ddot{q} = \dot{p} = -\beta p + mA + F \tag{3.319}$$

when one recalls that the average is made during a time large enough so that the average of the fluctuating force vanishes.

To see the meaning of Eqs. (3.316), (3.317), and (3.318) we first examine equilibrium states where all averages are independent of time, i.e., $d\langle f\rangle/dt = 0$. Equations (3.314), (3.315), and (3.316) yield respectively

$$\langle p\rangle = 0 \tag{3.320}$$

$$\langle F\rangle = 0 \tag{3.321}$$

$$\langle qp\rangle = 0 \tag{3.322}$$

This last equation means that q and p are uncorrelated; carrying this result into Eq. (3.317) gives

$$\frac{\langle p^2\rangle}{2m} = \tfrac{1}{2}\langle qF\rangle \tag{3.323}$$

which we recognize as the scalar VT. Further, since q and p are uncorrelated, it also follows that

$$\langle pF \rangle = 0 \tag{3.324}$$

and from Eq. (3.318)

$$d = \beta \langle p^2 \rangle \tag{3.325}$$

Recalling the definition of d, Eq. (3.309), we see that Eq. (3.325) is similar to Einstein's formula, Eq. (3.302), the first known form of the fluctuation–dissipation theorem.

Going back into the time-dependent case, we can rewrite Eq. (3.317) as

$$\frac{2\langle p^2 \rangle}{2m} = -\langle qF \rangle + \beta \langle qp \rangle + \frac{d\langle qp \rangle}{dt} \tag{3.326}$$

which is the dynamic VT, or the equation of motion for qp. The two first terms on the right are the usual virials; the difference from the time-independent case comes from the correlation between q and p. Incidentally, by going over to three dimensions, we get immediately the tensor form of the VT (cf. Section II.A.5):

$$m^{-1}\langle p_k p_l \rangle = -\langle q_k F_l \rangle + \beta \langle q_k p_l \rangle + \frac{d\langle q_k p_l \rangle}{dt} \tag{3.327}$$

Its trace is the scalar VT, Eq. (3.317), and its antisymmetric part is the equation of motion for the angular momentum.

Still further, from Eq. (3.318) we obtain the equation

$$d = \beta \langle p^2 \rangle - \langle pF \rangle + \frac{1}{2} \frac{d\langle p^2 \rangle}{dt} \tag{3.328}$$

which generalizes the fluctuation–dissipation theorem to the time-dependent case. We see that corrections come from both the correlation between p and F and the time dependence of $\langle p^2 \rangle$. It is interesting to observe (de la Peña,[250] 1980) that the fluctuation–dissipation theorem in the form given above may easily be derived by studying the evolution of the average energy of the "classical" system, defined by means of the Hamiltonian $H_c = p^2/2m + U(q)$, where U is the potential associated with the external force $F(q)$, that is, neglecting the friction and the stochastic forces. Using Eq. (3.312), one gets

$$\frac{d\langle H_c \rangle}{dt} = \frac{d}{m} - \frac{\beta \langle p^2 \rangle}{m} \tag{3.329}$$

At equilibrium $d\langle H_c\rangle/dt = 0$ and we recover the fluctuation–dissipation theorem in the form of Eq. (3.325). Equation (3.329) is clearly equivalent to Eq. (3.328); taken together they imply

$$\frac{d\langle U\rangle}{dt} = -\frac{\langle pF\rangle}{m} \tag{3.330}$$

Hence we see that whereas $\langle qF\rangle$ determines the average of the rate of change of the kinetic energy, $\langle pF\rangle$ determines the average of the rate of change of the potential energy.

A more general discussion of stochastic fields and the VT is given by Moore[251] (1980).

In conclusion, we stress again that the method of moments is by no means restricted to the use of variables q and p, nor to the Fokker–Planck equation. The importance of this method has been emphasized by S. Chandrasekhar,[18,252] and its applications are legion. We quote from Chandrasekhar's Lectures in Theoretical Physics:[252]

> A standard technique for treating the integro-differential equations of mathematical physics is to take the moments of the equations concerned and consider suitably truncated sets of the resulting equations. An advantage in considering such moment equations is that the equations of the lowest order have, often, simple physical interpretations; and, moreover, in many instances, their solutions, with suitable assumptions of "closure," suggest methods of obtaining approximate solutions of the exact equations in a systematic way.

Acknowledgments

G.M. wishes to express his appreciation to the Department of Chemistry, University of California at Los Angeles, for award of a Chancellor's Fellowship.

We have benefitted from numerous discussions with W. Douglas Harris and Paul Pau.

We are grateful to Miss Anne Birmingham for assistance with the bibliography, to Ms. Madelon Lopez for typing some early drafts, and to Mmes. Susan Rodrigues and Nancy McMillan for care and patience in preparing the manuscript.

APPENDIX

In this Appendix, detailed references are given to some other works on the VT, classified under the headings Mathematics; Relativistic Quantum Mechanics; Scattering; Low Temperatures; Surfaces; and Astrophysics, General Relativity, and Magnetohydrodynamics (these three last subjects are closely interconnected).

Mathematics

S. Albeverio, "On bound states in the continuum of N-body systems and the VT," *Ann. of Phys.* **71**, 167–276 (1972).

H. Kalf, "Non-existence of eigenvalues of Schrödinger operators," *Proc. Roy. Soc. Edinburgh* **79A**, 157–172 (1977).

L. Vasquez, "Classical nonlinear scalar fields," *Hadronic J.* **2**, 917–954 (1979).

T. Weidmann, "The VT and its application to the spectral theory of Schrödinger operators," *Bull. Amer. Math. Soc.* **73**, 452–456 (1967).

C. Von Westenholz, "Scale symmetry and VT," *Ann. Inst. Henri Poincaré, Sect. A (France)* **29**, 415–422 (1978).

Relativistic Quantum Mechanics

J. N. Bahcall, "VT for many-electron Dirac systems," *Phys. Rev.* **124**, 923–924 (1961).

M. Brack, "VT for relativistic spin-$\frac{1}{2}$ and spin-0 particles," *Phys. Rev. D* **27**, 1950–1953 (1983).

S. Gupta, "Equation for moments and VT in Dirac wave mechanics," *Z. Physik* **68**, 573–576 (1931); "Angular momentum and virial equations of the Dirac electron," *Indian Phys. Math. J.* **3**, 105–113 (1932).

N. H. March, "The VT for Dirac's equation," *Phys. Rev.* **92**, 481–482 (1953).

Y. V. Novozhilov, "Variational principle and VT for the continuous spectrum of the Dirac equation," *J. Exp. Theoret. Phys. USSR* **31**, 928–30 (1956); "Scale transformation and VT in quantum field theory, *ibid*. **32**, 171–173 (1957).

A. E. Oganyan, "The VT and the spectral shift function for the Dirac operator," *Dokl. Akad. Nauk. SSSR* **239**, 569–572 (1978); transl., *Sov. Phys. Dokl.* **23**, 185–187 (1978).

J. Rafelski, "VT and stability of localized solutions of relativistic classical interacting fields," *Phys. Rev. D* **16**, 1890–1899 (1977).

M. E. Rose and T. A. Welton, "The VT for a Dirac Particle," *Phys. Rev.* **86**, 432–433 (1952).

G. Rosen, "Existence of particle-like solutions to nonlinear field theory," *J. Math. Phys.* **7**, 2066–2070 (1966).

N. R. Sen, "Equations of the theory of electrons and Dirac wave mechanics," *Physik* **66**, 122–128 (1930).

Scattering

Y. N. Demkov, *Variational Principle in the Theory of Collisions*, Pergamon, 1963, and references therein.

M. Heaton and B. L. Moiseiwitsch, "VT for variational functions in collision theory," *J. Phys. B* **4**, 332–342 (1971).

J. O. Hirschfelder, "Similarity of Wigner's delay time to the VT for scattering by a central field," *Phys. Rev A* **19**, 2463 (1979).

M. I. Lomonosov, A. E. Oganyan, and K. P. Cherkavasova, "Variational principles and the VT in quantum scattering theory," *Teor. Mat. Fiz. USSR* **31**, 62–74 (1977); *Transl. Theor. Math. Phys.* **31**, 319–28 (1977).

B. L. Moiseiwitsch, "Recent progress in atomic collision theory," *Rep. Prog. Phys.* **40**, 843–904 (1977).

J. D. G. McWhirter and B. L. Moiseiwitsch, "Generalizations of the VT in collision theory," *J. Phys. B* **5**, 2439–2455 (1972); "The VT for elastic scattering of electrons and positrons by helium," *ibid.* **6**, 2489–2494 (1973); "VT and Kohn's variational principle for scattering amplitude," *ibid.* **7**, 229–235 (1974); "VT for scattering amplitude," *Acta Phys. Acad. Sci. Hungary* **35**, 37–45 (1974).

Y. V. Novozhilov: See under "Relativistic Quantum Mechanics" above.

R. E. Oganyan: See under "Relativistic Quantum Mechanics" above.

R. T. Pack and J. O. Hirschfelder, "VT for inelastic molecular collisions; atom–rigid rotor scattering," *J. Chem. Phys.* **73**, 3823–3830 (1980).

R. J. Weiss, "Relationship between the electronic potential energy of a crystal and X-ray scattering," *Phil. Mag. B* **38**, 289–293 (1978).

R. Yaris and P. Winkler, "Electron scattering resonances using dilatation transformations. I—Condition for dilatational stability; II—Restricted perturbation methods [P.W.]; III—Variational study of the stability of resonance eigenvalues [P.W. and R.Y.], *J. Phys. B* **11**, 1475–1480, 1481–1495, 4257–4270 (1978).

Low Temperatures

M. Rasetti and T. Regge, "Coherent states and VT for liquid helium," *J. Low Temp. Phys.* **13**, 249–259 (1973).

Surfaces

A. B. Bathia and N. H. March, "Statistical Mechanical Theory of Liquid Surface Tension," *Phys. and Chem. Liq.* **9**, 1–10 (1979).

S. Chandrasekhar, "The stability of a rotating liquid drop," *Proc. Roy. Soc. London A* **286**, 1–26 (1965).

C. E. Rosenkilde, "Surface-energy tensors," *J. Math. Phys.* **8**, 84–88; "Surface-energy tensors for ellipsoids," *ibid.* **8**, 88–97; "Stability of axisymmetric figures of equilibrium of a rotating charged liquid drop," *ibid.* **8**, 98–118 (1967).

Astrophysics, General Relativity, and Magnetohydrodynamics

J. Binney, "On the rotation of elliptical galaxies," *Mon. Not. Roy. Astron. Soc.* **183**, 501–514 (1978).

S. Bonazzola, "The VT in general relativity," *Astrophys. J.* **182**, 335–340 (1973).

S. Chandrasekhar, *Ellipsoidal Figures of Equilibrium*, Yale U. P., 1969 (Chandrasekhar gives a few selected references on the VT at the end of this book; we do not reproduce them here, and most of our references are within the last 10 years).

M. Davis, M. J. Geller, and J. Huchra, "The local mean mass density of the universe; new methods for studying galaxy clustering," *Astrophys. J.* **221**, 1–18 (1978).

A. S. Eddington, "The kinetic energy of a star cluster," *Monthly Not. Astron. Soc. London* **76**, 525–528 (1916).

A. Di Fazio and F. Occhionero, "Dynamical evolution of elliptical protogalaxies," *Astron. and Astroph. Germany* **62**, 349–354 (1978).

F. Ferrini, "Thermodynamics of gravitating fermions. I. Thermodynamic potentials," *Astrophys. and Space Sci. Netherlands* **62**, 263–277 (1979).

M. J. Geller and M. Davis, "On application of statistical VT," *Astrophys. J.* **225**, 1–6 (1978).

A. Georgiou, "A VT for general relativistic charged fluids," *J. Phys. A* **13**, 3751–3759 (1980).

E. A. Godfredsen, "Dynamical stability of the local group," *Astrophys. J.* **134**, 69–96 (1961).

M. V. Goldman and D. R. Nicholson, "VT of direct Langmuir collapse," *Phys. Rev. Lett.* **41**, 406–410 (1978).

J. Gott III, "*N*-body simulations and the value of omega," in *Large Scale Structure of the Universe*, IAU Symp. #79, 1978, pp. 63–70.

F. D. A. Hartwick, "The virial mass discrepancy in groups and clusters of galaxies," *Astrophys. J.* **219**, 345–351 (1978).

D. Layzer, "On the gravitation of self-gravitating clouds," *Astrophys. J.* **137**, 351–362 (1963).

F. R. Klinkhamer, "Scale covariant gravitation; virial masses of groups of galaxies," *Astron. and Astrophys. Germany* **87**, 354–356 (1980).

D. Koester, "VT, energy content and mass–radius relation for white dwarfs," *Astron. and Astrophys. Germany* **64**, 289–294 (1978).

D. Lynden-Bell and R. M. Lynden-Bell, "On the negative specific heat paradox," *Month. Not. Roy. Astron. Soc. London* **181**, 405–419 (1977).

B. J. McNamara and W. L. Sanders, "An upper limit to the mass and velocity dispersion of M67", *Astron. and Astrophys. Germany* **62**, 259–260 (1978).

J. E. Palmore, "Relative equilibria and the VT," *Celestial Mech. Netherlands* **19**, 167–171 (1979).

J. A. Rose, "The dynamical nature of compact groups of galaxies," *Astrophys. J.* **231**, 10–22 (1979).

H. Smith, Jr., "Application of the VT to clusters of galaxies with unseen mass," *Astrophys. J.* **241**, 63–66 (1980).

N. Spyrou, "Tensor-virial equations for post-newtonian relativistic stellar dynamics," *Gen. Relat, and Gravit.* **8**, 463–489 (1977).

V. Yu. Terebizh, "Estimation of the virial mass for systems of galaxies," *Astron. Zh. SSSR* **56**, 258–262 (1979); transl., *Sov. Astron.* **23**, 141–143 (1979).

C. Vilain, "VT in general relativity; consequences for stability of spherical symmetry," *Astrophys. J.* **227**, 307–318 (1979).

J. Weidenschilling, "A constraint on pre-main-sequence mass loss," *Moon and Planets (Netherlands)* **19**, 279–287 (1978).

P. S. Wesson, "Does the binding energy of binaries masquerade as missing mass," *Astron. and Astrophys. Germany* **90**, 1–7 (1980).

J. M. Whittaker, "Dynamics with variable masses," *Proc. Roy. Soc. London A* **372**, 485–487 (1980).

REFERENCES

1. A. Ore, *Fysik. Verden, Fra.* **12**, 191 (1950); **13**, 90 (1951); *Nature* **167**, 402, 728 (1951).
2. R. M. Schectman and R. H. Good, *Am. J. Phys.* **25**, 219 (1957).
3. J. O. Hirschfelder, in *Progr. Inst. Res. Thermodyn. Transport Properties 2nd Symp. Thermophys. Properties*, Princeton, N.J., 1962, p. 386.
4. J. Killingbeck, *Am. J. Phys.* **38**, 590 (1970).
5. E. Weislinger and G. Oliver, in *Inst. J. Quant. Chem.*, VIIIS, 389 (1974).
6. R. J. E. Clausius, *Phil. Mag.* **40**, 122 (1870); *C.R. Acad. Sci. Paris* **70**, 1314 (1870); *Ann. Phys.* **141**, 124 (1870).
7. E. A. Milne, *Phil. Mag.* **50**, 409 (1925).
8. R. Courant and D. Hilbert, *Methods of Mathematical Physics*, Vol. II, Interscience, 1962, p. 11.
9. J. R. Murdoch, *J. Am. Chem. Soc.* **104**, 588 (1982).
10. J. C. Maxwell, *Trans. Roy. Soc. Edin.* **xxvi**, 14 (1870).
11. J. C. Maxwell, *Nature* **x**, 477 (1874).
12. Lord Rayleigh, *Phil. Mag.* **50**, 210 (1900); **9**, 494 (1905); **10**, 364 (1905).
13. E. N. Parker, *Phys. Rev.* **96**, 1686 (1954).

14. W. D. MacMillan, *Dynamics of Rigid Bodies*, McGraw-Hill, 1936; Dover reprint, 1970.
15. H. Poincaré, *Hypothèses Cosmogoniques*, Herman, Paris, 1911, p. 90.
16. A. S. Eddington, *Monthly Not. Roy. Astron. Soc. London* **76**, 525 (1916).
17. J. O. Hirschfelder, *J. Chem. Phys.* **33**, 1462 (1960).
18. S. Chandrasekhar, *Ellipsoidal Figures of Equilibrium*, Yale U. P., 1969.
19. J. Schwinger, Unpublished Lecture Notes on Classical Electrodynamics, 1982, p. 23.
20. J. D. Jackson, *Classical Electrodynamics*, 2nd ed., Wiley, 1975, p. 265.
21. V. Fock and G. Krutkov, *Phys. Z. Sowjet.* **1**, 756 (1932).
22. P. Kleban, *Am. J. Phys.* **47**, 883 (1979).
23. R. J. Finkelstein, *Nonrelativistic Mechanics*, Benjamin, 1973, Section 3-4.
24. H. Goldstein, *Classical Mechanics*, 2nd ed., Addison-Wesley, 1980, Chapter 9.
25. E. Merzbacher, *Quantum Mechanics*, 2nd ed., Wiley, 1970, p. 349.
26. V. Fock, *Z. Phys.* **63**, 855 (1930).
27. Y. N. Demkov, *Variational Principles in the Theory of Collisions*, Pergamon, 1963.
28. P.-O. Löwdin, *J. Molec. Spect.* **3**, 46 (1959).
29. L. D. Landau and E. M. Lifshitz, *Nonrelativistic Quantum Mechanics*, 3rd revised English ed., Pergamon, 1977, p. 38.
30. P. Ehrenfest, *Z. Phys.* **45**, 455 (1927).
31. J. P. Vinti, *Phys. Rev.* **58**, 882 (1940).
32. L. Cohen, *J. Math. Phys.* **19**, 1838 (1978).
33. L. I. Schiff, *Quantum Mechanics*, 3rd ed., McGraw-Hill, 1968, p. 180.
34. E. Merzbacher, *Quantum Mechanics*, 2nd ed., Wiley, 1970, p. 168.
35. I. N. Levine, *Quantum Chemistry*, 2nd ed., Allyn and Bacon, 1974, Chapter 14.
36. W. J. Carr, *Phys. Rev.* **106**, 414 (1957).
37. J. O. Hirschfelder and C. A. Coulson, *J. Chem. Phys.* 36, 941 (1962).
38. S. T. Epstein and J. O. Hirschfelder, *Phys. Rev.* **123**, 1945 (1961).
39. A. D. McLean, *J. Chem. Phys.* **40**, 2774 (1964).
40. J. H. Epstein and S. T. Epstein, *Am. J. Phys.* **30**, 266 (1962).
41. M. Born, W. Heisenberg, and P. Jordan, *Z. Phys.* **35**, 557 (1925); reprinted in Sources of Quantum Mechanics, B. L. van der Waerden (ed.), Dover, 1968, p. 343.
42. B. N. Finkelstein, *Z. Phys.* **50**, 293 (1928).
43. J. C. Slater, *J. Chem. Phys.* **1**, 687 (1933).
44. T. L. Cottrell and S. Paterson, *Phil. Mag.* **42**, 391 (1951).
45. S. Srebrenik, R. F. W. Bader, and T. Tung Nguyen-Dang, *J. Chem. Phys.* **68**, 3667 (1978).
46. W. Pauli, in *Handbuch der Physik*, Springer, Berlin, 1953.
47. J. I. Musher, *Am. J. Phys.* **34**, 267 (1966).
48. H. Hellmann, *Einführung in die Quantenchemie*, Deuticke, Vienna, 1937.
49. R. P. Feynman, *Phys. Rev.* **56**, 340 (1939).
50. B. M. Deb (ed.), *The Force Concept in Chemistry*, Van Nostrand Reinhold, 1981; see also B. M. Deb, *Rev. Mod. Phys.* **45**, 22 (1973).
51. W. A. McKinley, *Am. J. Phys.* **39**, 905 (1971).
52. S. T. Epstein, *J. Chem. Phys.* **42**, 3813 (1965).

53. P. N. Argyres, *Int. J. Quant. Chem.* **1S**, 669 (1967).
54. J. de Boer, *Physica* **15**, 843 (1949).
55. N. H. March, *Phys. Rev.* **110**, 604 (1958).
56. H. Kanazawa, *J. Phys. Soc. Jap.* **23**, 476 (1967).
57. P. Ziesche, *J. Phys. C* **13**, 3625 (1980).
58. J. Vannimenus and H. F. Budd, *Phys. Rev. B* **15**, 5302 (1977).
59. J. Heinrichs, *Phys. Status Solidi B (Germany)* **92**, 185 (1979).
60. J. E. van Himbergen and R. Silbey, *Phys. Rev. B* **20**, 567 (1979).
61. H. Minkowski, in *The Principle of Relativity*, A. Einstein et al., Dover, 1951.
62. H. Goldstein, *Classical Mechanics*, 2nd ed., Addison-Wesley, 1980, p. 338.
63. L. D. Landau and E. M. Lifshitz, *The Classical Theory of Fields*, 4th revised English ed., Pergamon, 1979, p. 84; 1st ed., Moscow, 1941.
64. V. D. Shafranov, *Sov. Phys. Usp.* **22**, 368 (1979).
65. W. Pauli, *Z. Phys.* **43**, 601 (1927).
66. S. Gupta, *Indian Phys. Math. J.* **3**, 105 (1932).
67. M. E. Rose and T. A. Welton, *Phys. Rev.* **86**, 432 (1952).
68. N. H. March, *Phys. Rev.* **92**, 481 (1953).
69. M. Brack, *Phys. Rev. D* **27**, 1950 (1983).
70. V. F. Weisskopf, *Science* **187**, 605 (1975).
71. M. Kregar and V. F. Weisskopf, *Am. J. Phys.* **50**, 213 (1982).
72. F. Juttner, *Z. Physik* **47**, 542 (1928).
73. V. Danilow, *Z. Physik* **63**, 692 (1930).
74. J. Frenkel, *Z. Physik* **29**, 214 (1924); **50**, 234 (1928).
75. A. Fröman and P.-O. Löwdin, *J. Phys. Chem. Solids* **23**, 75 (1962).
76. A. Rothwarf, *Phys. Lett.* **29A**, 54 (1969).
77. E. A. Mason and S. T. Spurling, *The Virial Equation of State*, Pergamon, 1969.
78. J. E. Mayer, *J. Chem. Phys.* **18**, 1426 (1950).
79. W. G. McMillan and J. E. Mayer, *J. Chem. Phys.* **13**, 276 (1945).
80. T. L. Hill, *J. Chem. Phys.* **30**, 93 (1959).
81. M. Thiensen, *Ann. Phys.* **24**, 467 (1885).
82. H. Kammerlingh-Onnes, *Comm. Phys. Lab. Leiden*, Nos. 71, 74 (1901).
83. D. A. McQuarrie, *Statistical Mechanics*, Harper and Row, 1976.
84. S. Ono, *J. Chem. Phys.* **19**, 504 (1951).
85. J. E. Kilpatrick, *J. Chem. Phys.* **21**, 274 (1953).
86. J. G. Kirkwood, *Phys. Rev.* **44**, 31; **45**, 116 (1933).
87. H. S. Green, *Physica* **15**, 882 (1949); *J. Chem. Phys.* **18**, 1123 (1950).
88. M. Born and H. S. Green, *Proc. Roy. Soc. A* **143**, 168 (1947).
89. J. de Boer, Physica **15**, 843 (1949).
90. J. Yvon, *C.R. Acad. Sci. Paris* **227**, 763 (1948).
91. R. J. Riddell and G. E. Uhlenbeck, *J. Chem. Phys.* **18**, 1066 (1950).
92. R. W. Zwanzig, *J. Chem. Phys.* **18**, 1412 (1950).
93. W. B. Brown, *J. Chem. Phys.* **28**, 522 (1958).

94. E. M. Lifshitz and L. D. Pitaevski, *Statistical Physics*, 3rd revised and enlarged ed., Part I, Pergamon, 1980.

95. Don C. Kelly, *Am. J. Phys.* **31**, 827 (1963).

96. P. Debye and E. Hückel, *Phys. Z.* **24**, 185 (1923).

97. J. G. Kirkwood and J. Poirier, *J. Phys. Chem.* **58**, 591 (1954).

98. P. Gross and O. Halpern, *Physik Z.* **26**, 403 (1925).

99. E. O. Adams, *J. Am. Chem. Soc.* **48**, 621 (1926).

100. P. van Rysselberghe, *J. Chem. Phys.* **1**, 205 (1933).

101. R. M. Fuoss, *J. Chem. Phys.* **2**, 818 (1934).

102. B. N. Finkelstein, *Proc. Camb. Phil. Soc.* **31**, 281 (1935).

103. S. R. Milner, *Phil. Mag.* **23**, 551 (1912).

104. R. M. May, *Phys. Lett.* **25A**, 282 (1967).

105. G. Knorr, *Am. J. Phys.* **38**, 433 (1970).

106. J. O. Hirschfelder, C. F. Curtiss, and R. B. Bird, *Molecular Theory of Gases and Liquids*, Wiley, New York, 1954, p. 254.

107. D. H. Tsai, *J. Chem. Phys.* **70**, 1375 (1979).

108. M. Ross, *Phys. Rev. A* **179**, 612 (1969).

109. D. A. Liberman, *Phys. Rev. B* **3**, 2081 (1971).

110. J. F. Janak, *Phys. Rev. B* **9**, 3985 (1974).

111. R. M. More, *Phys. Rev. A* **19**, 1234 (1979).

112. L. J. Bartolotti and R. G. Parr, *J. Chem. Phys.* **72**, 1593 (1980).

113. D. G. Pettifor, *J. Phys. F* **1**, L62 (1971).

114. D. G. Pettifor, *J. Phys. F* **7**, 613; **8**, 219 (1977).

115. Y. Kakehashi, *J. Phys. Soc. Japan* **49**, 28, 1790, 2421 (1980).

116. W. G. McMillan and A. L. Latter, *J. Chem. Phys.* **29**, 15 (1958).

117. L. M. Libby and W. F. Libby, *Proc. Nat. Acad. Sci. U.S.A.* **69**, 3305 (1972).

118. H. Reiss et al., *J. Chem. Phys.* **34**, 2001; **35**, 820 (1961); **36**, 144 (1962).

119. C. S. Allen, *Astrophysical Quantities*, reprinted, Univ. of London, Athlone Press, 1964, p. 103.

120. H. A. Bethe and R. Jackiw, *Intermediate Quantum Mechanics* 2nd revised ed., Benjamin/Cummings, 1980.

121. D. R. Hartree, *Proc. Cambridge Phil. Soc.* **24**, 89, 111 (1928).

122. J. C. Slater, Phys. Rev. **35**, 210 (1930).

123. V. Fock, *Z. Physik* **61**, 126 (1930).

124. T. H. Koopmans, *Physica* **1**, 104 (1933).

125. J. C. Slater, *Phys. Rev.* **81**, 385 (1951).

126. D. R. Hartree, *The Calculation of Atomic Structures*, Wiley, New York, 1957.

127. W. Kohn and L. J. Sham, *Phys. Rev.* **140**, 1133 (1965).

128. R. Gaspar, *Acta Phys. Acad. Sci. Hung.* **3**, 263 (1954).

129. R. D. Cowan, A. C. Larson, D. Liberman, J. B. Mann, and J. Waber, *Phys. Rev.* **144**, 5 (1966).

130. B. Y. Tong and L. J. Sham. *Phys. Rev.* **144**, 1 (1966).

131. F. Herman and S. Skillman, *Atomic Structure Calculations*, Prentice-Hall, 1963.

132. R. D. Cowan, *Phys. Rev.* **163**, 54 (1967).
133. D. C. Griffin, R. D. Cowan, and K. L. Andrew, *Phys. Rev. A* **3**, 1233 (1971).
134. D. A. Liberman, *Phys. Rev.* **171**, 1 (1968); **B2**, 244 (1970).
135. E. L. Hill, *Phys. Rev.* **51**, 370 (1937).
136. M. Ross, *Phys. Rev.* **179**, 612 (1969).
137. L. J. Sham, *Phys. Rev. A* **1**, 969 (1970).
138. B. W. Holland, *Phys. Lett.* **79A**, 437 (1980).
139. M. Berrondo and O. Goscinski, *Phys. Rev.* **184**, 10 (1969).
140. J. C. Slater and J. H. Wood, *Int. J. Quant. Chem.* **S4**, 3 (1971).
141. J. C. Slater, *J. Chem. Phys.* **57**, 2389 (1972).
142. M. S. Gopinathan and M. A. Whitehead, *J. Chem. Phys.* **65**, 196 (1976).
143. D. A. Liberman, *Phys. Rev. B* **3**, 2081 (1971).
144. J. Ihm, A. Zunger, and M. L. Cohen, *J. Phys. C* **12**, 4409 (1979).
145. L. Szasz, *Z. Naturforsch.* **35A**, 628 (1980).
146. L. H. Thomas, *Proc. Cambridge Phil. Soc.* **23**, 542 (1927).
147. E. Fermi, *Z. Physik* **48**, 73; **49**, 550 (1928).
148. P. A. M. Dirac, *Proc. Cambridge Phil. Soc.* **26**, 376 (1930).
149. E. Lieb and B. Simon, *Phys. Rev. Lett.* **31**, 681 (1973); *Adv. Math.* **23**, 22 (1977); E. Lieb, *Rev. Mod. Phys.* **48**, 553 (1976).
150. E. U. Condon and H. Odabasi, *Atomic Structure*, Cambridge U.P., 1980.
151. J. C. Slater and H. M. Krutter, *Phys. Rev.* **47**, 559 (1935).
152. R. P. Feynman, N. C. Metropolis, and E. Teller, *Phys. Rev.* **75**, 1561 (1949).
153. P. Gombás, *Die Statistische Theorie des Atoms und Ihre Anwendungen*, Springer, Vienna, 1949.
154. C. F. von Weizsäcker, *Z. Physik* **96**, 431 (1935).
155. J. S. Plaskett, *Proc. Phys. Soc. London* **66A**, 178 (1953).
156. P. Gombás and K. Ladânyi, *Z. Physik* **158**, 261 (1960).
157. E. Teller, *Rev. Mod. Phys.* **34**, 627 (1962).
158. J. W. Sheldon, *Phys. Rev.* **99**, 1291 (1955).
159. M. K. Brachman, *Phys. Rev.* **84**, 1263 (1951).
160. R. D. Cowan and J. Ashkin, *Phys. Rev.* **105**, 144 (1957).
161. R. Latter, *J. Chem. Phys.* **24**, 280 (1956); *Phys. Rev.* **99**, 1854 (1955).
162. R. Latter, *Phys. Rev.* **99**, 510 (1955).
163. R. G. Parr, R. H. Donelly, M. Levy, and W. E. Palke, *J. Chem. Phys.* **68**, 3801 (1978).
164. S. Flügge, *Practical Quantum Mechanics*, Springer, 1974, Problem 174.
165. L. D. Landau and E. M. Lifshitz, *Quantum Mechanics* (*Non-relativistic Theory*), 3rd revised ed., Pergamon, 1977, p. 262.
166. L. A. Young, *Phys. Rev.* **34**, 1226 (1929).
167. A. J. Allard, *J. Phys. et Rad.* **9**, 225 (1948).
168. L. L. Foldy, *Phys. Rev.* **83**, 397 (1951).
169. J. M. C. Scott, *Phil. Mag.* **7, 43**, 859 (1952).
170. J. Schwinger, *Phys. Rev. A* **22**, 1827 (1980); **24**, 2353 (1981).

171. G. I. Plindov and I. K. Dmitrieva, *Phys. Lett.* **64A**, 348 (1978).

172. E. A. Hylleraas, *Z. Physik* **54**, 347 (1929).

173. R. J. Boyd, *J. Phys. B* **11**, L655 (1978).

174. M. Cohen, *J. Phys. B* **12**, L219 (1979).

175. J. R. Murdoch and D. E. Magnoli, *J. Chem. Phys.* **77**, 4558 (1982).

176. V. Fock, *Physik Z. Sowjet.* **1**, 747 (1932).

177. H. Jensen, *Z. Physik* **81**, 611 (1933).

178. R. J. Duffin, *Phys. Rev.* **47**, 421 (1935).

179. H. Jensen, *Z. Physik* **89**, 713 (1934).

180. N. H. March, *Phil. Mag.* **7**, **43**, 1042 (1952).

181. H. Jensen, *Z. Physik* **111**, 373 (1939).

182. N. H. March, *Phil. Mag.* **7**, **44**, 1193 (1953).

183. W. G. McMillan, *Phys. Rev.* **99**, 661 (1955).

184. W. G. McMillan, *Phys. Rev.* **111**, 479 (1958).

185. M. Born and R. Oppenheimer, *Ann. Phys.* **84**, 457 (1927).

186. I. N. Levine, *Quantum Chemistry*, 2nd ed., Allyn & Bacon, 1974, p. 292.

187. E. U. Condon, *Am. J. Phys.* **15**, 365 (1947).

188. H. Essén, *Int. J. Quant. Chem.* **XII**, 721 (1977).

189. G. Oliver and E. Weislinger, *C. R. Acad. Sci. Paris* **B284**, 131 (1977).

190. B. Nelander, *J. Chem. Phys.* **51**, 469 (1969).

191. F. W. King and S. M. Rothstein, *Phys. Rev. A* **21**, 1378 (1980).

192. D. E. Magnoli and J. R. Murdoch, *Int. J. Quant. Chem.* **XXII**, 1249 (1982).

193. W. Kolos and L. Wolniewicz, *J. Chem. Phys.* **41**, 3663 (1964).

194. C. Cohen-Tannoudji, B. Diu, and F. Laloë, *Mécanique Quantique*, Tome II, 2nd ed. revue et corrigée, Herman, 1977, p. 1181.

195. B. M. Deb (ed.), *The Force Concept in Chemistry*, Van Nostrand Reinhold, 1981.

196. J. Goodisman, *Diatomic Interaction Potential Theory*, Vol. 1, Academic, 1973, pp. 195–214; also in Ref. 195, pp. 246–283.

197. W. Moffitt, *Proc. Roy. Soc. London* **210A**, 245 (1951); **218A**, 486 (1953); *J. Chem. Phys.* **22**, 1820 (1954).

198. R. F. W. Bader and T. T. Nguyen-Dang, *Adv. Quantum Chem.* **14**, 63–123 (1981); Bader, in Ref. 195, Chapter 2.

199. S. Srebrenik, R. F. W. Bader, and T. Tung Nguyen-Dang, *J. Chem. Phys.* **68**, 3667 (1978).

200. I. N. Levine, *Molecular Spectroscopy*, Wiley, 1975, p. 143.

201. D. Steele, E. R. Lippincott, and J. T. Vanderslice, *Rev. Mod. Phys.* **34**, 239 (1962).

202. E. Fues, *Ann. Physik* **80**, 367 (1926).

203. R. G. Parr and R. F. Borkman, *J. Chem. Phys.* **49**, 1055; **48**, 1116 (1968).

204. R. G. Parr and R. J. White, *J. Chem. Phys.* **49**, 1059 (1968).

205. E. Bright Wilson, Jr., J. C. Decius, and P. C. Cross, *Molecular Vibrations*, McGraw-Hill, 1955 (Dover reprint, 1980), Appendix III.

206. F. M. Fernández and E. A. Castro, *Int. J. Quant. Chem.* **XIX**, 521, 533; *Phys. Rev. A* **24**, 2344 (1981).

207. J. Å Schweitz, *Int. J. Quant. Chem.* XVIII, 811 (1980); *J. Phys.* A10, 507, 517 (1977).

208. J. O. Hirschfelder and C. A. Coulson, *J. Chem. Phys.* **36**, 941 (1962), Sec. IV.
209. S. I. Vetchinkin, *J. Chem. Phys.* **41**, 1991 (1964).
210. C. A. Coulson, *Quart, J. Math.* **16**, 279 (1965).
211. P. D. Robinson, *Proc. Roy. Soc. London A* **283**, 229 (1965).
212. S. T. Epstein, *Theor. Chim. Acta* **52**, 89 (1979).
213. K. Banerjee, *Proc. Roy. Soc. London A* **363**, 147 (1978).
214. E. A. Castro, *Int. J. Quant. Chem.* XIII, 455; XIV, 231 (1978).
215. E. A. Castro and F. M. Fernàndez, *Lett. Nuovo Cimento* **28**, 484 (1980).
216. S. T. Epstein, *Am. J. Phys.* **44**, 251 (1976); S. T. Epstein, J. H. Epstein, and B. Kennedy, *J. Math. Phys.* **8**, 1747 (1967); J. H. Epstein and S. T. Epstein, *Am. J. Phys.* **30**, 266 (1962).
217. J. L. Friar, *Phys. Rev. C* **22**, 2636 (1980).
218. D. P. Chong and M. L. Benston, *Mol. Phys.* **12**, 487 (1967); *J. Chem. Phys.* **49**, 1302 (1968).
219. C. P. Yue and D. P. Chong, *Theor. Chim. Acta* **13**, 159 (1969).
220. C. J. Bradley and D. E. Hughes, *Int. J. Quant. Chem.* IS, 687 (1967); **3**, 699 (1969).
221. C. A. Coulson and J. C. Nash, *J. Phys. B* **5**, 421 (1972); **7**, 263 (1974).
222. E. Biemont and M. Godefroid, *Phys. Scripta (Sweden)* **18**, 323 (1978).
223. K. Banerjee, *Phys. Lett.* **63A**, 223 (1977).
224. K. Sakamoto and T. Terasaka, *Int. J. Quant. Chem.* XVI, 1357 (1979).
225. R. J. Swenson and S. H. Danforth, *J. Chem. Phys.* **57**, 1734 (1972).
226. S. Flügge, *Practical Quantum Mechanics*, Springer, 1974.
227. W. E. Caswell, *Ann. Phys.* **123**, 153 (1979).
228. L. S. Bartell, *J. Chem. Phys.* **38**, 1827 (1963).
229. R. A. Bonham and L. S. Su, *J. Chem. Phys.* **45**, 2827 (1966).
230. I. K. Dmitrieva and G. I. Plindov, *Phys. Lett.* **79A**, 47 (1980).
231. J. L. Richardson and R. Blankenbecler, *Phys. Rev. D* **19**, 496 (1979).
232. R. Blankenbecler, T. DeGrand, and R. L. Sugar, *Phys. Rev. D* **21**, 1055 (1980).
233. R. H. Tipping, *J. Chem. Phys.* **59**, 6433, 6443 (1973).
234. J. L. Richardson and R. Blankenbecler, *Phys. Rev. D* **20**, 1351 (1979).
235. A. Maduemezia, *J. Phys. A* **6**, 778 (1973); **7**, 1520 (1974).
236. J. Killingbeck, *Phys. Lett.* **65A**, 87 (1978).
237. M. Grant and C. S. Lai, *Phys. Rev. A* **20**, 718 (1979).
238. J. Killingbeck, *J. Phys. A* **14**, 1005 (1981).
239. R. Jackiw, *Phys. Rev.* **157**, 1220 (1967).
240. F. Hausdorff, *Ber. Sachs Ges. (Akad.) Wiss.* **58**, 19 (1906); see Ref. 211.
241. R. J. Swenson, *J. Chem. Phys.* **52**, 1012 (1970).
242. E. J. Robinson, *Phys. Rev. A* **16**, 2196 (1977).
243. S. T. Epstein, *J. Chem. Phys.* **58**, 5184 (1973).
244. A. Burnell and H. Caprasse, *Phys. Rev. D* **21**, 2000 (1980).
245. A. Einstein, *Ann. Physik* **17**, 549 (1905); **19**, 371 (1906).

246. P. Langevin, *C. R. Acad. Sci. Paris* **146**, 530 (1908).

247. E. M. Lifshitz and L. P. Pitaevskii, *Statistical Physics*, 3rd ed., Part I, Pergamon, 1980, p. 362.

248. S. Chandrasekhar, *Rev. Mod. Phys.* **15**, 1 (1943); reprinted in *Selected Papers on Noise and Stochastic Processes*, N. Wax (ed.), Dover, 1954.

249. J. Perrin, *Atoms*, Constable, London, 1916.

250. L. de la Peña, *Am. J. Phys.* **48**, 1080 (1980).

251. S. M. Moore, *J. Math. Phys.* **21**, 2102 (1980).

252. S. Chandrasekhar, in *Lectures in Theoretical Physics*, E. Brittin (ed.), 1964, Vol. 6, p. 1.

AUTHOR INDEX

SUBJECT INDEX